P9-BBT-599

Introduction to
Mathematical Statistics

A WILEY PUBLICATION
IN MATHEMATICAL STATISTICS

Introduction to Mathematical Statistics

FOURTH EDITION

PAUL G. HOEL

Professor of Mathematics
University of California
Los Angeles

JOHN WILEY & SONS, INC.

New York   London   Sydney   Toronto

Library of Congress Catalog Card Number: 70-139277
ISBN 0-471-40365-2

Printed in the United States of America

10 9 8 7 6 5 4 3 2 1

Preface

This edition differs from the third edition principally in the organization of material. The first six chapters have been rewritten to enable the instructor to postpone statistical inference until after the material on probability has been covered. This will permit a two-quarter (or two-semester) course to be given that will satisfy both the students who desire only a one-quarter course in probability and those who want a two-quarter course in probability and statistics. In order to organize a course in this manner it is necessary to postpone Chapter 4 and section 9 of Chapter 5 until after Chapter 6 has been completed.

Although Chapter 4, which is a gentle introduction to statistical inference, may be postponed until after the basic material on probability has been studied, I feel that so little time is required to cover it and so much insight into statistical reasoning is obtained by studying it that it is a mistake to omit it even for students who wish to learn only probability. It should be part of the knowledge acquired by any student who is interested in probability.

The reorganization of material required considerable rewriting of the text. This was particularly true in the early chapters on probability in which it was necessary to formalize the notation and treatment. The later chapters differ very little from those of the earlier edition except in the choice of topics. The exercises at the end of each chapter have been changed considerably by changing the data in many of the numerical problems (so that old student files on solutions will be worthless) and by adding new problems. The exercises have been labeled with the number of the section to which they belong, beginning with routine numerical problems and followed by theoretical ones. The more difficult problems occur at the end of each set.

Since this book is written for the student who has only an elementary calculus background in mathematics, much of the statistical theory must be accepted on faith. Some of the more important proofs are outlined in the appendix for the benefit of those students who possess more mathematical maturity than that normally obtained from an elementary calculus course.

From time to time I have received letters and reviews from some of the users of the earlier editions of this book with suggestions for its improvement.

I have always appreciated such letters and reviews even though I may not have included all those suggestions in the revision. I wish to thank those who have used my book over the years and especially those who have submitted ideas for its improvement.

PAUL G. HOEL

Los Angeles, California
January 1971

Contents

Introduction to
Mathematical Statistics

CHAPTER 1

Introduction

Mathematical statistics is the study of how to deal with data by means of probability models. It grew out of methods for treating data that were obtained by some repetitive operation such as those encountered in games of chance and in industrial processes. These methods soon found application in such diverse fields as medical research, insurance, marketing, agriculture, chemistry, and industrial experimentation.

In its broadest sense, statistical methods are often described as methods for making decisions in the face of uncertainty. The outcome of an experiment is usually uncertain but, hopefully, if it is repeated a number of times one may be able to construct a probability model for it and make decisions concerning the experimental process by means of it. Although probabilty can be applied to experiments or situations in which it is difficult to conceive of them as being of the repetitive type, the emphasis in this book will be on the repetitive type situation.

Experience indicates that many repetitive operations or experiments behave as though they occurred under essentially stable circumstances. Games of chance, such as coin tossing or dice rolling, usually exhibit this property. Many experiments and operations in the various branches of science and industry do likewise. Under such circumstances, it is often possible to construct a satisfactory mathematical model of the repetitive operation. This model can then be employed to study properties of the operation and to draw conclusions concerning it. Although mathematical models are especially useful devices for studying real-life problems when the model is realistic of the actual operation involved, it often happens that such models prove useful even though the operation is not highly stable.

The mathematical model that a statistician selects for a repetitive operation is usually one that enables him to make predictions about the frequency with which certain results can be expected to occur when the operation is repeated a number of times. For example, the model for studying the inheritance of color in the propagation of certain flowers might be one that predicted three times as many flowers of one color as of another color. In the investigation of the quality of manufactured parts the model might be one that predicts the percentage of defective parts that can be expected in the manufacturing process.

Because of the nature of statistical data and models, it is only natural that probability should be the fundamental tool in statistical theory. The statistician looks on probability as an idealization of the proportion of times that a certain result will occur in repeated trials of an experiment; consequently, a probability model is the type of mathematical model selected by him. Because probability is so important in the theory and applications of statistical methods, an introduction to probability is given before the study of statistical methods as such is taken up.

Although probability is being interpreted herein as an idealized relative frequency, it is treated as a measure of an individual's betting odds by those individuals who apply probability to a broader class of problems than those considered in this book. In applying probability techniques to such problems, however, it is necessary to realize that the reliability of a decision is heavily dependent on the realism of the individual's betting odds.

The idea of a mathematical model for assisting in the solution of real-life problems is a familiar one in the various sciences. For example, a physicist studying projectile motion often assumes that the simple laws of mechanics yield a satisfactory model, in spite of the complexity of the actual problem. For more refined work, he introduces a more complicated model. Since a model is only an idealization of the actual situation, the conclusions derived from it can be relied on only to the extent that the model chosen is a sufficiently good approximation to the actual situation being studied. In any given problem, therefore, it is essential to be well acquainted with the field of application in order to know what models are likely to be realistic. This is just as true for statistical models as for models in the various branches of science.

The science student will soon discover the similarity between certain of the statistical methods and certain scientific methods in which the scientist sets up a hypothesis, conducts an experiment, and then tests the hypothesis by means of his experimental data. Although statistical methods are applicable to all branches of science, they have been applied most actively in the biological and social sciences because the laboratory methods of the physical sciences have not been sufficiently broad to treat many of the problems of those other sciences. Problems in the biological and social sciences often involve undesired variables that cannot be controlled, as contrasted to the physical sciences in which such variables can often be controlled satisfactorily in the laboratory. Statistical theory is concerned not only with how to solve certain problems of the various sciences but also with how experiments in those sciences should be designed. Thus the science student should expect to learn statistical techniques to assist him in treating his experimental data and in designing his experiments in a more efficient manner.

The theory of statistics can be treated as a branch of mathematics in which probability is the basic tool; however, since the theory developed from an attempt to solve real-life problems, much of it would not be fully appreciated if it were removed from such applications. Therefore the theory and the applications are considered simultaneously throughout this book, although the emphasis is on the theory.

In the process of solving a real-life problem in statistics three steps may be recognized. First, a mathematical model is selected. Second, a check is made of the reasonableness of the model. Third, the proper conclusions are drawn from this model to solve the proposed problem. In this book the emphasis is on the first and third steps. In order to do justice to the second step, it would be necessary to be well acquainted with the field of application. It would also be necessary to know how the conclusions are affected by changes in the assumptions necessary for the model.

Students who have not had experience with applied science are sometimes disturbed by the readiness with which a statistician will accept certain of his model assumptions as being sufficiently well satisfied in a given problem to justify confidence in the validity of the conclusions. One of the striking features of much of statistical theory is that its field of application is much broader than the assumptions involved would seem to justify. The rapid development of, and interest in, statistical methods during the last few decades can be attributed in part to the highly successful application of statistical techniques to so many different branches of science and industry.

CHAPTER 2

Probability

1 INTRODUCTION

An individual's approach to probability depends on the nature of his interest in the subject. The pure mathematician usually prefers to treat probability from an axiomatic point of view, just as he does, say, the study of geometry. The applied statistician usually prefers to think of probability as the proportion of times that a certain event will occur if the experiment related to the event is repeated indefinitely. The approach to probability here is based on a blending of these two points of view.

The statistician is usually interested in probability only as it pertains to the possible outcomes of experiments. Furthermore, most statisticians are interested in only those experiments that are repetitive in nature or that can be conceived of as being so. Experiments such as tossing a coin, counting the number of defective parts in a box of parts, or reading the daily temperature on a thermometer are examples of simple repetitive experiments. An experiment in which several experimental animals are fed different rations may be performed only once with those same animals; nevertheless, the experiment may be thought of as the first in an unlimited number of similar experiments and therefore may be conceived of as being repetitive.

2 SAMPLE SPACE

Consider a simple experiment such as tossing a coin. In this experiment there are but two possible outcomes, a head and a tail. It is convenient to represent the possible outcomes of such an experiment, and experiments in general, by points on a line or by points in higher dimensions. Here it would be convenient to represent a head by the point 1 on the x axis and a tail by the point 0. This choice is convenient because the number corresponds to the number of heads obtained in the toss. If the experiment had consisted of

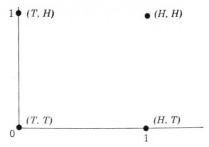

Fig. 1. A simple sample space.

tossing the coin twice, there would have been four possible outcomes, namely *HH*, *HT*, *TH*, and *TT*. For reasons of symmetry, it would be desirable to represent these outcomes by the points $(1, 1)$, $(1, 0)$, $(0, 1)$, and $(0, 0)$ in the x, y plane. Figure 1 illustrates this choice of points to represent the possible outcomes of the experiment.

If the coin were tossed three times, it would be convenient to use three dimensions to represent the possible experimental outcomes. This representation, of course, is merely a convenience, and if desired one could just as well mark off any eight points on the x axis to represent the eight possible outcomes.

In the experiment of rolling two dice, there are 36 possible outcomes, which have been listed in Table 1. The first number of each pair denotes

TABLE 1

11	21	31	41	51	61
12	22	32	42	52	62
13	23	33	43	53	63
14	24	34	44	54	64
15	25	35	45	55	65
16	26	36	46	56	66

the number that came up on one of the dice and the second number denotes the number that came up on the other die. It is assumed that the two dice are distinguishable or are rolled in order. For this experiment a natural set of points to represent the possible outcomes are the 36 points in the x, y plane whose coordinates are the corresponding number pairs of Table 1. This choice is shown in Fig. 2.

An experiment that consists of reading the temperature of a patient in a hospital has a very large number of possible outcomes depending on the degree of accuracy with which the thermometer is read. For such an experiment it is convenient to assume that the patient's temperature can assume

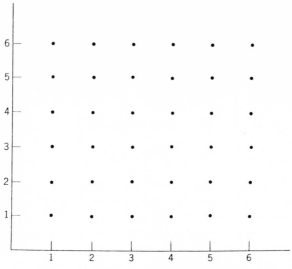

Fig. 2. A sample space for rolling two dice.

any value between, say, 95° and 110°; therefore the possible outcomes would be represented in a natural way by the points inside the interval from 95 to 110 on the x axis. This, of course, is a convenient idealization and ignores the impossibility of reading a thermometer to unlimited accuracy.

DEFINITION: *The set of points representing the possible outcomes of an experiment is called the sample space of the experiment.*

The idea of a sample space is introduced because it is a convenient mathematical device for developing the theory of probability as it pertains to the outcomes of experiments.

3 EVENTS

Consider an experiment such that whatever the outcome of the experiment it can be decided whether an event A has occurred. This means that each sample point can be classified as one for which A will occur or as one for which A will not occur. Thus, if A is the event of getting exactly one head and one tail in tossing a coin twice, the two sample points (H, T) and (T, H) of Fig. 1 correspond to the occurrence of A. If A is the event of getting a total of seven points in rolling two dice, then A is associated with the six sample points $(1, 6)$, $(2, 5)$, $(3, 4)$, $(4, 3)$, $(5, 2)$, and $(6, 1)$ of Fig. 2. If A is the event that a patient's temperature will be at least as high as 102, then A will consist of the interval of points from 102 to 110 on the x axis.

DEFINITION: *An event is a subset of a sample space.*

Since a subset of a set of points is understood to include the possibility that the subset is the entire set of points or that it contains none of the points of the set, this definition includes an event that is certain to occur or one that cannot possibly occur when the experiment is performed.

In view of the correspondence between events and sets of points the study of the relationship between various events is reduced to the study of the relationship between the corresponding sets. For this purpose it is convenient to represent the sample space, whatever its dimension or whatever the number of points in it, by a set of points inside a rectangle in a plane. An event A, which is therefore a subset of the points in this rectangle, is represented by the points lying inside a closed curve contained in the rectangle. If B is some other event of interest, it will be represented by the points inside some other closed curve in the rectangle. This representation is shown in Fig. 3. No attempt has been made to indicate whether the number of points is finite or infinite because a representation for both types of sample spaces is desired.

If A and B are two events associated with an experiment, one may be interested in knowing whether at least one of the events A and B will occur when the experiment is performed. Now the set of points that consists of all points that belong to A, or B, or both A and B is called the union of A and B and is denoted by the symbol $A \cup B$. This set of points, which is shown as the shaded region in Fig. 3, therefore represents the event that at least one of the events A and B will occur.

As an illustration, if A is the event of getting a six on the first die when rolling two dice and B is the event of getting a six on the second die, then $A \cup B$ is the event of getting at least one six in rolling two dice. The event A consists of the six points found in the last column of points in the sample space shown in Fig. 2 and the event B consists of the six points found in the last row of points in that sample space. The event $A \cup B$ is then the set of eleven points found in the union of the last column and last row of points.

Another event of possible interest is that of knowing whether both events A and B will occur when the experiment is performed. The set of points that consists of all points that belong to both A and B is called the intersection of

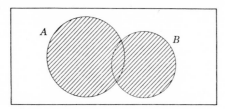

Fig. 3. Representation of $A \cup B$.

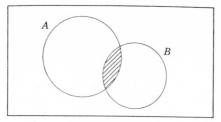

Fig. 4.   Representation of $A \cap B$.

A and B and is denoted by the symbol $A \cap B$. This set of points, which is shown as the shaded region in Fig. 4, therefore represents the event that both A and B will occur when the experiment is performed.

In the preceding illustration concerning the two dice, $A \cap B$ is the event that both dice will show a six. It is represented by the single point $(6, 6)$, which is the intersection of the last column, and last row, sets of points.

Corresponding to any event A there is an associated event, denoted by $\bar{A}$, which states that A will not occur when the experiment is performed. It is represented by all the points of the rectangle not found in A and is shown as the shaded region in Fig. 5. The set $\bar{A}$ is called the complement of the set A relative to the sample space.

If two sets, A and B, have no points in common they are said to be disjoint sets. In the language of events, such events are called *disjoint events*, but they are also called *mutually exclusive events*, because the occurrence of one of those events excludes the possible occurrence of the other.

4   PROBABILITY

The familiar functions of calculus are what are known as point functions. The function defined by the formula $f(x) = x^2$ is an example of a point function, for to each point on the x axis this formula assigns the value of the function. The notion of function is much broader than this, however, and permits the elements of the domain of the function to be sets of points rather than individual points. In this case the function is called a *set function*. As

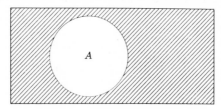

Fig. 5.   Representation of $\bar{A}$.

an illustration, the domain might consist of all intervals on the x axis and the function might be the function that gives the length of the interval. Set functions are introduced here because they are needed for defining probability.

Probability is an idealization of the proportion of times that a certain result will occur in repeated trials of an experiment; therefore a probability model for events should be one for which the probability that an event A will occur, which is denoted by $P\{A\}$, is equal to the proportion of times that the event A would be expected to occur in repeated trials of the experiment. Since $P\{A\}$ is a function defined on sets, such as A, it is a set function.

If an experiment could be repeated a large number of times the results could be used to assign a value to $P\{A\}$. However, there is no requirement that the experiment be performed at all before such a probability is assigned. Thus, if A is the event of getting two sixes in rolling two dice, symmetry considerations would suggest the value $\frac{1}{36}$ for the probability that this event will occur. The experimenter is free to assign any value he desires, but if his assignment is unrealistic from the long run proportion point of view his probability model is not likely to prove very useful to him for making predictions about future experiments.

If probabilities of events are to be interpreted as models for the proportion of times those events will occur in repetitions of the experiment, such probabilities should possess the essential properties of proportions. Thus, a probability should be a number between 0 and 1 because a proportion is that kind of number. Further, the probability of the event S, where S is the sample space, should be 1 because some one of the possible outcomes is certain to occur when the experiment is performed. Finally, if two events A and B are disjoint, the probability of the union of those events should be equal to the sum of the probabilities of the two events because for such events the proportion of times that A or B occurs will be equal to the proportion of times that A occurs plus the proportion of times that B occurs. It has been found that any other reasonable properties of probabilities that might be desired will be satisfied if the following three conditions, which are based on the preceding discussion, are satisfied. These are called the axioms, or postulates, of probability. They place restrictions on the type of set function P that can be used to calculate probabilities of events. Such a function is called a probability measure.

AXIOMS OF PROBABILITY: *A probability measure P is a real valued set function defined on a sample space S that satisfies*

(1) $0 \leq P\{A\} \leq 1$ *for every event A*

(2) $P\{S\} = 1$

(3) $P\{A_1 \cup A_2 \cup \cdots\} = P\{A_1\} + P\{A_2\} + \cdots$ *for every finite or infinite sequence of disjoint events $A_1, A_2, \cdots$.*

It would be very difficult to find a function P that gives numbers corresponding to expected proportions for every possible subset A of a sample space because the number of such subsets is extremely large even for a sample space containing only a few points. Fortunately, for sample spaces containing only a finite number, or an infinite sequence, of sample points it suffices to assign probabilities to each of the sample points. The value of $P\{A\}$ for any subset A is then easily determined from the probabilities assigned to the individual sample points by means of axiom (3).

For a sample space such as that proposed for the experiment of reading a patient's temperature and which consists of the points inside the interval from 95 to 110 on the x axis, the determination of a function P that will satisfy the axioms appears to be a formidable task. Difficulties can arise unless one restricts the class of events slightly. With this restriction it turns out that it suffices to define P for intervals only. The function so determined can then be used to calculate the probability for any desired event of the restricted type. The restriction needed here is so mild that it plays no role in applications and is of mathematical interest only. As will be seen later, finding a function that is satisfactory for intervals is not particularly difficult.

5 THE ADDITION RULE

Applications of probability are often concerned with a number of related events rather than with just one event. For simplicity, consider two such events, A_1 and A_2, associated with an experiment. Then one is often interested in knowing whether both A_1 and A_2 will occur when the experiment is performed, or whether at least one of the events A_1 and A_2 will occur. From the notation that was developed in section 3, the first of these two composite events is represented by $A_1 \cap A_2$ and the second by $A_1 \cup A_2$. To answer questions concerning the relative frequency with which these composite events are expected to occur it is necessary to know the values of their probabilities, namely $P\{A_1 \cap A_2\}$ and $P\{A_1 \cup A_2\}$. The purpose of this section is to derive a useful formula for $P\{A_1 \cup A_2\}$.

Let the sample space for an experiment be represented by the points in the rectangle of Fig 6 and let the sample points corresponding to the occurrence of A_1 and A_2 be the points inside the regions labeled A_1 and A_2, respectively. Then, as it may be recalled from Fig. 3, the event $A_1 \cup A_2$ consists of all points lying inside these two regions. The derivation of a formula for $P\{A_1 \cup A_2\}$ is based on expressing $A_1 \cup A_2$ as the union of disjoint events and then applying axiom (3). From Fig. 6 it will be seen that $A_1 \cup A_2$ is the union of the three disjoint sets $A_1 \cap \bar{A}_2$, $A_2 \cap \bar{A}_1$, and $A_1 \cap A_2$; therefore

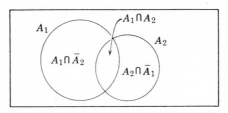

Fig. 6.   Disjoint sets for $A_1 \cup A_2$.

by axiom (3)

(1) $P\{A_1 \cup A_2\} = P\{A_1 \cap \bar{A}_2\} + P\{A_2 \cap \bar{A}_1\} + P\{A_1 \cap A_2\}$.

Since, as will be observed from Fig. 6, A_1 is the union of the two disjoint sets $A_1 \cap \bar{A}_2$ and $A_1 \cap A_2$, it follows from axiom (3) that

$$P\{A_1\} = P\{A_1 \cap \bar{A}_2\} + P\{A_1 \cap A_2\} .$$

In the same manner,

$$P\{A_2\} = P\{A_2 \cap \bar{A}_1\} + P\{A_1 \cap A_2\} .$$

Using these last two relations to eliminate $P\{A_1 \cap \bar{A}_2\}$ and $P\{A_2 \cap \bar{A}_1\}$ from (1) will yield the desired formula for $P\{A_1 \cup A_2\}$, which is known as the addition rule of probability.

(2)   ADDITION RULE: $P\{A_1 \cup A_2\} = P\{A_1\} + P\{A_2\} - P\{A_1 \cap A_2\}$.

Two events A_1 and A_2 may have no sample points in common. As stated earlier, the events are then said to be disjoint or mutually exclusive. Formula (2) then reduces to the following one:

(3) $P\{A_1 \cup A_2\} = P\{A_1\} + P\{A_2\}$ *when A_1 and A_2 are disjoint* .

This formula is, of course, also a direct consequence of axiom (3). Formula (2) can be generalized to more than two events; however, since the generalization is not needed in later work, it is not considered here. The generalization of formula (3) is needed but it is already available in axiom (3).

The preceding rules are applicable to any type of sample space for which a probability measure P is given. Their use, of course, depends upon knowing the values of the probabilities on the right side of the formulas. As suggested earlier, the calculation of probabilities of events of this type is especially simple when the sample space contains only a finite number of points; therefore the discussion will now be restricted to such sample spaces.

The first step in calculating the probability of an event A for a finite sample space is to assign a probability to each of the sample points. These probabilities must be assigned in such a manner as to conform to the first two axioms of probability. That is, they must be non-negative numbers that sum to 1. If the probability model is to be useful for making predictions, the probability assigned to a particular sample point should correspond to the proportion of times that that particular sample point would be expected to be obtained in a large number of repetitions of the experiment. This expectation could be based on one's experience, or on outside information, or on symmetry considerations, or on a mixture of them. Thus, it would be realistic to assign the probability of $\frac{1}{36}$ to each sample point of the sample space shown in Fig. 2, whether or not one had had experience in rolling dice.

Let the total number of sample points be denoted by n and let $p_1, p_2, \cdots,$ p_n be the probabilities assigned to the respective sample points. Each point represents a possible outcome, which in turn is an event. Events of this type are often called *simple events*. These events will be labeled $e_1, e_2, \cdots, e_n$. It is clear that they are disjoint. Now any event A is a set of sample points, and therefore is the union of the corresponding simple events. Application of axiom (3) therefore gives

$$(4) \qquad P\{A\} = \sum_A P\{e_i\} = \sum_A p_i$$

where the summation is over those p's associated with the sample points lying in A.

For many games of chance, which were the original impetus for the development of a theory of probability, the sample space is not only finite but the sample points are assigned the same probability. This would be true, for example, for the game of craps for which the sample space of Fig. 2 is appropriate. Each of those sample points would be assigned the probability $\frac{1}{36}$. If n denotes the total number of sample points and $N(A)$ denotes the number of those sample points in the set A, then, since it is being assumed that $p_i = 1/n$ for all $i = 1, 2, \cdots, n$, formula (4) will reduce to

$$(5) \qquad P\{A\} = \frac{N(A)}{n}.$$

As an illustration, consider once more the experiment of rolling two dice for which Fig. 2 is the sample space. Let A be the event of getting a total of seven points on the two dice. Then if each sample point is assigned the probability $\frac{1}{36}$, formula (5) will give $P\{A\} = \frac{6}{36}$ because $n = 36$ and there are six sample points corresponding to the occurrence of A, namely $(1, 6)$, $(2, 5)$, $(3, 4)$, $(4, 3)$, $(5, 2)$, and $(6, 1)$. In a similar manner, if B is the event of getting a total of at least seven points, then $P\{B\} = \frac{21}{36}$ because the 21 sample points lying on or below the diagonal extending from the point $(1, 6)$ to the point $(6, 1)$ correspond to the occurrence of B.

As an illustration for which formula (4) must be used because the p's are not all equal, consider a pair of modified dice in which each one-spot has been changed to a two-spot. As a result, each die will possess two 2's but no 1. In order to compensate for this alteration in Table 1, it is necessary to replace each 1 by a 2 in that table. The first two rows, and also the first two columns, will then become identical. If similar expressions are combined, the possible outcomes for this experiment are those listed in Table 2.

TABLE 2

22(4)	32(2)	42(2)	52(2)	62(2)
23(2)	33(1)	43(1)	53(1)	63(1)
24(2)	34(1)	44(1)	54(1)	64(1)
25(2)	35(1)	45(1)	55(1)	65(1)
26(2)	36(1)	46(1)	56(1)	66(1)

The numbers in parentheses following the outcomes give the number of outcomes in Table 1 that produced the outcomes in Table 2. Thus a (4) follows the outcome 22 because the outcomes 11, 12, 21, and 22 of Table 1 all reduce to 22 after each 1 is replaced by a 2. In view of the earlier assumption that each of the 36 possible outcomes of Table 1 will occur with the same relative frequency, the natural probabilities to assign the possible outcomes listed in Table 2 are those obtained by multiplying $\frac{1}{36}$ by the numbers in parentheses.

Now if A is the event of getting a total of seven points in the experiment of rolling the two altered dice it will follow from Table 2 and formula (4) that

$$P\{A\} = \tfrac{2}{36} + \tfrac{1}{36} + \tfrac{1}{36} + \tfrac{2}{36} = \tfrac{6}{36}$$

because these numbers are the probabilities assigned to the four points in A, namely $(2, 5)$, $(3, 4)$, $(4, 3)$, and $(5, 2)$. This result is the same as that for the earlier experiment when two normal dice were rolled. If B is the event of getting a total of at least seven points for the experiment of rolling the altered dice, then from Table 2 it follows that $P\{B\} = \frac{23}{36}$. This is obtained by adding the probabilities attached to the sample points lying on and below the diagonal extending from $(2, 5)$ to $(5, 2)$. This result is not the same as that obtained earlier when two normal dice were rolled. For that experiment the probability was $\frac{21}{36}$.

6 THE MULTIPLICATION RULE

Suppose that one is interested in knowing whether an event A_2 will occur subject to the condition that A_1 is certain to occur. For the purpose of discussing probabilities associated with such events, assume for the present that the sample space contains only a finite number of points. Since A_1 is certain to occur only when the sample space is restricted to those points lying inside the region labeled A_1 in Fig. 6, it is necessary to consider how probabilities should be assigned to the points of this new smaller sample space. If originally a sample point in A_1 had been assigned, say, twice as large a probability as another point in A_1, then it should be assigned twice as large a probability in the new sample space also, because ignoring experimental outcomes that do not yield the event A_1 should not affect the two-to-one ratio of expected frequencies for those two sample points. It is merely necessary therefore to multiply the original probabilities assigned to points in A_1 by a constant factor c such that the sum of the new probabilities will be 1. Thus if π_i denotes the new probability corresponding to p_i in the original assignment, one should choose $\pi_i = cp_i$, where

$$1 = \sum_{A_1} \pi_i = c \sum_{A_1} p_i = cP\{A_1\} .$$

As a result, $c = 1/P\{A_1\}$ and therefore

$$\pi_i = \frac{p_i}{P\{A_1\}} .$$

Now that the new sample space probabilities have been assigned, one can calculate probabilities in the usual manner by merely applying formula (4). All such probabilities will be conditional probabilities, subject to the occurrence of A_1. If the probability that A_2 will occur, subject to the restriction that A_1 must occur, is denoted by $P\{A_2 \mid A_1\}$, then it follows from Fig. 6, formula (4), and the expression for π_i that

$$P\{A_2 \mid A_1\} = \sum_{A_1 \cap A_2} \pi_i = \frac{\sum\limits_{A_1 \cap A_2} p_i}{P\{A_1\}} .$$

The first sum is over those π_i corresponding to sample points lying inside $A_1 \cap A_2$ because they are the only sample points inside A_1 associated with the occurrence of A_2. Since the numerator sum in the last expression is the one that defines $P\{A_1 \cap A_2\}$, it follows that the formula for conditional probability reduces to

(6) $$P\{A_2 \mid A_1\} = \frac{P\{A_1 \cap A_2\}}{P\{A_1\}} .$$

It is assumed here that A_1 is an event for which $P\{A_1\} > 0$. In this derivation it was also assumed that the sample space contains only a finite number of points. For more general sample spaces this formula will be taken as the definition of conditional probability. It is easily shown that the three axioms of probability are satisfied by $P\{A_2 \mid A_1\}$; therefore it is legitimate to define conditional probability in this manner regardless of the type of sample space.

This formula when written in product form yields the fundamental multiplication rule for probabilities.

(7)   MULTIPLICATION RULE: $P\{A_1 \cap A_2\} = P\{A_1\}P\{A_2 \mid A_1\}$.

Although formula (6) holds only when $P\{A_1\} \neq 0$, formula (7) may be treated as holding in general if it is agreed to give the right side the value 0 when the factor $P\{A_1\}$ is equal to 0. If the order of the two events is interchanged, formula (7) becomes

(8) $P\{A_1 \cap A_2\} = P\{A_2\}P\{A_1 \mid A_2\}$.

Now, suppose that A_1 and A_2 are two events such that $P\{A_2 \mid A_1\} = P\{A_2\}$ and such that $P\{A_1\}P\{A_2\} > 0$. Then the event A_2 is said to be independent in a probability sense, or more briefly, independent, of the event A_1. This name follows from the property that the probability of A_2 occurring is not affected by adding the condition that A_1 must occur. When A_2 is independent of A_1, (7) reduces to

(9) $P\{A_1 \cap A_2\} = P\{A_1\}P\{A_2\}$.

Conversely, when (9) is true, it follows from comparing (9) and (7) that A_2 is independent of A_1. If the right members of (8) and (9) are equated, it will be seen that $P\{A_1 \mid A_2\} = P\{A_1\}$. But this states that the event A_1 is independent of the event A_2. Thus, if A_2 is independent of A_1, it follows that A_1 must be independent of A_2. Because of this mutual independence and because (9) implies this independence, it is customary to define independence in the following manner:

(10) DEFINITION: *Two events, A_1 and A_2, are said to be independent if* $P\{A_1 \cap A_2\} = P\{A_1\}P\{A_2\}$.

Formulas (7) and (10) can be generalized in an obvious manner for more than two events by always combining events into two groups.

7   ILLUSTRATIONS

As illustrations of the application of the preceding rules of probability, consider a few simple problems related to games of chance. From symmetry considerations, it is usually assumed in such games that all possible outcomes should be assigned the same probability. This was done, for example, in discussing the probability of events in connection with Table 1. It was not done in connection with Table 2 because symmetry was missing in the experiment that gave rise to Table 2. In the following illustrations it is assumed that symmetry is present and therefore that probabilities may be calculated by means of formula (5).

(a) If two dice are rolled, what is the probability of getting either a total of 7 or a total of 11 points? Let A_1 and A_2 denote the events of getting a total of 7 and 11 points, respectively. Since these events are mutually exclusive, formula (3) may be applied. From Fig. 1 there are six sample points giving rise to event A_1 and two giving rise to A_2; consequently, application of formula (5) gives $P\{A_1\} = \frac{6}{36}$ and $P\{A_2\} = \frac{2}{36}$. Formula (3) then yields

$$P\{A_1 \cup A_2\} = \frac{6}{36} + \frac{2}{36} = \frac{8}{36}.$$

This result is, of course, the same as that obtained by counting favorable and total outcomes in Fig. 1 and applying formula (5) directly.

(b) If two dice are rolled, what is the probability that each of them will show at least five points? Let A_1 denote the event of getting a 5 or 6 on the first die and A_2 the event of getting a 5 or 6 on the second die. If the dice are rolled properly, events A_1 and A_2 may be assumed to be independent; therefore formula (10) may be applied. Now one can treat the experiment of rolling two dice as composed of two consecutive independent experiments in which one die is rolled first and then the second die is rolled. From this point of view, the event A_1 is concerned with the first experiment only for which there are six sample points, two of which correspond to the occurrence of A_1. Under the symmetry assumption, it therefore follows that $P\{A_1\} = \frac{2}{6}$. The event A_2 plays the same role with respect to the second experiment as A_1 does in the first; hence $P\{A_2\} = \frac{2}{6}$. Formula (10) now yields the desired result, namely,

$$P\{A_1 \cap A_2\} = \frac{2}{6} \cdot \frac{2}{6} = \frac{4}{36}.$$

This result could also have been obtained directly by counting sample points in the sample space of Fig. 1 for the complete experiment. Since there are 36 sample points and the four sample points given by the outcomes 55, 56, 65, 66 correspond to the occurrence of $A_1 \cap A_2$, it follows from (5) that $P\{A_1 \cap A_2\} = \frac{4}{36}$. The advantage of using formula (10) here is that it enables one to work with simpler sample spaces than the original sample space. The real purpose of these illustrations, however, is to develop familiarity with the formulas and not to simplify calculations, because in many problems the experimental sample space is not available for a direct application of formula (5).

(c) Two cards are drawn from an ordinary deck of 52 cards but the first card drawn is replaced before the second card is drawn. What is the probability that at least one of the cards will be a spade? Let A_1 denote the event of drawing a spade on the first draw and A_2 the event of drawing a spade on the second draw. The problem then is to calculate $P\{A_1 \cup A_2\}$ by means of formula (2). As in the preceding illustration, the complete experiment can be broken down into two consecutive independent experiments. Here $P\{A_1\} = \frac{13}{52} = \frac{1}{4}$ because A_1 is concerned with the first drawing only and there are 52 sample points, 13 of which are favorable (spades), in that experiment. Similarly, $P\{A_2\} = \frac{1}{4}$. Because of the independence of A_1 and A_2, it follows from formula (10) that $P\{A_1 \cap A_2\} = \frac{1}{4} \cdot \frac{1}{4} = \frac{1}{16}$. Application of formula (2) now gives

$$P\{A_1 \cup A_2\} = \frac{1}{4} + \frac{1}{4} - \frac{1}{16} = \frac{7}{16}.$$

This problem can also be solved indirectly by first calculating the probability that neither card drawn will be a spade, which is $\frac{39}{52} \cdot \frac{39}{52}$, and then subtracting this result from 1. The reasoning here is that the opposite of "neither card will be a spade" is "at least one card will be a spade." Of course, one could also solve this problem by counting sample points and applying formula (5). The total number of sample points is $52 \cdot 52$, whereas the favorable number is $52 \cdot 52 - 39 \cdot 39$ because only those sample points corresponding to a pair of nonspades are unfavorable.

(d) Two cards are drawn from a deck of cards. What is the probability that both cards will be spades? As in the preceding illustration, let A_1 and A_2 denote the events of getting a spade on the first and second drawings, respectively. Since the first card drawn is not replaced before the second drawing, these events are certainly not independent; therefore formula (7) must be used. As before, the complete experiment may be treated as composed of two consecutive experiments; however, here they are not independent experiments. In the first experiment there are 52 sample points, of which 13 correspond to the occurrence of A_1; therefore $P\{A_1\} = \frac{13}{52}$. In calculating $P\{A_2 \mid A_1\}$ it is necessary to consider only that part of the original sample space for which A_1 is certain to occur. It will contain $13 \cdot 51$ sample points because to each of the 13 possible spades that may be obtained on the first drawing there are always 51 remaining cards that may be obtained on the second drawing. There are $13 \cdot 12$ points in this sample space that correspond to the occurrence of A_2, because to each of the 13 possible spades that may be obtained on the first drawing there are always 12 remaining spades that may be obtained on the second drawing to give $13 \cdot 12$ spade pairs. As a result, $P\{A_2 \mid A_1\} = 13 \cdot 12/13 \cdot 51 = \frac{12}{51}$. By using symmetry this computation could have been simplified considerably. Since the conditional probability of getting a spade on the second drawing, given that a spade was obtained on the first drawing, should be equal to the corresponding probability when a particular spade is known to have been obtained on the first drawing, one could just as well have worked with the sample space in which a particular spade is obtained on the first drawing. This reduced sample space contains only 51 points, 12 of which are favorable. The calculation of $P\{A_2 \mid A_1\}$ now becomes $\frac{12}{51}$. The application of formula (7) can now be made and yields the result

$$P\{A_1 \cap A_2\} = \tfrac{13}{52} \cdot \tfrac{12}{51} = \tfrac{1}{17}.$$

Hereafter, formulas (2) and (7) will be applied without discussing the nature of the various sample spaces involved in the computations. Furthermore, symmetry considerations such as those used to simplify the preceding computations will be used whenever they are advantageous. If one is not certain in a given problem that his intuition is correct in choosing simple sample spaces, he should go back to the original sample space.

(e) This last illustration is a somewhat more complicated exercise in the manipulation of formulas (3) and (7). One box contains two red balls. A second box of identical appearance contains one red and one white ball. If a box is selected by chance and one ball is drawn from it, what is the probability that the first box was the selected one, if the drawn ball turns out to be red? Let A_1 denote the event of selecting the first box and $\bar{A}_1$ that of selecting the second box. Let A_2 denote the event of drawing a red ball and $\bar{A}_2$ that of drawing a white ball. Then the problem is to calculate the conditional probability $P\{A_1 \mid A_2\}$. Interchanging A_1 and A_2 in (6) will give

$$P\{A_1 \mid A_2\} = \frac{P\{A_1 \cap A_2\}}{P\{A_2\}}.$$

The numerator probability may be calculated by using formula (7) directly, with the understanding that to select a box by chance means that the probability of selecting, say, the first box is $\frac{1}{2}$. Thus

$$P\{A_1 \cap A_2\} = P\{A_1\}P\{A_2 \mid A_1\} = \tfrac{1}{2} \cdot 1 = \tfrac{1}{2}.$$

The denominator probability may be calculated by considering the event A_2 in conjunction with the selection of a box. Now A_2 will occur if, and only if, one of the two mutually exclusive events $A_1 \cap A_2$ and $\bar{A}_1 \cap A_2$ occurs. Thus, by formula (3),

(11) $$P\{A_2\} = P\{A_1 \cap A_2\} + P\{\bar{A}_1 \cap A_2\}.$$

But

$$P\{\bar{A}_1 \cap A_2\} = P\{\bar{A}_1\}P\{A_2 \mid \bar{A}_1\} = \tfrac{1}{2} \cdot \tfrac{1}{2} = \tfrac{1}{4}.$$

Since $P\{A_1 \cap A_2\}$ was found to be equal to $\frac{1}{2}$, it follows from (11) and this last result that $P\{A_2\} = \frac{3}{4}$, and hence that

$$P\{A_1 \mid A_2\} = \frac{\frac{1}{2}}{\frac{3}{4}} = \frac{2}{3}.$$

This problem was designed to give practice in the use of the two basic probability formulas; however, it could have been solved more simply by looking at the sample space and applying formula (4). A sample space consisting of the four points I1, I2, II1, II2, would be quite natural here. The Roman numeral denotes the box number, and the Arabic numeral the ball number. One would assign equal probabilities to these four points. The condition A_2 restricts the sample space to the first three sample points if ball number 2 in the second box is understood to be the white one. Thus each of these three sample points must be assigned the probability $\frac{1}{3}$. Now, the first two sample points correspond to the occurrence of A_1; hence $P\{A_1 \mid A_2\} = \frac{2}{3}$.

8 BAYES' FORMULA

Illustration (e) of the preceding section is typical of problems in which one looks at the outcome of an experiment and then asks for the probability that the outcome was due to a particular one of the possible "causes" of the outcome. Thus in illustration (e) there are two possible causes, or ways, for a red ball to be obtained, and the problem is to calculate the probability that it was due to the first one. Although the solution to the problem was obtained by merely applying the two rules of probability in the proper sequence, the computations are sufficiently extensive to make it worthwhile to derive a formula for treating such problems systematically.

For the purpose of obtaining a formula, let the sample space be divided into k disjoint sets whose union is the sample space. This is called a *partition* of the sample space. These k sets will be denoted by $H_1, H_2, \cdots, H_k$. They represent the k possible causes of an experimental outcome. Next, let A be an event that occurred when the experiment was performed and consider the problem of calculating the probability that H_i was the cause of the occurrence of A. The geometry of a partitioning and its relation to an event A is shown in Fig. 7 in which S is partitioned into four subsets.

The problem is to calculate the conditional probability $P\{H_i \mid A\}$. From

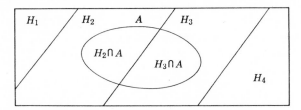

Fig. 7. A partitioning of a sample space.

formula (6) this is given by

(12)
$$P\{H_i \mid A\} = \frac{P\{H_i \cap A\}}{P\{A\}}.$$

Formula (7) gives

(13)
$$P\{H_i \cap A\} = P\{H_i\}P\{A \mid H_i\}.$$

Hence (12) can be written in the form

(14)
$$P\{H_i \mid A\} = \frac{P\{H_i\}\, P\{A \mid H_i\}}{P\{A\}}.$$

From Fig. 7 it is clear that the event A is the union of the disjoint events $H_1 \cap A,\ H_2 \cap A,\ \cdots,\ H_k \cap A$. Some of these sets may, of course, be empty as is true in Fig. 7. Axiom (3) for disjoint events therefore gives

$$P\{A\} = P\{H_1 \cap A\} + P\{H_2 \cap A\} + \cdots + P\{H_k \cap A\}.$$

If formula (13) is applied to each term on the right, this last relation may be expressed in the following form:

$$P\{A\} = \sum_{j=1}^{k} P\{H_j\}\, P\{A \mid H_j\}.$$

The substitution of this expression in (14) yields the desired formula for calculating probabilities of causes. The result may be summarized as follows:

BAYES' FORMULA:

$$P\{H_i \mid A\} = \frac{P\{H_i\}\, P\{A \mid H_i\}}{\displaystyle\sum_{j=1}^{k} P\{H_j\}\, P\{A \mid H_j\}}, \qquad i = 1, 2, \cdots, k.$$

Illustration (e) of the preceding section, which was solved by the expeditious use of the two rules of probability, is solved here to illustrate the direct use of Bayes' formula. Let H_1 and H_2 correspond to the events of getting box number 1 and box number 2, respectively, and let A be the event of getting a red ball. Since a box is selected by chance, $P\{H_1\} = P\{H_2\} = \frac{1}{2}$. Further, it is clear from the contents of the two boxes that $P\{A \mid H_1\} = 1$ and $P\{A \mid H_2\} = \frac{1}{2}$. Bayes' formula then yields

$$P\{H_1 \mid A\} = \frac{\frac{1}{2} \cdot 1}{\frac{1}{2} \cdot 1 + \frac{1}{2} \cdot \frac{1}{2}} = \frac{2}{3}.$$

9   COUNTING TECHNIQUES

The simplest problems on which to develop facility in applying the addition and multiplication rules of probability are some of those related to games of chance. For many such problems, however, the counting of sample points corresponding to various events becomes tedious unless compact counting methods are developed. A few of the formulas that yield such methods are derived in this section.

9.1   Trees

If an experiment is such that it can be treated as a two-stage experiment, the problem of counting sample points can be considerably simplified by the use of a tree diagram. For example, suppose a box contains two red, two black, and one green ball and that two balls are to be drawn. This is a two-stage experiment for which the various possibilities that can occur may be represented by a horizontal tree such as that shown in Fig. 8.

Each stage of a multi-stage experiment has as many branches as there are possibilities at that stage. Here there are three main branches for the first stage and three secondary branches at each of the second stages, except for the last one where there are only two branches because it is impossible to obtain a green ball at the second drawing if a green ball is obtained on the first drawing. The total number of terminating branches in such a tree gives the total number of possible outcomes in the compound experiment and

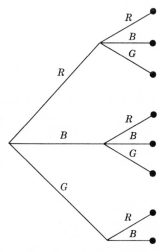

Fig. 8.   A counting tree.

therefore the end points of those branches may be treated as the sample points of a sample space corresponding to the experiment.

If there are several stages to an experiment and several possibilities at each stage, the tree associated with the experiment would become too large to be manageable. For such problems the counting of sample points is simplified by means of algebraic formulas. Toward the objective of deriving such formulas, consider a two-stage experiment for which there are r possibilities at the first stage and s possibilities at the second stage corresponding to each of the first stage possibilities. A tree to represent this experiment would have r main branches and s secondary branches emanating from each of the r main branches; consequently the total number of terminating branches would be rs. If a third stage with t possibilities were added, the total number would become rst. This can be extended in an obvious manner to any number of stages. The counting principle obtained here by means of a tree, and which actually has been used in some of the earlier illustrative examples, will now be used to derive some useful algebraic counting formulas.

9.2 Permutations

Consider a set of n different objects, such as n blocks having different numbers. Let r of the n objects be selected and arranged in a line. Such an arrangement is called a *permutation* of the r objects. If two of the r objects are interchanged in their respective positions, a different permutation results. In order to count the total number of permutations, it suffices to consider the r positions on the line as fixed and then count the number of ways in which blocks can be selected to be placed in the r positions.

This is essentially an r-stage experiment to which the counting technique of the preceding section may be applied. Starting from the position farthest to the left, any one of the n blocks may be chosen to fill that position. Thus, there are n possibilities at the first stage. After the first position has been filled, there will be only $n - 1$ blocks left to choose from to fill the second position. For each choice for the first position there are therefore $n - 1$ choices for the second position and hence $n - 1$ possibilities for the second stage of the experiment. Thus, by the tree counting principle of the preceding section there are $n(n - 1)$ total choices for the two positions. If this selection procedure is continued, there will be $n - r + 1$ blocks left to choose from for the rth position. If the total number of such permutations is denoted by $_nP_r$, it therefore follows that

(15) $$_nP_r = n(n - 1) \cdots (n - r + 1).$$

The symbol $_nP_r$ is usually called the number of permutations of n things taken r at a time.

As an illustration, suppose one is given the four letters a, b, c, d. The number of permutations of these four letters taken two at a time is given by $_4P_2 = 4 \cdot 3 = 12$. These permutations are easily enumerated as follows: ab, ba, ac, ca, ad, da, bc, cb, bd, db, cd, dc.

If r is chosen equal to n, (15) reduces to

$$(16) \qquad\qquad _nP_n = n(n-1) \cdots (1) = n! \,.$$

The symbol $n!$ is called n factorial. In order to permit formulas that involve factorials to be correct even when $n = 0$, it is necessary to define $0! = 1$. This is consistent for $n = 1$ with the factorial property that $(n-1)! = n!/n$.

9.3   Combinations

If one is interested only in what particular objects are selected when r objects are chosen from n objects, without regard to their arrangement in a line, then the unordered selection is called a *combination*. Thus, if two letters are chosen from the four letters a, b, c, d, the combination ab is the same combination as ba, but of course it differs from the combination ac. The total number of combinations possible in selecting r objects from n different objects is denoted by the symbol $\binom{n}{r}$. This symbol is usually called the number of combinations of n things taken r at a time.

In order to derive a formula for $\binom{n}{r}$, it suffices to compare the total number of permutations and total number of combinations possible. Since a permutation is obtained by first selecting r objects and then arranging them in some order, whereas a combination is obtained by performing only the first step, it follows that the total number of permutations is obtained by taking every possible combination, the total number of which is $\binom{n}{r}$, and arranging them in all possible ways. But from (16) the total number of arrangements of r objects in r places is $r!$; hence the total number of permutations is given by multiplying the number of combinations, $\binom{n}{r}$, by $r!$. Thus $_nP_r = \binom{n}{r} \cdot r!$. Using formula (15), it therefore follows that

$$\binom{n}{r} = \frac{n(n-1) \cdots (n-r+1)}{r!}\,.$$

Since $n(n-1) \cdots (n-r+1) = n!/(n-r)!$, this formula may be written in the following more compact form:

(17) $$\binom{n}{r} = \frac{n!}{r!\,(n-r!)}.$$

As an illustration, the number of combinations of two letters selected from the four letters a, b, c, d is given by $\binom{4}{2} = 4!/2!\,2! = 6$. The actual combinations are ab, ac, ad, bc, bd, cd.

9.4 Permutations When Some Elements Are Alike

In the preceding derivations it has been assumed that all the n objects are different. It sometimes happens, however, that the n objects contain a number of similar objects. Thus one might have five colored balls of which three are white and two black, instead of five distinct colors. Now suppose that there are only k distinct kinds of objects and that there are n_1 of the first kind, n_2 of the second kind, $\cdots$, and n_k of the kth kind, where $n_1 + n_2 + \cdots + n_k = n$. The total number of different permutations of these n objects arranged in a line is obviously less than $n!$. In order to find the total number of distinct permutations, it suffices to compare the number of permutations now, which is denoted by P, with the number that would be obtained if the like objects were given marks to distinguish them. The comparison is similar to that made between $\binom{n}{r}$ and $_nP_r$ in deriving fromula (17). Each permutation in the problem under consideration gives rise to additional permutations when the like objects are made different by markings. For example, if the n_1 similar objects in a permutation are made different, they can be rearranged in their positions in $n_1!$ ways. Since this is true for each of the P permutations, there will be $n_1!$ times as many permutations when the n_1 similar objects are made different as before. In the same manner, the n_2 similar objects may be made different to give $n_2!$ times as many permutations as before. Continuing this procedure, the total number of permutations after all similar objects have been made different will be $n_1!\,n_2!\cdots n_k!$ times as large as the number of permutations before the similar objects were made different; hence the total number after these changes will be $Pn_1!\,n_2!\cdots n_k!$. But after all similar objects have been made different, the total number of permutations will be the number of permutations of n different things taken n at a time, which is $n!$. Equating these two results and solving for P, one obtains

(18) $$\frac{n!}{n_1!\,n_2!\cdots n_k!}$$

for the total number of permutations of n things in which there are n_1 alike, n_2 alike, $\cdots$, n_k alike. As an illustration, consider the number of permutations of the five letters a, a, a, b, b. Formula (18) yields $5!/3!\,2! = 10$. These permutations are easily written down: *aaabb*, *aabab*, *abaab*, *baaab*, *aabba*, *ababa*, *baaba*, *abbaa*, *babaa*, *bbaaa*.

9.5 Illustrations of the Use of Combinatorial Formulas

(a) Consider a bridge hand consisting of 13 cards chosen from an ordinary deck. What is the probability that such a hand will contain exactly seven spades? Since a bridge hand is not concerned with the order in which the various cards are obtained, the total number of possible bridge hands is equal to the number of ways of choosing 13 objects from 52 objects, or $\binom{52}{13}$. This is therefore the total number of sample points in the sample space. The number of hands containing exactly seven spades is equal to the number of ways of choosing 7 spades from 13 spades, or $\binom{13}{7}$, multiplied by the number of ways of choosing 6 nonspades from 39 nonspades, or $\binom{39}{6}$. Hence the desired probability is given by

$$\frac{\binom{13}{7}\binom{39}{6}}{\binom{52}{13}} = \frac{13!\,39!\,13!\,39!}{7!\,6!\,6!\,33!\,52!}.$$

(b) What is the probability that a bridge hand will contain at most one ace? The total number of bridge hands containing at most one ace consists of those with one ace and those with no ace. The number of hands with one ace is given by $\binom{4}{1}\binom{48}{12}$, whereas the number with no ace is given by $\binom{48}{13}$; consequently, the total number of favorable hands is

$$\binom{4}{1}\binom{48}{12} + \binom{48}{13}.$$

Since the total number of possible bridge hands was found earlier to be $\binom{52}{13}$, the desired probability is given by the ratio

$$\frac{\binom{4}{1}\binom{48}{12} + \binom{48}{13}}{\binom{52}{13}}.$$

(c) If you know that a bridge hand contains at most one ace, what is the probability that it contains no ace? The number of sample points in this sample space is given by the numerator of the preceding result. The number of those corresponding to the desired event, namely bridge hands with no ace, is given by $\binom{48}{13}$; consequently, the desired probability is given by the ratio

$$\frac{\binom{48}{13}}{\binom{4}{1}\binom{48}{12} + \binom{48}{13}}.$$

10 RANDOM VARIABLES

Consider a sample space corresponding to the tossing of two coins and suppose that interest is centered on the total number of heads that will be obtained. For the purpose of calculating the probabilities of the various possible totals, it is convenient to introduce a symbol X to represent the total number of heads that will be obtained. If the sample space displayed in Fig. 1 is used, X will assume the value 0 at the sample point $(0, 0)$, the value 1 at the two sample points $(1, 0)$ and $(0, 1)$, and the value 2 at the sample point $(1, 1)$. This association of X values for this sample space is shown in Fig. 9. A numerical valued variable X such as this is an example of what is known as a random variable.

As another illustration, if X represents the sum of the points obtained in rolling two dice, then X is a random variable that can assume integer values from 2 to 12, inclusive. As a third illustration, if X represents the distance a dart thrown at a circular target of radius 20 inches is from the center of the target, and complete misses are ignored, then X is a random variable that can assume any value between 0 and 20.

In all of these illustrations X is numerical valued and its value depends upon the sample point. Thus, X is a function whose domain of definition is the set of sample points and whose range is a set of real numbers.

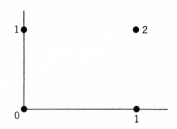

Fig. 9. Random variable values.

DEFINITION: *A random variable is a real valued function defined on a sample space.*

The name random, or chance, is given to the variable X in these illustrations and definition because it is defined on a sample space associated with a physical experiment in which the outcome of any one experiment is uncertain and is therefore said to depend upon chance.

Since the objective of this section is to study random variables and calculate probabilities related to them, it is necessary that a probability measure be assigned to the sample space associated with the random variable X. Therefore it is assumed without mentioning the fact that any sample space being studied has a probability measure associated with it.

11   DISCRETE RANDOM VARIABLES

After a random variable X has been defined on a sample space, interest usually centers on determining the probability that X will assume specified values in its range of possible values. For example, if X represents the sum of the points obtained in rolling two dice, then in the game of craps it is of interest to calculate the probability that X will assume the value 7. Or if X represents the distance a dart will land from the center of a target, it may be of interest to calculate the probability that X will assume some value less than 5.

The calculation of $P\{X = 7\}$ in the first illustration is much simpler than the calculation of $P\{X < 5\}$ in the second illustration because it is much simpler to work with a sample space consisting of a finite number of points than with one consisting of an interval of points on the x axis. A sample space that consists of a finite number, or an infinite sequence, of points is called a *discrete sample space* whereas one that consists of one or more intervals of points is called a *continuous sample space*. The word interval is understood to include higher dimensional intervals as well as the familiar one-dimensional interval. Thus, a rectangle is a two-dimensional interval. Permitting a space to have an infinite sequence of points rather than only a finite number does not complicate the theory needed for such spaces; therefore it is convenient theoretically to associate such spaces with the finite ones under the name discrete.

Since the theory for discrete sample spaces is much simpler than that for continuous spaces, the discussion in the next few sections will be restricted to discrete sample spaces. There cannot be more values assumed by a random variable than there are sample points; therefore a random variable defined on a discrete sample space can assume only a finite number, or an infinite sequence, of values. Such a random variable is called a *discrete random variable*. As might be anticipated, a *continuous random variable* is one that can assume any value in some interval of values and for which the probability is zero that it will assume any particular value in the interval.

Return now to the problem of calculating probabilities for random variables defined on discrete sample spaces. Formula (4) for calculating the probability of an event *A* was derived on the assumption that the sample space contained only a finite number of points; however, the same derivation applies if the sample space consists of an infinite sequence of points. This formula is therefore applicable to any discrete sample space. If x is a particular value of the random variable X defined on a discrete sample space, this formula may be used to calculate the probability of the event that X will assume the value x. This probability is therefore given by

$$(19) \qquad\qquad P\{X = x\} = \sum_{X=x} p_i .$$

Here the summation is over all those sample points for which the random variable has the value x.

As an illustration, if X represents the sum of the points obtained in rolling two dice, for which Fig. 2 is the sample space and for which all sample points are assigned the probability $\frac{1}{36}$, then $P\{X = 7\} = \frac{6}{36}$. This result arises from the fact that there are six sample points for which the value of X is 7.

12 DENSITY FUNCTIONS

For the purpose of calculating probabilities related to discrete random variables it is convenient to introduce a function called a *discrete probability density function*, or more briefly a *density*.

DEFINITION: *Let X be a discrete random variable. Then the function f defined by $f(x) = P\{X = x\}$ is called the discrete density function of X.*

In calculus the word density is used in connection with a continuous distribution of matter along a line or in a plane; therefore the use of it here in connection with a discrete distribution of probability may seem irregular. Since it is desirable to use the same name for this type of function whether the random variable is discrete or continuous and since the word density is traditional for continuous variables, the name density will be used here also.

A discrete density function often consists of merely a table of values. Thus, if two coins are tossed and if X represents the total number of heads that will be obtained, it suffices to define f by means of the following set of values $f(0) = \frac{1}{4}, f(1) = \frac{1}{2},$ and $f(2) = \frac{1}{4}.$

In order to judge how a random variable is distributed, that is, how its probability changes as the variable changes, it is helpful to graph the density function by means of a line chart. As an illustration of such a graph, let X represent the sum of the points that will be obtained when rolling two dice. Attaching the proper value of X to each of the 36 sample points of the sample space shown in Fig. 2 and assuming equal probabilities for the sample points, application of formula (19) will show that $f(2) = f(12) = \frac{1}{36}, f(3) = f(11) = \frac{2}{36}, f(4) = f(10) = \frac{3}{36}, f(5) = f(9) = \frac{4}{36}, f(6) = f(8) = \frac{5}{36},$ and $f(7) = \frac{6}{36}.$ A line chart of $f(x)$ is given in Fig. 10.

The purpose of introducing a density function is to simplify the notation and calculation of probabilities related to random variables. After a density function has been determined, it is no longer necessary to calculate probabilities of random variable events by summing sample point probabilities as in (19). Instead one can look upon the points of the x axis where $f(x) > 0$ as a new discrete sample space with the probabilities given by $f(x)$ attached to those points. For example, if X represents the sum of the points in rolling two dice, the new sample space will consist of the eleven points 2, 3, $\cdots$, 12 and the probabilities attached to those points will be the values of $f(x)$ that were calculated in the preceding paragraph. This new sample space is shown in Fig. 11. It is merely a sample space representation of the information conveyed by Fig. 10.

Now consider the calculation of the probability of the event $X \in R$, where R is some set of points on the x axis and where $X \in R$ is a symbol to represent the event that X will assume some value in the set R of values. In terms of the new sample space generated by X and $f(x)$ a direct application of formula (4) gives the result

$$(20) \qquad P\{X \in R\} = \sum_{x \in R} f(x)$$

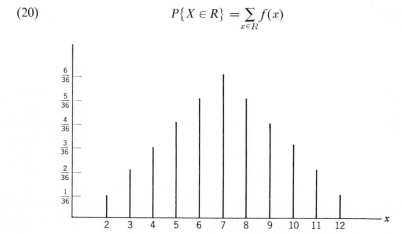

Fig. 10.   Line chart for a density function.

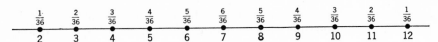

Fig. 11. A random variable sample space.

where the summation is over those x values in R for which $f(x) > 0$. If R contains points not belonging to the new sample space, there is no harm done in including them in the summation because $f(x) = 0$ for such points. The calculation of this probability in this manner is usually much simpler than the calculation that is based on the original sample space of the experiment.

Using the dice illustration again, suppose one wished to calculate the probability that the sum of the points will exceed 7. In terms of the new sample space shown in Fig. 11 this probability is given by

$$P\{X > 7\} = \sum_{x=8}^{12} f(x) = \tfrac{5}{36} + \tfrac{4}{36} + \tfrac{3}{36} + \tfrac{2}{36} + \tfrac{1}{36} = \tfrac{15}{36}.$$

If this probability were to be calculated using the original sample space shown in Fig. 2, it would be necessary to sum the probabilities of all the sample points lying below the diagonal extending from $(1, 6)$ to $(6, 1)$. In this problem there is no particular advantage of using the new sample space; however for more complicated sample spaces the advantage can be considerable.

A function closely related to the density function f is the corresponding distribution function F. It is defined by the relation

$$F(x) = P\{X \le x\} = \sum_{t \le x} f(t)$$

where the summation is over all those values of the random variable that are

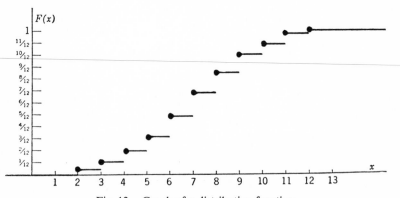

Fig. 12. Graph of a distribution function.

less than or equal to the specified value x. Thus $F(x)$ gives the probability that X will assume a value less than or equal to x as contrasted to $f(x)$ which gives the probability that X will assume the particular value x. The graph of $F(x)$ corresponding to the $f(x)$ of Fig. 10 is given in Fig. 12. It should be noted that the value of $F(x)$ for x an integer is the upper horizontal line value rather than the lower. This has been indicated by using dots at the start of each line segment. The distribution function is very useful for later theoretical work and will be discussed more thoroughly then.

13 JOINT DENSITY FUNCTIONS

Many experiments involve several random variables rather than just one. For simplicity, consider two discrete random variables X and Y. A mathematical model for these two variables is a function that gives the probability that X will assume a particular value x while at the same time Y will assume a particular value y. A function $f(x, y)$ that gives such probabilities is called a joint density function of the two random variables X and Y. The adjective joint is often omitted because there is little possibility of confusing a function of two variables with a function of one variable.

As an illustration, let X represent the number of spades obtained in drawing one card from an ordinary deck and let Y represent the number of spades obtained in drawing a second card from the deck, without the first card being replaced. Here X and Y can assume the values 0 and 1 only. Then $f(x, y)$ is defined by the following table of values: $f(0, 0) = \frac{39}{52} \cdot \frac{38}{51}$, $f(1, 0) = \frac{13}{52} \cdot \frac{39}{51}$, $f(0, 1) = \frac{39}{52} \cdot \frac{13}{51}$, and $f(1, 1) = \frac{13}{52} \cdot \frac{12}{51}$. The graph of $f(x, y)$ as a line chart is given in Fig. 13.

As a second illustration, let X and Y denote the number of red and white balls, respectively, obtained in drawing two balls from a bag containing two red, two white, and two black balls. Here the joint density function is given by the formula

$$(21) \quad f(x, y) = \frac{\binom{2}{x}\binom{2}{y}\binom{2}{2 - x - y}}{\binom{6}{2}}, \qquad x, y = 0, 1, 2, \quad 0 \le x + y \le 2.$$

Although this density function is defined by an explicit formula, it could have been defined by means of a table of values as in the first illustration. It is understood in this formula that $f(x, y) = 0$ if x and y are values for which the combination symbols as defined by (17) do not have meaning. The numerator in (21) is obtained by realizing that the x red balls must come

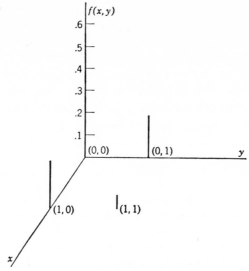

Fig. 13.   Graph of a joint density function.

from the two red balls, the y white balls must come from the two white balls, and the remaining $2 - (x + y)$ balls must come from the two black balls.

A joint density, such as the preceding one, can be used to calculate various probability statements about X and Y by summing $f(x, y)$ over the appropriate values of x and y, just as probability statements about X were calculated by summing $f(x)$ over the appropriate values of x. As in the one-dimensional problem, it is useful to work in a new sample space that consists of the points in the x, y plane where $f(x, y) > 0$ and whose probabilities are the values given by $f(x, y)$. For example, the probability of getting at least one red ball in the experiment related to (21) can be expressed in the form

$$P\{X \geq 1\} = \sum_{x=1}^{2} \sum_{y=0}^{2-x} f(x, y).$$

In evaluating this double sum it is necessary to realize that $f(x, y) = 0$ for those integer values of x and y in the sum that do not satisfy the restriction $0 \leq x + y \leq 2$.

In much of the statistical theory that is developed in the following chapters, the variables are unrelated in a probability sense. In the first illustration, the variables X and Y would have been such variables if the first card drawn had been replaced before the second card was drawn. It would seem appropriate to call two variables that are unrelated in this sense independent variables. Since the word independent has been defined for events by (10), a corresponding definition for random variables should conform to that definition. In terms of events such as $X \in A$ and $Y \in B$, where A and B are any two sets in the domains of X and Y, respectively, the definition of independence for events would require that

(22) $P\{X \in A \text{ and } Y \in B\} = P\{X \in A\} P\{Y \in B\}$ for all A and B.

This is often taken as the definition of independence for two random variables; however, since the emphasis in this section is on density functions, an equivalent definition based on densities will be given instead.

Let X and Y possess the joint density $f(x, y)$ and let $g(x)$ and $h(y)$ be their individual densities. Then by choosing the sets A and B to be the single points x and y, respectively, (22) will reduce to

$P\{X = x \text{ and } Y = y\} = P\{X = x\} P\{Y = y\}$ for all x and y.

But in terms of densities this is $f(x, y) = g(x)h(y)$ for all x and y. Conversely, if $f(x, y) = g(x)h(y)$ for all x and y, then working in the sample space of the two random variables, it follows by analogy with (20) that

$$P\{X \in A \text{ and } Y \in B\} = \sum_{x \in A} \sum_{y \in B} f(x, y) = \sum_{x \in A} \sum_{y \in B} g(x)h(y)$$

$$= \sum_{x \in A} g(x) \sum_{y \in B} h(y) = P\{X \in A\} P\{Y \in B\}.$$

This shows that (22) will hold for all A and B if and only if $f(x, y) = g(x)h(y)$ holds for all x and y. Since the latter relationship is in terms of densities and is more useful than (22), it will be used to define independence.

(23) DEFINITION: *The random variables X and Y whose joint density function is $f(x, y)$ and whose individual density functions are $g(x)$ and $h(y)$ are said to be independent if and only if*

$$f(x, y) = g(x)h(y) \quad \text{for all } x \text{ and } y.$$

The preceding definition can be generalized in an obvious manner to define independence for n random variables. Thus,

(24) DEFINITION: *The random variables $X_1, X_2, \cdots, X_n$ whose joint density function is $f(x_1, x_2, \cdots, x_n)$ and whose individual density functions are $f_1(x_1), f_2(x_2), \cdots, f_n(x_n)$ are said to be independent if and only if*

$$f(x_1, x_2, \cdots, x_n) = f_1(x_1)f_2(x_2) \cdots f_n(x_n).$$

As an illustration, consider the first of the two preceding illustrations, modified to the extent that the first card drawn is replaced before the second card is drawn. The density function of X, which is denoted by $f_1(x)$, is given by the formula

$$f_1(x) = \frac{\binom{13}{x}\binom{39}{1-x}}{\binom{52}{1}}.$$

Since the first card is returned to the deck before the second drawing, the second drawing does not differ from the first drawing in properties; consequently, the density function of Y, which is denoted by $f_2(y)$, is given by the same formula with x replaced by y. Since X and Y are obviously independent here,

$$f(x, y) = \frac{\binom{13}{x}\binom{39}{1-x}}{\binom{52}{1}} \frac{\binom{13}{y}\binom{39}{1-y}}{\binom{52}{1}} = f_1(x)f_2(y).$$

As an example in which (23) does not hold, consider the second of the two preceding illustrations for which the joint density function is given by (21). If $f_1(x)$ denotes the density function of X alone, then

$$f_1(x) = \frac{\binom{2}{x}\binom{4}{2-x}}{\binom{6}{2}}.$$

Similarly, if $f_2(y)$ denotes the density function of Y alone, then

$$f_2(y) = \frac{\binom{2}{y}\binom{4}{2-y}}{\binom{6}{2}}.$$

If (23) were to hold here, which means that $f(x, y)$ as given by (21) would have to be equal to $f_1(x)f_2(y)$, then it would be necessary that

$$\binom{4}{2-x}\binom{4}{2-y} = \binom{6}{2}\binom{2}{2-x-y}.$$

This relationship would be required to hold for all experimentally possible values of x and y. It obviously does not hold for $x = 1$ and $y = 1$. As a matter of fact, it does not hold for a single pair of possible values.

14   MARGINAL AND CONDITIONAL DISTRIBUTIONS

In the preceding section it was necessary to obtain the density functions of the individual variables before one could decide whether the variables of a joint density function were independent. Since it is important to know whether a set of variables is independent, it would be desirable to have a systematic way of finding the density functions of the individual variables. Such a method is readily obtained by means of formula (7) for the case of two variables. Although one can easily extend the method so that it will apply to more than two variables, there is no need for the extension in later work; therefore the discussion is limited to two variables.

Consider an experiment for which A_1 is the event that a random variable X will assume the value x and A_2 is the event that a second random variable Y will assume the value y. The multiplication formula

$$(25) \qquad P\{A_1 \cap A_2\} = P\{A_1\}P\{A_2 \mid A_1\}$$

can then be expressed in terms of density functions. Since $P\{A_1 \cap A_2\}$ now gives the probability that the two random variables will assume the values x and y, respectively, it is equivalent to $f(x, y)$, the value of the joint density function at the point (x, y). Similarly, $P\{A_1\}$ is the probability that X will assume the value x; therefore it is equivalent to $f(x)$. Since $P\{A_2 \mid A_1\}$ gives the probability that Y will assume the value y, given that X is known to have the value x, it may be treated as the value of a conditional density function, which is denoted by $f(y \mid x)$. Formula (25) then becomes

$$(26) \qquad f(x, y) = f(x)f(y \mid x) .$$

Since $f(y \mid x)$ gives the conditional probability that Y will assume the value y when X has the fixed value x, the sum of $f(y \mid x)$ over all possible values of y for this fixed value of x must equal 1. Hence, if both sides of (26) are summed over all possible values of y, the formula for $f(x)$ given below in (27) will be obtained. In connection with the joint density function $f(x, y)$, the function $f(x)$ is called the X *marginal density function;* however, it is merely the density function of X. This result may be expressed as follows:

$$(27) \qquad \text{MARGINAL DISTRIBUTION:} \quad f(x) = \sum_y f(x, y) .$$

In a similar manner, the Y marginal density function, $g(y)$, can be obtained by summing $f(x, y)$ over all values of x with y held fixed. Thus $g(y) = \sum_x f(x, y)$. These results show that if one has the joint density function of two random variables and if one desires the density function of one of them, it is merely necessary to sum the joint density function over all values of the other variable.

The conditional density function $f(y \mid x)$ gives the distribution of Y when X is held fixed. Because of (26), if $f(x) \neq 0$, one may therefore write

(28)   CONDITIONAL DISTRIBUTION: $f(y \mid x) = \dfrac{f(x, y)}{f(x)}$.

The conditional distribution for X with Y held fixed is given by an analogous formula. This shows that if one has the joint density function of two variables and desires the conditional density function for one of them when the other is held fixed, it is merely necessary to divide the joint density function by the density function of the fixed variable.

For the purpose of illustrating the preceding ideas, suppose that a bag contains two white and four black balls and that two balls are drawn from the bag. Let X and Y represent the results of the two drawings, 0 corresponding to a black ball and 1 corresponding to a white ball. Then every possible result will be represented by one of the four points in the x, y plane shown in Fig. 14. From the contents of the bag and the order in which the drawings are made, it follows directly from formula (26) that

$$f(0, 0) = f(0)f(0 \mid 0) = \tfrac{4}{6} \cdot \tfrac{3}{5} = \tfrac{6}{15}$$
$$f(0, 1) = f(0)f(1 \mid 0) = \tfrac{4}{6} \cdot \tfrac{2}{5} = \tfrac{4}{15}$$
$$f(1, 0) = f(1)f(0 \mid 1) = \tfrac{2}{6} \cdot \tfrac{4}{5} = \tfrac{4}{15}$$
$$f(1, 1) = f(1)f(1 \mid 1) = \tfrac{2}{6} \cdot \tfrac{1}{5} = \tfrac{1}{15} .$$

The values of $f(x, y)$ have been graphed in Fig. 14 by means of a simple line chart.

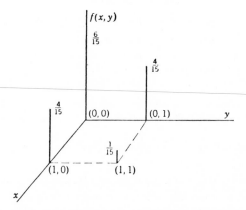

Fig. 14. Theoretical distribution for two discrete variables.

In order to illustrate the method of obtaining a marginal distribution and a conditional distribution from the joint distribution, assume now that only the final values of $f(x, y)$ just calculated are known. Thus the only information available is that given in Fig. 14. One should erase from his mind how these numbers were obtained.

The X marginal density function can be obtained by applying formula (27). Thus

$$f(0) = f(0, 0) + f(0, 1) = \tfrac{6}{15} + \tfrac{4}{15} = \tfrac{2}{3}$$
$$f(1) = f(1, 0) + f(1, 1) = \tfrac{4}{15} + \tfrac{1}{15} = \tfrac{1}{3}.$$

If the four points in the x, y plane are thought of as mass points whose total mass is 1, then the X marginal distribution represents the distribution of mass along the x axis after the mass points in the x, y plane have been projected on the x axis.

The conditional density function of Y for x fixed can be obtained by applying formula (28) and using the results just obtained. Thus, if x is assigned the value $x = 1$,

$$f(0 \mid 1) = \frac{f(1, 0)}{f(1)} = \frac{\tfrac{4}{15}}{\tfrac{1}{3}} = \tfrac{4}{5}$$

$$f(1 \mid 1) = \frac{f(1, 1)}{f(1)} = \frac{\tfrac{1}{15}}{\tfrac{1}{3}} = \tfrac{1}{5}.$$

Geometrically, $f(y \mid 1)$ represents the distribution of probability mass along the line $x = 1$ when the two points on this line have had their probability masses multiplied by a number, $1/f(1)$, to make the sum of their masses equal 1.

As a second illustration, in which the joint density function is given directly, consider the density function defined by the formula

$$f(x, y) = \tfrac{1}{27}(x + 2y)$$

where x and y can assume only the integer values 0, 1, or 2. The sample space with its probabilities calculated by means of this formula is shown in Fig. 15.

From formula (27), the marginal density function of X is given by

$$f(x) = \sum_y \tfrac{1}{27}(x + 2y).$$

In carrying out the summation it is clear from Fig. 15 that y may range over all its possible values regardless of the value of x that was fixed; hence

(29) $$f(x) = \sum_{y=0}^{2} \tfrac{1}{27}(x + 2y) = \tfrac{1}{9}(x + 2).$$

Similarly, summing over all x values with y fixed gives

$$g(y) = \tfrac{1}{9}(2y + 1) \, .$$

It is clear that $f(x, y)$ is not equal to the product of its marginal density functions here and therefore that X and Y are not independent random variables.

From formula (28) and the result given in (29), it follows that the conditional density function of Y for X fixed is given by

$$f(y \mid x) = \frac{\tfrac{1}{27}(x + 2y)}{\tfrac{1}{9}(x + 2)} = \frac{x + 2y}{3(x + 2)} \, .$$

If x is assigned the value 2, for example,

$$f(y \mid 2) = \tfrac{1}{6}(y + 1) \, .$$

This function would be useful if one wished to calculate probabilities for various values of Y when it is known that X has the value 2. It can easily be checked that this is a probability density function by summing the three probabilities obtained from this formula by letting $y = 0$, 1, and 2 and verifying that the sum is 1.

It was a simple matter to find the marginal density function in the preceding problem because the sum over y in formula (27) was over all possible values of y regardless of the fixed value of x. The problem would have been somewhat more difficult if the density function had been given by the formula

$$f(x, y) = \tfrac{1}{16}(x + 2y)$$

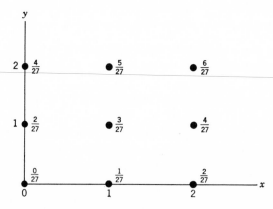

Fig. 15.   Sample space for $f(x, y) = \tfrac{1}{27}(x + 2y)$.

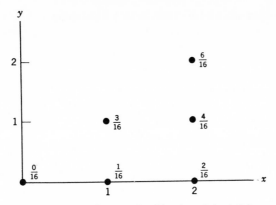

Fig. 16. Sample space for $f(x, y) = \frac{1}{16}(x + 2y)$.

and the sample space had been the one shown in Fig. 16. This sample space differs from the one in Fig. 15 in that Y is not permitted to exceed X in value. Now if one wanted the marginal value $f(1)$, the sum in (27) would become

$$f(1) = \sum_{y=0}^{1} \tfrac{1}{16}(1 + 2y) = \tfrac{4}{16}.$$

However, if one wanted the marginal value $f(2)$, the sum would become

$$f(2) = \sum_{y=0}^{2} \tfrac{1}{16}(2 + 2y) = \tfrac{12}{16}.$$

Thus the values of y over which the sum is to be taken depend upon what marginal value of x is desired. Although the simplest procedure here is to perform the summation separately for each value of x, one can sum for a general x. Calculations for a general x will show that the marginal function can be expressed by means of the formula

$$f(x) = \tfrac{1}{8}x(x + 1).$$

15 CONTINUOUS RANDOM VARIABLES

The last few sections have been concerned with discrete random variables and their density functions. In this section density functions for continuous variables will be introduced. Although a formal definition of such a function can be written down immediately, the definition will be motivated by considering what happens to a discrete density when the sample space becomes more dense.

PROBABILITY

Continuous random variables are usually associated with experiments that involve some type of measurement. Thus, such variables as weights, lengths, temperatures, and velocities are considered to be continuous. In reality all such variables are discrete because whatever measuring device is used it is of limited reading accuracy; however it is mathematically convenient to assume that no such limitation exists.

For the purpose of discussing properties of continuous variables, consider a particular continuous random variable X that represents the thickness of a metal washer obtained from a certain machine turning out washers. If the machine were permitted to turn out, say, 100 washers, and if the thicknesses of these 100 washers were measured to the nearest .001 inch, there would be available 100 values of X with which to study the behavior of the machine. If these 100 values were collected and represented in table form, one might find a table of values such as that displayed in Table 3 that gives the absolute frequency f with which various values of X occurred. The word "frequency" usually implies the ratio of the observed number of values of X to the total number of observational values; however, it is also used to denote the numerator of this ratio. If throughout the subsequent chapters there is any question which meaning is being used, the words "relative frequency" and "absolute frequency" will be employed. Absolute frequencies are recorded in Table 3.

For the purpose of displaying these results graphically, a graph called a *histogram* is used. A histogram is a graph of the type shown in Fig. 17, in which areas are used to represent observed frequencies, particularly relative frequencies. Thus the area of the rectangle that is centered at $x = .234$ should equal the relative frequency .18; however, in practice it is customary to choose any convenient unit on the y axis, with the result that the areas of the rectangles may be only proportional to the corresponding frequencies rather than equal to them. The histogram shown in Fig. 17 for the data of Table 3 has been constructed with such a convenient choice of units; hence areas there are only proportional to frequencies.

If the histogram is to be constructed so that areas will be equal to relative frequencies, then the total area of the histogram must equal 1 because the sum of the relative frequencies must equal 1. If h denotes the distance between consecutive x values, the height of the rectangle centered at, say, x_i will be f_i/Nh, where f_i denotes the absolute frequency of x_i and N denotes the number

TABLE 3

x	.231	.232	.233	.234	.235	.236	.237	.238	.239
f	1	2	8	18	28	24	13	4	2

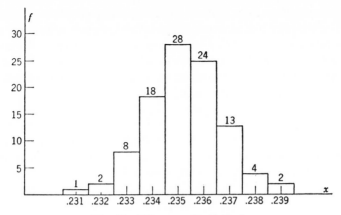

Fig. 17.   Histogram for Table 3.

of observations. This result is obvious when it is realized that this ordinate when multiplied by the base h must equal the relative frequency f_i/N.

The histogram of Fig. 17 indicates the frequency with which various values of X were obtained for 100 runs of the experiment. If 200 runs had been made, the resulting histogram would have been twice as large as that based on 100 runs. In order to compare histograms based on different numbers of experiments, it is necessary to choose units on the y axis, as discussed in the preceding paragraph, in such a manner that the area of the histogram will always be equal to one. With this choice of units, the histogram would be expected to approach a fixed histogram as the number of runs of the experiment is increased indefinitely. Furthermore, if it is assumed that X can be measured as accurately as desired so that the unit on the x axis, h, can be made as small as desired, then the histogram would be expected to smooth out and approximate a continuous curve as the number of runs of the experiment is increased indefinitely and h is chosen very small. Such a curve is thought of as an idealization for the relative frequency with which different values of X would be expected to be obtained for runs of the actual experiment.

When the area of the histogram is made equal to 1, it follows from the preceding discussion that the sum of the areas of several neighboring rectangles is equal to the relative frequency with which the value of X was observed to lie in the interval that forms the base of those rectangles. Since this property will continue to hold as the number of runs of the experiment increases indefinitely, the area under the expected limiting, or idealized, curve between any two given values of x should be equal to the relative frequency with which X would be expected to lie in the interval determined by those values of x. The function $f(x)$ whose graph is conceived as being the limiting form of the histogram is treated as the mathematical model for the continuous random variable X and is called the density function of the variable. Since relative frequency in the case of a histogram is replaced by probability in the case of a mathematical model, the definition of a density function for a continuous variable may be stated in the following form:

(30) DEFINITION: *A density function for a continuous random variable X is a function f that possesses the following properties:*

(i) $$f(x) \geq 0$$

(ii) $$\int_{-\infty}^{\infty} f(x)\, dx = 1$$

(iii) $$\int_{a}^{b} f(x)\, dx = P\{a < X < b\}$$

where a and b are any two values of x satisfying $a < b$.

Property (i) is obviously necessary since probability must be non-negative. Property (ii) corresponds to the requirement that the probability of an event that is certain to occur should be equal to one. Here X is certain to assume some real value when an observation of it is made.

In dealing with continuous random variables, one asks only for the probability that the variable will lie in some interval or intervals. As a result, probabilities for continuous variables are always given by integrals, whereas those for discrete variables are given by sums. If the domain of X is not the entire real line, it is assumed that $f(x)$ is defined to be equal to 0 for those values outside the specified domain of the variable.

As an illustration, consider the possibility of using $f(x) = ke^{-x}$ as a density function for X where k is some constant. From (i) it is clear that k must be positive. Since the integral of e^{-x} from $-\infty$ to $+\infty$ is infinite, it follows that X must be restricted; hence assume, for example, that X can take on only non-negative values. Then $f(x)$ will be defined to be 0 for negative values and to be given by the formula for non-negative values. From (ii) it then follows that k must be equal to 1 because the integral of e^{-x} from 0 to ∞ is equal to 1. The calculation of, say, $P\{1 < X < 2\}$ would then become

$$\int_{1}^{2} e^{-x}\, dx = e^{-1} - e^{-2} = .23 \ .$$

The graph of this density function and the representation of $P\{1 < X < 2\}$ as an area is given in Fig. 18.

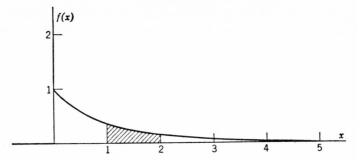

Fig. 18.   Graph of a density function for a continuous variable.

Although $f(x)$ may be chosen at will in any given problem, a choice for which the resulting probabilities are not approximated well by observed relative frequencies is not likely to be a useful choice. As in the case of discrete variables, there are particular density functions that have proved very useful in statistical work and whose explicit formulas are considered later.

The *distribution function*, F, for the continuous variable X is defined by

(31) $$F(x) = P(X \le x) = \int_{-\infty}^{x} f(t)\, dt \ .$$

The graph of $F(x)$ for the preceding illustration is given in Fig. 19. It should be noted that $P\{1 < X < 2\}$ is now given by $F(2) - F(1)$, that is, by the difference of the ordinates on the graph of $F(x)$. Here the graph was constructed by first determining $F(x)$ from definition (31). Thus

$$F(x) = \int_{0}^{x} e^{-t}\, dt = 1 - e^{-x}, \qquad x \ge 0$$
$$= 0, \qquad x < 0 \ .$$

The density function is the one commonly used in the applications of statistical theory; however, the distribution function is also very useful in

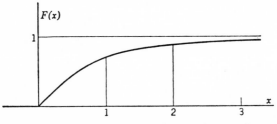

Fig. 19.   Graph of the distribution function for a continuous variable.

deriving some of that theory. For example, it is often easier to find the distribution function of a random variable than it is to find the density function. But after the distribution function has been found, the density function can be obtained by differentiating the distribution function, since, by employing a familiar calculus formula for differentiating an integral with respect to its variable upper limit, it follows from (31) that

$$\frac{dF(x)}{dx} = f(x)$$

provided that $f(x)$ is a continuous function of x. This technique, of course, cannot be used on discrete variable distributions. For such distributions it is necessary to take differences of $F(x)$ values to obtain $f(x)$ values.

16 JOINT CONTINUOUS DENSITY FUNCTIONS

A density function for two or more continuous random variables is the natural generalization of a density function for one variable. Thus, a density function for two continuous random variables X and Y is denoted by $f(x, y)$ and is represented geometrically by a surface in three dimensions, just as a density function $f(x)$ of one variable is represented by a curve in two dimensions. If integrals of $f(x, y)$ are to yield probabilities it is necessary that the total volume under this surface be equal to one and that the volume under this surface lying above a region R in the x, y plane give the probability that the random variables X and Y will assume values corresponding to a point inside this region. These essential properties for a density function of two variables may be formalized as follows:

DEFINITION: *A joint density function for a pair of continuous random variables X and Y is a function f that possesses the following properties*

(i) $$f(x, y) \geq 0$$

(ii) $$\int_{-\infty}^{\infty} \int_{-\infty}^{\infty} f(x, y) \, dx \, dy = 1$$

(iii) $$\iint_R f(x, y) \, dx \, dy = P\{X, Y \in R\} \qquad \text{for all regions } R \,.$$

It is assumed here that the region R is such that the integral of $f(x, y)$ over that region exists. Very often the region R will be a rectangle of the type $a < x < b$ and $c < y < d$, in which case condition (iii) becomes

(32) $$\int_a^b \int_c^d f(x, y) \, dy \, dx = P\{a < X < b \text{ and } c < Y < d\} \,.$$

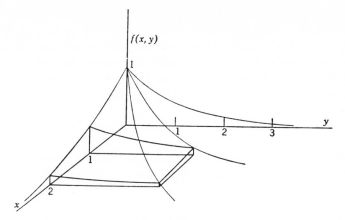

Fig. 20. Graph of a joint density function for two continuous variables.

As an illustration, consider the function $f(x, y) = e^{-(x+y)}$ which is a two-dimensional generalization of the example used in the preceding section. If $f(x, y)$ is defined to be zero for negative values of x or y, it will be observed that (i) and (ii) are satisfied. From (32) the calculation of, say, $P\{1 < X < 2$ and $0 < Y < 2\}$ will then be given by

$$\int_1^2 \int_0^2 e^{-x} e^{-y} \, dy \, dx = (e^{-1} - e^{-2})(e^0 - e^{-2}) \doteq .20 \, .$$

The graph of $f(x, y)$ and the representation of $P\{1 < X < 2$ and $0 < Y < 2\}$ as a volume under the graphed surface is given in Fig. 20.

Two continuous random variables that are unrelated in a probability sense are said to be independently distributed, just as in the case of discrete variables. The definition of independence given in (22) is appropriate whether the two variables are discrete or continuous. Under the assumption that X and Y are discrete variables it was demonstrated that definition (22) is equivalent to definition (23). Although that proof is not applicable to continuous variables, the same equivalence can be shown to hold for continuous variables; therefore definition (23) will be used for both continuous and discrete variables. That is the reason why the word discrete was purposely omitted from definition (23).

Just as in the case of discrete variables, this two-variable definition can be generalized to define independence for n continuous random variables and is identical to the definition given in (24). In anticipation of this fact, the word discrete was also omitted in definition (24).

The density function whose graph is given in Fig. 20 is an illustration of a joint density function of two independent random variables. In the present notation $g(x) = e^{-x}$ and $h(y) = e^{-y}$.

It should be noted that in writing probability statements for continuous variables, such as in (32), it is irrelevant whether, say, $a < x < b$ or $a \leq x \leq b$ is used to determine the desired region because the integral is the same for the two cases. This property does not hold, of course, for discrete variables.

By using integrals in place of sums, formulas can be derived for marginal and conditional distributions just as in the case of discrete variables. Since the derivations are somewhat sophisticated at this stage of the theory and are not needed for some time, they are postponed to a later chapter.

EXERCISES

Sec. 5

1. A die has two of its sides painted red, two black, and two yellow. If the die is rolled twice, describe a two-dimensional sample space for the experiment. What probabilities would you assign to the various sample points?

2. A box contains 4 white and 2 black balls. One ball is drawn from it. Describe a sample space and assign probabilities to the sample points when (a) the balls are also numbered, (b) like colored balls cannot be distinguished.

3. A box contains 3 white and 2 black balls. Two balls are to be drawn from the box. Describe a two-dimensional sample space and assign probabilities to the sample points when (a) the balls are also numbered, (b) like colored balls cannot be distinguished.

4. Show that for the events A, B, C, the probability that at least one of the events will occur is given by $P\{A\} + P\{B\} + P\{C\} - P\{A \cap B\} - P\{A \cap C\} - P\{B \cap C\} + P\{A \cap B \cap C\}$.

Sec. 6

5. If the die of problem 1 is rolled until a red side comes up, describe a sample space for the experiment. What probabilities would you assign to the various sample points?

Sec. 7

6. Two balls are drawn from an urn containing 1 white, 3 black, and 4 green balls. (a) What is the probability that the first is white and the second is black? (b) What is this probability if the first ball is replaced before the second drawing?

7. One urn contains 2 white and 2 black balls; a second urn contains 2 white and 3 black balls. (a) If one ball is chosen from each urn, what is the probability that they will be the same color? (b) If an urn is selected by chance and one ball drawn from it what is the probability the ball will be white? (c) If an urn is selected by chance and two balls drawn from it, what is the probability that they will be the same color?

8. Compare the chances of rolling a 5 with one die and rolling a total of 10 with two dice.

9. Assuming that the ratio of male children is $\frac{1}{2}$, find the probability that in a family of 5 children (a) all children will be of the same sex, (b) the three oldest will be boys and the two youngest will be girls, (c) three of them will be boys and two will be girls.

10. Successive drawings of a card from an ordinary deck are made with replacement each time. How many drawings are necessary before the probability is at least .9 that a spade will be obtained at least once?

11. A box contains two honest coins and one coin that has two heads. A coin is selected at random from the box and tossed twice. (a) What is the probability it will show two heads? (b) If it shows three heads, what is the probability that it is the dishonest coin?

12. Two boxes contain 1 black and 2 white balls, and 2 black and 2 white balls, respectively. One ball is transferred from the first to the second box, after which a ball is drawn from the second box. What is the probability that it is white?

13. A coin is tossed once. If it comes up heads a die is rolled and you are paid the number showing in dollars. If it comes up tails two dice are rolled and you are paid in dollars the sum of the numbers showing. What is the probability that you will be paid at most 5 dollars?

14. A, B, and C in order toss a coin. The first one to throw a head wins. What are their respective chances of winning? Note that the game may continue indefinitely.

15. Fourteen quarters and one five-dollar gold piece are in one purse and 15 quarters are in another purse. Five coins are taken from the first purse and placed in the second, and then 5 coins are taken from the second and placed in the first. What is the probability that after these transactions the gold coin will be found in the first purse?

16. A card is drawn from an ordinary deck. What is the probability that it is a king, given that it is a face card?

17. Two dice are rolled. What is the probability that the sum of the faces exceeds 8, given that one or more of the faces is a 6?

18. A box contains 2 red tickets numbered 1 and 2, and 3 green tickets numbered 1, 2, and 3. If two tickets are drawn from the box, what is the probability that both will be red, given that one of them is known to be (a) red, (b) the red ticket numbered 1?

19. A bag contains 2 black balls and 2 white balls. A ball is drawn and replaced by a ball of the opposite color. Then another ball is drawn from the bag. Find the conditional probability that the first ball drawn was white, given that the second ball drawn was white.

Sec. 8

20. A group of businessmen consists of 40 percent Democrats and 60 percent Republicans. If 30 percent of the Democrats and 50 percent of the Republicans

smoke cigars, what is the probability that a cigar-smoking businessman is a Republican?

21. A test for detecting cancer which appears promising has been developed. Suppose it was found that 97 percent of the cancer patients in a large hospital reacted positively to the test whereas only 5 percent of those not having cancer did so. If 2 percent of the patients in the hospital have cancer, what is the probability that a patient selected by chance who reacts positively to the test will actually have cancer?

22. Each of 3 boxes has 2 drawers. One box contains a gold coin in each drawer, another contains a silver coin in each drawer, and the third contains a gold coin in one drawer and a silver coin in the other. A box is chosen, a drawer is opened and found to contain a gold coin. What is the probability that the coin in the other drawer is also gold?

Sec. 9

23. If a player is dealt three spades and two hearts and is permitted to draw two cards from the rest of the deck, what is the probability that the two cards will be spades?

24. If a bridge player has two aces in his hand, what is the probability that his partner has (a) one ace? (b) two aces?

25. If a poker hand of 5 cards is drawn from a deck, what is the probability that it will contain 3 aces?

26. What is the probability that a bridge hand will contain 13 cards of the same suit?

27. If a box contains 40 good and 10 defective fuses and 7 fuses are selected, what is the probability that all of them will be good?

28. From a group of 40 people, 3 are to be chosen. Find the probability that none of 10 particular people in the group will be chosen.

29. If you hold 2 tickets to a lottery for which n tickets were sold and 5 prizes are to be given, what is the probability that you will win at least one prize?

30. What is the probability that the bridge hands of north and south together contain exactly 2 aces?

31. If a bridge player and his partner have 9 spades between them, what is the probability that the 4 spades held by their opponents will be split 3 and 1?

32. Find the probability that a poker hand of 5 cards will contain only black cards, (a) given that it contains at least 4 black cards, (b) given that it contains at at least 4 spades.

33. Find the probability that a poker hand of 5 cards will contain no card smaller than 8, given that it contains at least 3 cards over 10, where aces are treated as low cards.

34. Let X and Y denote the respective number of heads that will be obtained in tossing 2 coins twice. Calculate the probability that $Y - X$ will be less than 1.

35. A tosses 3 coins and B tosses 2 coins, simultaneously. The one with the

greater number of heads wins. (a) What is the probability that A will win? (b) What is this probability if the experiment is repeated whenever a tie occurs?

36. Show that $\begin{pmatrix} n \\ r-1 \end{pmatrix} + \begin{pmatrix} n \\ r \end{pmatrix} = \begin{pmatrix} n+1 \\ r \end{pmatrix}.$

Sec. 12

37. Given the discrete density $f(x) = e^{-1}/x!$, $x = 0, 1, 2, \cdots$, (a) calculate $P\{X = 2\}$, (b) calculate $P\{X < 2\}$, (c) show that e^{-1} is the proper constant for this density.

38. A coin is tossed until a head appears. (a) What is the probability that a head will first appear on the fourth toss? (b) What is the probability, $f(x)$, that x tosses will be required to produce a head? (c) Graph the density function $f(x)$.

39. If the probability is $\frac{1}{2}$ that a finesse in bridge will be successful, (a) what is the probability that 2 out of 4 such finesses will be successful? (b) What is the probability, $f(x)$, that x out of 4 finesses will be successful? (c) Graph the density function $f(x)$ as a line chart.

40. Graph the distribution function $F(x)$ for the density obtained in problem 38.

41. Graph the distribution function $F(x)$ for the density obtained in problem 39.

42. Two dice are rolled. Let X be the difference of the face numbers showing, the higher minus the lower, and 0 for ties. Find the density function of X.

43. A box contains 4 red and 2 black balls. Two balls are drawn. Let X be the number of red balls obtained. Find the density function of X and also its distribution function.

44. A die is tossed once. If a 5 or 6 comes up, let X equal the number showing. If a 1, 2, 3, or 4 comes up, toss the die again and let X equal the sum of the two numbers obtained. Find the density function of X.

45. In the game of odd man wins, 3 people toss coins. The game continues until someone has an outcome different from the other two. The individual with the different outcome wins. Let X equal the number of games needed before a decision is reached. Find the density function of X.

46. There are N tickets numbered $1, 2, \cdots, N$, from which n are chosen. Let X equal the smallest number appearing on the ticket drawn. Find the density function of X.

Sec. 13

47. Let X and Y denote the respective number of heads obtained in tossing a coin twice. Find an expression for the density function $f(x, y)$.

48. Let X and Y denote the respective number of heads obtained in tossing a pair of coins twice. Find an expression for the density function $f(x, y)$.

49. Five dice are rolled. Let X denote the number of 1's and Y the number of 2's that show. Find an expression for the density function $f(x, y)$.

50. Four cards are drawn from a deck. Let X denote the number of aces and Y the number of kings that show. Find an expression for the density function $f(x, y)$.

51. Let X and Y denote the number of spades and hearts, respectively, in a bridge hand of 13 cards. Find the density $f(x, y)$.

Sec. 14

52. For the first illustration of section 13, calculate the values of (a) $f(1)$, (b) $g(0)$, (c) $f(y \mid 1)$, (d) $g(x \mid 0)$. Here f and g are the densities of X and Y.

53. For the distribution given by (21) in section 13, find $f(0)$ and $f(y \mid 0)$.

54. For the density of problem 47 find $f(x)$ and $f(y \mid x)$. Comment.

55. For the distribution of problem 48, find the conditional distribution density $g(x \mid y)$.

56. Calculate the marginal values $f(2)$ and $g(4)$ for problem 49.

57. Use a result from problem 56 to obtain an expression for the conditional density $f(y \mid 2)$ for the distribution of problem 49.

58. Consider a deck of cards consisting of the ace, king, queen, and jack of each of the four suits. If 2 cards are drawn from this deck, and X and Y denote the number of spades and hearts obtained, find (a) the joint density $f(x, y)$, (b) the marginal density $f(x)$, (c) the conditional density, $f(y \mid 1)$, of Y for $X = 1$.

59. Given $f(x, y) = c(x + y)$ at the points $(1, 1)$, $(2, 1)$, $(2, 2)$, $(3, 1)$, and zero elsewhere, (a) evaluate c, (b) find $f(x)$, (c) find $f(y \mid x)$.

60. Explain why two variables X and Y cannot be independently distributed, regardless of the nature of $f(x, y)$, if the region in the x, y plane where $f(x, y)$ is positive is the triangular region of Fig. 16.

61. If $f(x) = e^{-\mu} \mu^x / x!$, $x = 0, 1, 2, \cdots$, and $f(y \mid x) = \binom{x}{y} p^y (1-p)^{x-y}$, $y = 0, 1, 2, \cdots, x$, show that the marginal density of Y is given by $g(y) = e^{-\mu p} (\mu p)^y / y!$.

Sec. 15

62. Given the continuous density $f(x) = cxe^{-x}$, $x > 0$, (a) determine the value of c, (b) calculate $P\{X < 2\}$, (c) calculate $P\{2 < X < 3\}$.

63. Given the continuous density $f(x) = c$, $1 < x < 3$, (a) determine the value of c, (b) calculate $P\{X < 2\}$, (c) calculate $P\{X > 1.5\}$.

64. If $f(x)$ is the density of the random variable X, find, and graph, the distribution function $F(x)$ in each case.

(a) $f(x) = \frac{1}{3}$, $x = 0, 1, 2$

(b) $f(x) = 3(1 - x)^2$, $0 < x < 1$

(c) $f(x) = 1/x^2$, $x > 1$.

65. Find the distribution function $F(x)$ and graph it if the density of X is (a) $f(x) = \frac{1}{2}$, $0 \leq x \leq 2$, (b) $f(x) = x$ for $0 \leq x \leq 1$ and $f(x) = -x + 2$ for $1 < x \leq 2$, (c) $f(x) = [\pi(1 + x^2)]^{-1}$.

66. If $f(x) = \frac{1}{2}e^{-x/2}$, $x > 0$, find a number x_0 such that $P\{X > x_0\} = \frac{1}{2}$.

67. Suppose the life in hours, X, of a type of radio tube has the density $f(x) = c/x^2$, $x > 100$. (a) Evaluate c. (b) Find $F(x)$. (c) Calculate $P\{X > 500\}$.

68. Suppose the probability that an atom of a radioactive material will disintegrate in time t is given by $1 - e^{-at}$, where a is a constant depending on the material. Find the density function of X, the length of life of such an atom.

69. If the radioactive material of problem 68 will disintegrate in 1000 units of time, calculate the probability that the life of an atom of this material will exceed 2000 units of time.

Sec. 16

70. Given the density $f(x, y) = 8xy$, $0 < x < 1$, $0 < y < x$, calculate (a) $P\{X < .5, Y < .5\}$, (b) $P\{X < .5\}$, (c) $P\{Y < .5\}$. (d) From the preceding calculations, what can you say about the independence of X and Y?

71. Given the density $f(x, y) = xye^{-(x+y)}$ $x > 0$, $y > 0$, calculate $P\{X > 1, Y < 1\}$.

72. If X and Y are independent random variables with the same continuous distribution function F, find an expression for $P\{X < t, Y < t\}$. Use this to find the distribution function, $G(z)$, of the variable $Z = \max\{X, Y\}$.

73. Give an example of two random variables X and Y that are not independent but such that X^2 and Y^2 are independent.

CHAPTER 3

Some Particular Probability Distributions

1 INTRODUCTION

The purpose of this chapter is to introduce a few of the probability distributions that have proved particularly useful as models for real-life phenomena. Some of the distributions are for discrete variables, others are for continuous variables. In every case the distribution will be specified by presenting the probability density function of the random variable.

In many problems it suffices to consider only certain properties of a distribution rather than to study the entire distribution. In particular, it often suffices to be given the low order moments of a distribution. This chapter, therefore, will be concerned with the moments of particular distributions as well as with their densities.

2 DISCRETE VARIABLES

Most of the discrete variables that occur in repetitive type experiments are the counting type. For example, the random variable might be the number of accidents a car owner has per year, or the number of insects surviving a spraying, or the number of counts recorded by a Geiger counter in a given time interval. Such random variables assume only non-negative integer values. Although the discrete variables to be presented in this chapter are of the counting type, the notation employed will be sufficiently general to include other types of discrete random variables as well.

2.1 Expectation

Before considering particular densities for discrete variables, a brief discussion of moments of a distribution will be given so that the moments of those particular distributions may be computed. The moment of mass and the moment of inertia are two moments with which the student should already be familiar from his calculus course. Rather than restrict the discussion to moments, however, a more general concept called *expected value* will be introduced and defined. It includes moments as a special case, but it also serves as a very useful tool for studying other properties of a distribution.

Consider a game in which you toss three honest coins and receive one dollar for each head that shows. How much money should you expect to get if you were permitted to play this game once? If X denotes the amount that will be won, then X must assume one of the values 0, 1, 2, or 3 with corresponding probabilities $\frac{1}{8}$, $\frac{3}{8}$, $\frac{3}{8}$, and $\frac{1}{8}$. Intuitively, therefore, you should expect to get 0 dollars $\frac{1}{8}$ of the time, 1 dollar $\frac{3}{8}$ of the time, 2 dollars $\frac{3}{8}$ of the time, and 3 dollars $\frac{1}{8}$ of the time, if the game were played a large number of times. You should therefore expect to average the amount

$$0 \cdot \tfrac{1}{8} + 1 \cdot \tfrac{3}{8} + 2 \cdot \tfrac{3}{8} + 3 \cdot \tfrac{1}{8} = 1\tfrac{1}{2} .$$

This amount, namely \$1.50, is what is commonly called the expected amount to be won if the game is played once. This example is an illustration of the general concept of expected value that follows.

Suppose the discrete random variable X must assume one of the values x_1, x_2, $\cdots$, x_k and that the probabilities associated with those values are $f(x_1)$, $f(x_2)$, $\cdots$, $f(x_k)$. Then the expected value of X is defined by the formula

$$(1) \qquad\qquad E[X] = \sum_{i=1}^{k} x_i f(x_i) .$$

In the preceding illustration the random variable X is the amount of money in dollars to be won in tossing three coins, the possible values of X are given by $x_1 = 0$, $x_2 = 1$, $x_3 = 2$, and $x_4 = 3$, and the corresponding probabilities are $\frac{1}{8}$, $\frac{3}{8}$, $\frac{3}{8}$ and $\frac{1}{8}$.

Now suppose the preceding game is altered so that you win the amount $g(x_i)$ instead of x_i when the value x_i is obtained. For example, $g(x)$ might be chosen to be the function $g(x) = x^2$, which means that you would receive in dollars the square of the number of heads that show. Then the expected value of the game to you would be given by the formula

$$(2) \qquad\qquad E[g(X)] = \sum_{i=1}^{k} g(x_i) f(x_i) .$$

Both of these formulas assume that there are only a finite number of possible values for the random variable X. To eliminate this restriction, the following more general definition which is applicable to any discrete random variable is introduced.

(3) DEFINITION: *The expected value of the function g(X) of the discrete random variable X, whose density function is f, is given by*

$$E[g(X)] = \sum_{i=1}^{\infty} g(x_i)f(x_i) \,.$$

It is understood in this definition that x_1, x_2, $\cdots$ are the possible values of X and $f(x_1), f(x_2), \cdots$ are the corresponding probabilities.

The expected value of the random variable X is usually called the mean of the random variable, or the mean of the distribution of the random variable X. If X can assume only a finite number, say k, of possible values the upper index in this sum will, of course, be k rather than ∞. When X possesses an infinite number of possible values with positive probabilities, it is necessary to assume that the infinite series in (3) converges. To avoid mathematical difficulties, the slightly stronger assumption that the series converges absolutely is made.

2.2 Moments

The moments of the distribution of a discrete random variable X are readily defined in terms of expected values as follows.

(4) DEFINITION: *The kth moment about the origin of the distribution of the discrete random variable X whose density is f is given by*

$$\mu_k' = E[X^k] = \sum_{i=1}^{\infty} x_i^k f(x_i) \,.$$

The kth moment of a distribution is also commonly called the kth moment of the random variable X whose distribution is being studied. Thus, one may speak of μ_k' as being the kth moment of X or as the kth moment of the distribution of X.

The first moment μ_1', which is $E[X]$, occurs so often that it is given the special symbol μ.

Since moments about the mean are used extensively, they also need to be defined. In terms of the symbol μ, moments about the mean are defined as follows:

(5) DEFINITION: *The kth moment about the mean of the distribution of the discrete random variable X whose density is f is given by*

$$\mu_k = E[(X - \mu)^k] = \sum_{i=1}^{\infty} (x_i - \mu)^k f(x_i) \,.$$

The moments of a distribution are very useful for assisting in the description of a distribution when the density function is not available. A famous problem in mathematics, called the moment problem, is that of determining when a knowledge of all the moments of a distribution uniquely specifies the distribution.

A large share of the problems involving moments will be concerned with only the first two moments because they often suffice to describe two useful properties of the distribution, namely where the distribution is centered and the degree to which the distribution is concentrated about this center. The first moment, μ, is used to determine where the distribution is centered and the second moment about the mean, μ_2, is used to determine the degree of concentration of the distribution about μ. Because the second moment about the mean is used so frequently in this connection it is denoted by the special symbol σ^2 and is called the *variance* of the distribution. The positive square root of the variance, namely σ, is called the *standard deviation* of the distribution. It is employed in place of the variance when a measure of concentration in the same units as the random variable is desired.

For the purpose of observing how μ and σ^2 help to describe a discrete probability distribution, let the probabilities $f(x_1), \cdots, f(x_k)$ associated with the values $x_1, \cdots, x_k$ of the random variable X be represented geometrically as point masses, totaling one, located at the points $x_1, \cdots, x_k$ on the x axis. This representation is shown in Fig. 1. The first moment, μ, then gives the center of gravity of this set of point masses and therefore may serve as a measure of where the distribution is located, or centered.

The second moment about the mean, σ^2, will tend to be small if most of the masses are concentrated near μ and there are no masses located far removed from μ. The more dispersed the masses are about the mean the larger σ^2 is likely to become. Thus, the variance of a distribution may serve as a measure of the degree to which the distribution is concentrated about the mean. It is easy to give illustrations of distributions for which it is clear that σ^2 is a poor measure of the concentration of the distribution about its mean; nevertheless, for a large share of the more common distributions it does serve this role satisfactorily.

As an illustration for which σ^2 seems appropriate as a relative measure of concentration, consider the two probability distributions shown in Figs. 2

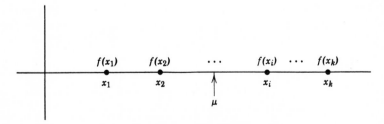

Fig. 1. Distribution of probability masses.

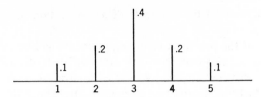

Fig. 2.   A probability distribution for which $\sigma^2 = 1.2$.

and 3. The probabilities are represented by vertical lines whose heights are proportional to the probabilities. For both distributions $\mu = 3$. Calculations based on formula (5) with $k = 2$ will show that σ^2 has the respective values 1.2 and 2.0 for the two distributions, indicating that the first distribution is more highly concentrated about its mean than is the second distribution.

As an illustration where the value of σ^2 does not seem to indicate the nature of the concentration of the distribution about its mean, consider the distribution shown in Fig. 4. Calculations give the value $\sigma^2 = 5.1$, which when compared to the values obtained for Figs. 2 and 3 would indicate poor concentration of the distribution about its mean, and yet most of the distribution is concentrated near the mean. By placing a very small amount of probability mass sufficiently far away from the center of the distribution, the value of σ^2 can be made as large as desired.

In spite of the limitations of σ^2 as a measure of the concentration of a distribution about its mean, as indicated by the preceding illustration, the mean and variance have proved to be very useful quantities in treating the particular distributions that are to be presented in this chapter.

In evaluating σ^2 it is usually more convenient to evaluate the first two moments about the origin and then calculate σ^2 from them rather than evaluate it directly from definition (5). This is accomplished by expanding the binomial in (5) for $k = 2$ in the following manner.

$$\mu_2 = \sum_{i=1}^{\infty} (x_i - \mu)^2 f(x_i)$$

$$= \sum_{i=1}^{\infty} x_i^2 f(x_i) - 2\mu \sum_{i=1}^{\infty} x_i f(x_i) + \mu^2 \sum_{i=1}^{\infty} f(x_i) .$$

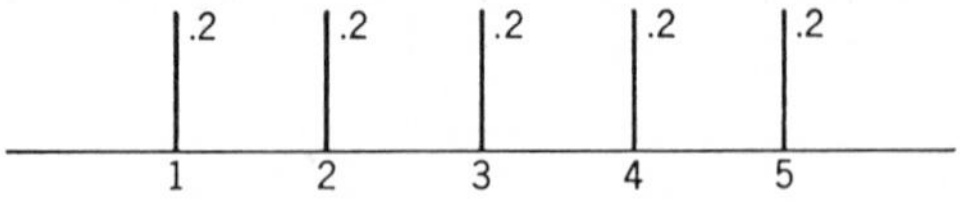

Fig. 3.   A probability distribution for which $\sigma^2 = 2.0$.

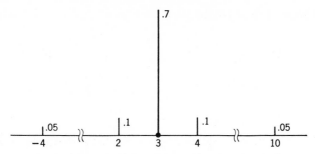

Fig. 4.   A probability distribution for which $\sigma^2 = 5.1$.

But from (4) this may be written

$$\mu_2 = \mu_2' - 2\mu\mu + \mu^2.$$

Combining terms gives the desired formula, namely

(6) $$\mu_2 = \mu_2' - \mu^2.$$

2.3   Moment Generating Function

Even though the direct computation of theoretical moments from definition (4) may be easy, it is convenient for later theory to be able to calculate such moments indirectly by another method. This method is introduced here and used throughout several chapters for proving theorems. It involves what is known as the moment generating function. As the name implies, the moment generating function is a function that generates moments. It is defined as follows:

(7)   DEFINITION: *The moment generating function of the discrete random variable X whose density is f is given by*

$$M_X(\theta) = E[e^{\theta X}] = \sum_{i=1}^{\infty} e^{\theta x_i} f(x_i).$$

This series is a function of the parameter θ only, but the subscript is placed on $M(\theta)$ to show what variable is being considered. The parameter θ has no real meaning here; it is merely a mathematical device introduced to assist in the determination of moments.

In order to see how $M_X(\theta)$ does produce moments, assume that f is a density for which the series in (7) converges. Now expand $e^{\theta x_i}$ in a power series and sum term by term. Since the power series for e^z is

$$e^z = 1 + z + \frac{z^2}{2!} + \frac{z^3}{3!} + \cdots$$

it follows from (7) and (4) that

$$M_X(\theta) = \sum_{i=1}^{\infty} \left[1 + \theta x_i + \frac{\theta^2 x_i^2}{2!} + \frac{\theta^3 x_i^3}{3!} + \cdots \right] f(x_i)$$

$$= \sum_{i=1}^{\infty} f(x_i) + \theta \sum_{i=1}^{\infty} x_i f(x_i) + \frac{\theta^2}{2!} \sum_{i=1}^{\infty} x_i^2 f(x_i) + \cdots$$

$$(8) \qquad = 1 + \theta \mu_1' + \frac{\theta^2}{2!} \mu_2' + \frac{\theta^3}{3!} \mu_3' + \cdots .$$

It will be observed that the coefficient of $\theta^k/k!$ in this expansion is the kth moment about the origin; consequently, if the moment generating function can be found for a variable X and can be expanded into a power series in θ, the moments of the variable can be obtained by merely inspecting the expansion. If a particular moment is desired, it may be more convenient to evaluate it by computing the proper derivative of $M_X(\theta)$ at $\theta = 0$, since repeated differentiation of (8) will show that

$$(9) \qquad \mu_k' = \frac{d^k M}{d\theta^k} \bigg|_{\theta=0} .$$

Applications of the preceding definitions are begun in the next section.

2.4 Binomial Distribution

Consider an experiment of the repetitive type in which only the occurrence or nonoccurrence of an event is recorded. Suppose the probability that the event will occur when the experiment is performed is p. Let $q = 1 - p$ denote the probability that it will fail to occur. If the event occurs at a given trial of the experiment, it will be called a success, otherwise a failure. Let n independent trials be made and denote by X the number of successes that will be obtained in the n trials. Then consider the problem of determining the probability of obtaining precisely x successes in n trials of the experiment. A formula for this probability would be needed, for example, if one knew that the probability of a marksman hitting a target is $\frac{1}{10}$ and if one wished to calculate the probability of getting at least two hits in taking 20 shots at the target.

For the purpose of deriving the desired formula, first determine the probability of obtaining x consecutive successes followed by $n - x$ consecutive failures. These n events are independent; therefore, by (10), Chapter 2, this probability is

$$\overbrace{p \cdot p \cdots p}^{x} \cdot \overbrace{q \cdot q \cdots q}^{n-x} = p^x q^{n-x} .$$

The probability of obtaining precisely x successes and $n - x$ failures in some other order of occurrence is the same as in this particular sequence because the p's and q's are merely rearranged to correspond to the other order. To solve the problem, it is therefore necessary to count the number of orders.

The number of orders is the number of permutations possible with n letters of which x are alike (p's) and the remaining $n - x$ are alike (q's). But by formula (18), Chapter 2, the number of such permutations is equal to

$$(10) \qquad \frac{n!}{x!\,(n - x)!} .$$

Now, by (3), Chapter 2, the probability that one or the other of a set of mutually exclusive events will occur is the sum of their separate probabilities; consequently it is necessary to add $p^x q^{n-x}$ as many times as there are different orders in which the desired result can occur. Since (10) gives the number of such orders, the probability of obtaining x successes in some order is therefore given by multiplying $p^x q^{n-x}$ by the quantity in (10). The resulting probability, which is that of obtaining x successes in n independent trials of an experiment for which p is the probability of success in a single trial, defines what is known as the binomial distribution.

$$(11) \qquad \text{BINOMIAL DISTRIBUTION:} \quad f(x) = \frac{n!}{x!\,(n - x)!}\, p^x q^{n-x} .$$

The name binomial comes from the relationship of (11) to the following binomial expansion:

$$(12) \qquad (q + p)^n = q^n + nq^{n-1}p + \frac{n(n - 1)}{2}\, q^{n-2}p^2 + \cdots + p^n .$$

The term in this expansion involving p^x, where x is an integer, is given by

$$\frac{n(n - 1) \cdots (n - x + 1)}{1 \cdot 2 \cdots x}\, q^{n-x}p^x = \frac{n!}{x!\,(n - x)!}\, p^x q^{n-x} .$$

As a result, one may write

$$(q + p)^n = \sum_{x=0}^{n} \frac{n!}{x!\,(n - x)!}\, p^x q^{n-x} .$$

But from (11) it will be observed that the general term in this expansion is precisely the value of $f(x)$ given in (11) and therefore that (12) may be written in the form

$$(q + p)^n = \sum_{x=0}^{n} f(x) .$$

Thus the various terms in the binomial expansion of $(q + p)^n$ give the probabilities of the various possible results in their natural order.

The binomial density is an example of a mathematical model that can be applied to many real-life problems involving a discrete variable. In any given application it is necessary to know or to estimate the values of the two parameters p and n before (11) can be used.

2.4.1 Illustrations. As illustrations of the direct application of formula (11), first consider two impractical problems related to the rolling of a die. If a true die is rolled five times, what is the probability that precisely two of the rolls will show 1's? Here success consists in obtaining a 1; hence $p = \frac{1}{6}, q = \frac{5}{6}$, and $n = 5$. When (11) is applied, the solution is

$$f(2) = \frac{5!}{2!\,3!}\left(\frac{1}{6}\right)^2\left(\frac{5}{6}\right)^3 = .16 .$$

It the die is rolled five times, what is the probability of obtaining at most two 1's? To answer this question it is necessary to compute the probabilities of obtaining precisely no 1's, one 1, and two 1's. Applying (11),

$$f(0) = \frac{5!}{0!\,5!}\left(\frac{1}{6}\right)^0\left(\frac{5}{6}\right)^5 = .40$$

$$f(1) = \frac{5!}{1!\,4}\left(\frac{1}{6}\right)^1\left(\frac{5}{6}\right)^4 = .40 .$$

Since these three possibilities are mutually exclusive events, it follows that

$$P\{X \le 2\} = f(0) + f(1) + f(2) = .96 .$$

As a somewhat more earthy problem, consider the one mentioned just before the derivation of (11), namely, that of calculating the probability of getting at least two hits on a target in taking 20 shots at it if the probability of a hit for a single shot is $\frac{1}{10}$. Here $p = \frac{1}{10}$ and $n = 20$; hence

$$P\{X \ge 2\} = 1 - f(0) - f(1)$$
$$= 1 - \left(\frac{9}{10}\right)^{20} - 20\left(\frac{1}{10}\right)^1\left(\frac{9}{10}\right)^{19}$$
$$= 1 - .122 - .270$$
$$= .608 .$$

The validity of using the binomial model in the last illustration is not so obvious as it is in the first two illustrations. The derivation of the binomial formula was based on independent trials with p constant from trial to trial. If the same man takes repeated shots at the same target, it might be expected that his chances of making a hit would increase somewhat with practice.

If a different man were used each time, p would undoubtedly change from trial to trial. Possible deviations in the basic assumptions should be taken into account when interpreting a resulting probability such as .608.

2.4.2 Binomial Moments. The first two moments of the binomial distribution will be needed shortly; therefore consider their computation. In order to illustrate the two methods for computing moments, these moments are calculated directly from definition and indirectly by means of the moment generating function.

If (1) is applied to (11) and if a few algebraic manipulations are made, it will be seen that

$$(13) \qquad E[X] = \mu = \sum_{x=0}^{n} x \frac{n!}{x!\,(n-x)!} p^x q^{n-x}$$

$$= \sum_{x=1}^{n} x \frac{n!}{x!\,(n-x)!} p^x q^{n-x}$$

$$= \sum_{x=1}^{n} \frac{n!}{(x-1)!\,(n-x)!} p^x q^{n-x} .$$

If n and p are factored out, this becomes

$$\mu = np \sum_{x=1}^{n} \frac{(n-1)!}{(x-1)!\,(n-x)!} p^{x-1} q^{n-x} .$$

Letting $y = x - 1$, the right side can be written

$$\mu = np \sum_{y=0}^{n-1} \frac{(n-1)!}{y!\,(n-1-y)!} p^y q^{n-1-y} .$$

But by (11) the quantity being summed is the binomial probability of y successes in $n - 1$ trials. Since the sum is over all possible values of y, the sum must equal one; hence $\mu = np$.

The second moment is calculated in a similar manner by using the identity $x^2 = x(x - 1) + x$. From (4) and (13), it follows that

$$\mu_2' = \sum_{x=0}^{n} x^2 \frac{n!}{x!\,(n-x)!} p^x q^{n-x}$$

$$= \sum_{x=0}^{n} [x(x-1) + x] \frac{n!}{x!\,(n-x)!} p^x q^{n-x}$$

$$= \sum_{x=0}^{n} x(x-1) \frac{n!}{x!\,(n-x)} p^x q^{n-x} + \mu .$$

Since the terms for $x = 0$ and $x = 1$ are equal to 0 because of the factor

$x(x - 1)$, the summation can begin with $x = 2$; hence

$$\mu_2' = \sum_{x=2}^{n} x(x - 1) \frac{n!}{x!\,(n - x)!}\, p^x q^{n-x} + \mu$$

$$= \sum_{x=2}^{n} \frac{n!}{(x - 2)!\,(n - x)!}\, p^x q^{n-x} + \mu .$$

If $n(n - 1)p^2$ is factored out, this becomes

$$\mu_2' = n(n - 1)p^2 \sum_{x=2}^{n} \frac{(n - 2)!}{(x - 2)!\,(n - x)}\, p^{x-2} q^{n-x} + \mu .$$

Letting $z = x - 2$, the right side can be written as

$$\mu_2' = n(n - 1)p^2 \sum_{z=0}^{n-2} \frac{(n - 2)!}{z!\,(n - 2 - z)!}\, p^z q^{n-2-z} + \mu .$$

The quantity being summed is the probability of z successes in $n - 2$ trials. Since the sum is over all possible values of z, its value must be one. Using this result and the earlier result that $\mu = np$, μ_2' reduces to

(14) $$\mu_2' = n(n - 1)p^2 + np .$$

If formula (6) is applied to the results just obtained for the binomial distribution,

$$\mu_2 = n(n - 1)p^2 + np - n^2 p^2$$
$$= -np^2 + np$$
$$= npq .$$

These calculations show that the mean and the standard deviation of a binomial distribution are given by the formulas

(15) $$\mu = np$$
$$\sigma = \sqrt{npq} .$$

Now consider the computation of these moments by means of the moment generating function. If (7) is applied to (11),

$$M_X(\theta) = \sum_{x=0}^{n} e^{\theta x} \frac{n!}{x!\,(n - x)!}\, p^x q^{n-x}$$

$$= \sum_{x=0}^{n} \frac{n!}{x!\,(n - x)!}\, (pe^\theta)^x q^{n-x} .$$

But from (12) this sum can be written as a binomial raised to the nth power because the expansion is purely algebraic and need not be interpreted in terms of probabilities. Hence

(16) $$M_X(\theta) = (q + pe^\theta)^n .$$

The desired moments may be obtained by applying (9). If (16) is differentiated twice with respect to θ and terms are combined,

$$M_X'(\theta) = npe^\theta(q + pe^\theta)^{n-1}$$

and

$$M_X''(\theta) = npe^\theta(q + pe^\theta)^{n-2}(q + npe^\theta) .$$

The values of these derivatives at $\theta = 0$ are np and $np(q + np)$, respectively; hence they are the values of μ and μ_2' respectively. If q is replaced by $1 - p$, it will be observed that μ_2' here agrees with the value obtained in (14). For this problem the moments are easier to obtain indirectly by means of the moment generating function than directly from definition.

2.5 Poisson Distribution

If the number of trials n is large, the computations involved in using formula (11) become quite lengthy; therefore, a convenient approximation to the binomial distribution would be very useful. It turns out that for large n there are two well-known density functions that give good approximations to the binomial density function: one when p is very small and the other when this is not the case. The approximation that applies when p is very small is known as the Poisson density function and is defined as follows:

(17)   POISSON DISTRIBUTION: $$f(x) = \frac{e^{-\mu}\mu^x}{x!} .$$

It will presently be seen that the parameter μ is the mean of the distribution; hence, it is proper to label it μ. Although the Poisson distribution is being introduced here as an approximation to the binomial distribution, it is a well-known and useful distribution in its own right and therefore should not be regarded as merely an approximation for the binomial distribution. It has been named after a pioneer in the theory of probability.

2.5.1 Poisson Approximation to the Binomial. In order to verify the fact that (17) does serve as a good approximation to the binomial distribution for very large n and very small p, consider what happens to the binomial density function when n becomes infinite and p approaches zero in such a manner that the mean $\mu = np$ remains fixed.

First, rewrite (11) as follows:

$$f(x) = \frac{n(n-1)\cdots(n-x+1)}{x!} p^x(1-p)^{n-x} .$$

If numerator and denominator are multiplied by n^x and the indicated algebraic manipulations are performed,

$$(18) \qquad f(x) = \frac{n(n-1)\cdots(n-x+1)}{n^x x!}(np)^x(1-p)^{n-x}$$

$$= \frac{n(n-1)\cdots(n-x+1)}{n \cdot n \cdots n}\frac{\mu^x}{x!}(1-p)^{n-x}$$

$$= \left(1-\frac{1}{n}\right)\left(1-\frac{2}{n}\right)\cdots\left(1-\frac{x-1}{n}\right)\frac{\mu^x}{x!}(1-p)^{n-x}$$

$$= \frac{\left(1-\frac{1}{n}\right)\left(1-\frac{2}{n}\right)\cdots\left(1-\frac{x-1}{n}\right)}{(1-p)^x}\frac{\mu^x}{x!}(1-p)^n .$$

Next, express $(1-p)^n$ in the form

$$(1-p)^n = [(1-p)^{-\frac{1}{p}}]^{-np} = [(1-p)^{-\frac{1}{p}}]^{-\mu} .$$

Now, from the definition of e,

$$\lim_{z \to 0}(1+z)^{\frac{1}{z}} = e ;$$

hence, letting $z = -p$,

$$\lim_{p \to 0}[(1-p)^{-\frac{1}{p}}]^{-\mu} = e^{-\mu} .$$

Furthermore,

$$\lim_{n \to \infty}\frac{\left(1-\frac{1}{n}\right)\left(1-\frac{2}{n}\right)\cdots\left(1-\frac{x-1}{n}\right)}{(1-p)^x} = 1$$

because $p \to 0$ as $n \to \infty$ when $np = \mu$ is fixed. By applying these two results to the right side of (18), it will be seen that

$$\lim_{n \to \infty}f(x) = \frac{e^{-\mu}\mu^x}{x!} .$$

This result may be expressed as a theorem.

THEOREM 1: *If the probability of success in a single trial p approaches 0 while the number of trials n becomes infinite in such a manner that the mean* $\mu = np$ *remains fixed, then the binomial distribution will approach the Poisson distribution with mean* μ.

Figures 5 and 6 were constructed to indicate how rapidly the binomial distribution approaches the Poisson distribution. The broken lines represent the fixed Poisson distribution for μ chosen equal to 4 and the solid lines the

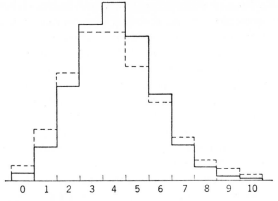

Fig. 5. Binomial (——) and Poisson (– – –) distribution for $\mu = 4$ and $p = \frac{1}{3}$.

binomial distribution for $p = \frac{1}{3}$ and $p = \frac{1}{24}$, respectively. It would appear from inspecting these graphs that the Poisson approximation should be sufficiently accurate for most applications if $n \geq 100$ and $p \leq .05$.

Since the Poisson distribution was obtained from the binomial distribution by holding $np = \mu$ fixed and allowing n to become infinite, it follows that the moments of the Poisson distribution can be obtained from the moments of the binomial distribution by calculating the limiting values of the latter. Thus, the mean of the Poisson distribution must be μ and the variance must be the limiting value of npq. But with $np = \mu$ fixed, p will approach zero as n becomes infinite; therefore

$$\lim_{n \to \infty} npq = \lim_{n \to \infty} \mu q = \lim_{p \to 0} \mu(1 - p) = \mu .$$

This gives the interesting result that the variance of a Poisson variable is equal to its mean.

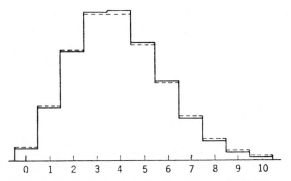

Fig. 6. Binomial (——) and Poisson (– – –) distributions for $\mu = 4$ and $p = \frac{1}{24}$.

Although the Poisson distribution has been introduced by means of its approximating property for the binomial distribution, it is a very useful model for treating certain types of problems unrelated to the binomial distribution. For example, the Poisson distribution has been found to be a satisfactory model for the number of disintegrating atoms from a radioactive material, for the number of telephone calls on a line in a fixed time interval, for the number of meteorites found on an acre of desert land, or for the number of flaws on a sheet of metal. These problems are the type in which a random variable is distributed over time or space. If it is assumed that the number of events occurring in nonoverlapping time intervals are independent, that the probability of a single event occurring in a small time interval is approximately proportional to the size of the interval, and that the probability of more than one event occurring in a small interval is negligible in comparison with the probability of one event occurring in that interval, then it can be shown by calculus techniques, when the preceding assumptions have been made mathematically precise, that the number of occurrences in any fixed size time interval will possess a Poisson distribution. The same derivation can be applied when time intervals are replaced by spatial intervals; therefore the number of occurrences in any fixed size region in space will also follow a Poisson distribution. Such a spatial interval may be one, two, or three dimensional. Thus, the number of flaws in a given length of wire, or in a given area of cloth, or in a given size block of concrete are all examples of spatial problems to which the Poisson distribution may be applied.

An experiment in which observations are taken over successive time or space intervals and for which the preceding assumptions are satisfied so that the occurrences possess a Poisson distribution is called a *Poisson process*. Thus, the Poisson distribution is a useful distribution for treating Poisson process problems and does not depend upon its approximating role for the binomial distribution for justification.

2.5.2 Applications. As an illustration of the use of the Poisson distribution as an approximation to the binomial distribution, consider the problem of calculating the probability that at most five defective fuses will be found in a box of 200 fuses if experience shows that 2 per cent of such fuses are defective. Here $\mu = np = 200(.02) = 4$; hence, using (17), the approximate answer is given by

$$P\{X \leq 5\} = \sum_{x=0}^{5} \frac{e^{-4}4^x}{x!}$$

$$= e^{-4}\left(1 + 4 + \frac{4^2}{2} + \frac{4^3}{6} + \frac{4^4}{24} + \frac{4^5}{120}\right)$$

$$= .785 .$$

This answer can be obtained directly from Table IX in the appendix, which is a table that sums Poisson probabilities for small values of μ. Lengthy calculations using (11) yield the answer .788; hence the approximation is very good here.

As an illustration of the use of the Poisson distribution as a model for a Poisson process, consider the following problem. Experience has shown that the mean number of telephone calls arriving at a switchboard during one minute is 5. If the switchboard can handle a maximum of 8 calls per minute, what is the probability that it will be unable to handle all the calls that come in during a period of one minute? The desired probability can be obtained by calculating the probability of receiving 8 or fewer calls and then subtracting this probability from 1. Using $\mu = 5$ in the Poisson density, and employing Table IX, gives

$$P\{X \le 8\} = \sum_{x=0}^{8} \frac{e^{-5} \cdot 5^x}{x!} = .932 .$$

Consequently the probability that the switchboard will be overtaxed is given by $P\{X > 8\} = .07$.

Because of the nature of the assumptions that led to the Poisson distribution for problems of this type, it is legitimate to choose any desired length time interval, or space interval, and calculate the probability of occurrences for this interval by means of the Poisson distribution with μ adjusted for that interval. For example, the solution of the problem of calculating the probability that at most 6 calls will be received in a period of 2 minutes when the mean number of calls per minute is 5 is obtained by choosing $\mu = 10$ and evaluating the following sum by means of Table IX.

$$P\{X \le 6\} = \sum_{x=0}^{6} \frac{e^{-10} \cdot 10^x}{x!} = .13 .$$

A solution based on treating this as a two-stage experiment with each minute of time as a stage would give rise to lengthy calculations. Students should be encouraged to solve the problem in this manner so that they will better appreciate the preceding property of the Poisson distribution.

2.6 Hypergeometric Distribution

The binomial distribution was derived on the basis of n independent trials of an experiment; however, if the experiment consists of selecting individuals from a finite population of individuals, the trials will not be independent. For example, if 20 students are to be chosen from a group of 100 students for the purpose of studying the extent to which students work part time, it is clear that the probability of selecting a student who works part time need not remain fixed as successive individuals are selected for the sample. For large finite populations the error arising from assuming that p is constant and the trials are independent, when sampling the population, is very small and it may be ignored, in which case the binomial model is satisfactory. However, for problems in which the population is so small that a serious error will be introduced in using the binomial distribution it is necessary to apply a more appropriate distribution known as the *hypergeometric distribution*. It can be derived as follows.

Let N denote the size of the population from which a set of n individuals is to be drawn. Let the proportion of individuals in this finite population who possess, say, property A be denoted by p. If X is the random variable corresponding to the number of individuals in the set of n who possess property A, then the problem is to find the density function of X, that is, to find a formula for $P\{X = x\}$.

Since the x individuals must come from the Np individuals in the population with property A and the remaining $n - x$ individuals must come from the $N - Np$ who do not possess the property, it follows from the methods illustrated in section 2.5, Chapter 2, that the desired density function will be given by the following formula.

(19)   HYPERGEOMETRIC DISTRIBUTION: $f(x) = \dfrac{\dbinom{Np}{x}\dbinom{N - Np}{n - x}}{\dbinom{N}{n}}.$

Calculations with this formula will show that when n is only a small percentage of N the value of N must be quite small before there will be any appreciable difference between the values given by this formula and the binomial formula in (11). As an illustration, suppose a population consists of 100 individuals, of whom 10 percent have high blood pressure. Then calculations will show, for example, that the probability of getting at most two individuals with high blood pressure when choosing ten individuals is

$$P\{X \leq 2\} = \sum_{x=0}^{2} \frac{\dbinom{10}{x}\dbinom{90}{10 - x}}{\dbinom{100}{10}} = .94 .$$

If the binomial formula (11) is used, additional calculations will show that one then obtains

$$P\{X \leq 2\} \doteq \sum_{x=0}^{2} \frac{10!}{x! \, (10 - x)!} \left(\frac{1}{10}\right)^{x} \left(\frac{9}{10}\right)^{10-x} = .93 .$$

The mean and variance of the hypergeometric distribution will be computed for the purpose of comparing them with the corresponding parameters of the binomial distribution. The computations are considerably simplified if the following two combination formulas are employed.

$$\binom{Np}{x} = \frac{Np}{x}\binom{Np-1}{x-1} \quad \text{and} \quad \binom{Np}{x} = \frac{(Np)(Np-1)}{x(x-1)}\binom{Np-2}{x-2}.$$

These formulas are readily verified if the combination symbols are written out in terms of factorials. Use of the first formula will give

$$E[X] = \sum_{x=0}^{n} x \frac{\binom{Np}{x}\binom{N-Np}{n-x}}{\binom{N}{n}} = Np \sum_{x=1}^{n} \frac{\binom{Np-1}{x-1}\binom{N-Np}{n-x}}{\binom{N}{n}}.$$

The second sum begins with $x = 1$ rather than $x = 0$ because the quantity being summed vanishes for $x = 0$. By letting $y = x - 1$ and using the relation

$$\binom{N}{n} = \frac{N}{n}\binom{N-1}{n-1}$$

in the denominator, the expression for $E[X]$ can be written in the form

$$E[X] = Np \frac{n}{N} \sum_{y=0}^{n-1} \frac{\binom{Np-1}{y}\binom{(N-1)-(Np-1)}{(n-1)-y}}{\binom{N-1}{n-1}}.$$

But this sum is merely the sum of the hypergeometric probabilities when a set of $n - 1$ individuals is extracted from a population of $N - 1$ individuals of whom $Np - 1$ possess property A; hence the sum is equal to 1 and

$$E[X] = np.$$

Thus, the mean of the hypergeometric distribution is the same as the mean of the corresponding binomial distribution. This implies, for example, that the mean number of spades that one would obtain in being dealt five cards from an ordinary deck of cards is the same as the mean number when each card received is returned to the deck before the next card is dealt and only a record kept of the five cards received.

The technique used to calculate the variance of a binomial variable is also effective for calculating the variance of a hypergeometric variable if the second of the foregoing formulas is employed. Thus, one first calculates

$$E[X(X-1)] = \sum_{x=0}^{n} x(x-1) \frac{\binom{Np}{x}\binom{N-Np}{n-x}}{\binom{N}{n}}$$

$$= (Np)(Np-1)\sum_{x=2}^{n} \frac{\binom{Np-2}{x-2}\binom{N-Np}{n-x}}{\binom{N}{n}}.$$

By letting $z = x - 2$ and using the relation $\binom{N}{n} = \frac{N(N-1)}{n(n-1)}\binom{N-2}{n-2}$ in the denominator, this can be expressed in the form

$$E[X(X-1)] = \frac{(Np)(Np-1)n(n-1)}{N(N-1)}$$

$$\times \sum_{z=0}^{n-2} \frac{\binom{Np-2}{z}\binom{(N-2)-(Np-2)}{(n-2)-z}}{\binom{N-2}{n-2}}.$$

It will be observed that this sum is the sum of the hypergeometric probabilities when a set of $n-2$ individuals is taken from a population of $N-2$ individuals of whom $Np-2$ possess property A; hence

$$E[X(X-1)] = \frac{(Np)(Np-1)n(n-1)}{N(N-1)} = \frac{np(n-1)(Np-1)}{N-1}.$$

Now

$$E[X(X-1)] = E[X^2] - E[X].$$

Application of this formula and formula (6) to the preceding result then gives

$$\sigma^2 = E[X^2] - \mu^2 = E[X(X-1)] + E[X] - \mu^2$$

$$= \frac{np(n-1)(Np-1)}{N-1} + np - n^2 p^2.$$

Some algebra will reduce this to

$$\sigma^2 = npq\frac{N-n}{N-1}.$$

This shows that the variance of the hypergeometric distribution is smaller than that for the corresponding binomial distribution by the factor $(N - n)/(N - 1)$. For N relatively large compared to n this factor will be close to one, which was to be expected because then it makes little difference whether n individuals are selected with or without replacement.

3 CONTINUOUS VARIABLES

In the preceding sections three particular discrete random variables were studied. In the next few sections three useful continuous random variables will be presented. Since it will be necessary to calculate the moments of those random variables, the definition of the kth moment of a continuous random variable is considered first.

3.1 Moments

Let $f(x)$ be the density of a continuous random variable X which is zero outside some finite interval (a, b). Figure 7 gives the graph of such a function. Let the interval (a, b) be divided into n equal subintervals and let x_i be the midpoint of the ith subinterval. Form the sum

$$(20) \qquad \sum_{i=1}^{n} x_i^k f(x_i) \, \Delta x$$

where Δx is the width of a subinterval. The quantity $f(x_i) \, \Delta x$ is the area of the shaded rectangle; hence $x_i^k f(x_i) \, \Delta x$ represents the approximate kth moment of this rectangular area about the origin and (20) represents the sum of such approximate kth moments of area. Since the rectangles approximate the area under the curve, the natural procedure is to define the kth moment of $f(x)$ as the limit of this sum as the width of the subinterval approaches zero. This limit, when it exists, is an integral and leads to the following definition.

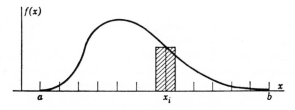

Fig. 7. A continuous density function.

(21) DEFINITION: *The kth moment about the origin of the distribution of the continuous random variable X whose density is f is given by*

$$\mu'_k = \int_a^b x^k f(x) \, dx \, .$$

By analogy with the definition given for a discrete variable, the kth moment about the mean, denoted by μ_k, is given by replacing x^k by $(x - \mu)^k$ in the preceding integral. As in the case of discrete variables, it is desirable to introduce the more general concept of expected value and treat moments as a special case of expected value. The desired definition follows by analogy with the one given for discrete variables and is as follows:

(22) DEFINITION: *The expected value of the function g(X) of the continuous random variable X whose density is f is given by*

$$E[g(X)] = \int_a^b g(x) f(x) \, dx \, .$$

Although the motivation for definition (21) assumed that $f(x)$ was zero outside the interval (a, b), no such restriction will be placed on the foregoing definitions. Thus, a may be $-\infty$ and b may be $+\infty$. When the interval of integration is infinite it is assumed that the integral converges absolutely. For convenience of notation, the limits $-\infty$ and $+\infty$ will be used hereafter whether a and b are infinite or finite.

Since the expected value operator E was designed to produce mean values of random variables, the question naturally arises whether the expected value of a function of a random variable X, say $g(X)$, is really the mean value of that function. Let the random variable $g(X)$ be denoted by Y. Then knowing the density $f(x)$ of X it is theoretically possible to find the density $h(y)$ of Y. The expected value of $g(X)$ is the same as the expected value of Y; therefore if $h(y)$ is available the latter expected value can be expressed in the form

$$E[Y] = \int_{-\infty}^{\infty} y h(y) \, dy \, .$$

By using change of variable techniques of calculus, it can be shown that this value is the same as the value given by the formula of (22). The advantage of (22) lies in the fact that (22) does not require one to find the density function h from a knowledge of the density function f before calculating the expected value of $Y = g(X)$. Change of variable techniques will be studied in Chapter 10.

3.2   Moment Generating Function

The moment generating function for a continuous variable is defined by analogy with (7) to be

$$(23) \qquad M_X(\theta) = E[e^{\theta X}] = \int_{-\infty}^{\infty} e^{\theta x} f(x)\, dx \,.$$

If $e^{\theta x}$ is expanded in a power series and if the integration is performed term by term, it will be found that $M_X(\theta)$ will assume the same expanded form as in (8); hence (23) generates moments in the same manner as (7) does.

In order to be able to generate moments of a function $g(X)$ of the continuous random variable X, it is necessary to generalize the definition of the moment generating function. From the manner in which $M_X(\theta)$ generates moments, it is clear that moments of $g(X)$ will be generated if $e^{\theta x}$ is replaced by $e^{\theta g(x)}$ in (23). The desired definition is the following:

(24) DEFINITION: *The moment generating function of the function $g(X)$ of the continuous random variable X, whose density function is f, is given by*

$$M_{g(X)}(\theta) = E[e^{\theta g(X)}] = \int_{-\infty}^{\infty} e^{\theta g(x)} f(x)\, dx \,.$$

This generalized form of the moment generating function is used to prove a number of theorems, but in such proofs two properties of moment generating functions are needed; therefore, we consider those properties now.

Let c be any constant and let $h(X)$ be a function of X for which the moment generating function exists. Then, since $g(X)$ in (24) represents an arbitrary function, $g(X)$ may be chosen as $g(X) = ch(X)$; consequently,

$$M_{ch}(\theta) = \int_{-\infty}^{\infty} e^{\theta ch(x)} f(x)\, dx = M_h(c\theta) \,.$$

The second property is obtained by choosing $g(X) = h(X) + c$. Then

$$\begin{aligned}
M_{h+c}(\theta) &= \int_{-\infty}^{\infty} e^{\theta[h(x)+c]} f(x)\, dx \\
&= e^{\theta c} \int_{-\infty}^{\infty} e^{\theta h(x)} f(x)\, dx \\
&= e^{\theta c} M_h(\theta) \,.
\end{aligned}$$

For the purpose of expressing these results in terms of the same notation as that used in (24), $h(x)$ is replaced by $g(x)$ in the preceding results to give the following two important formulas.

(25) PROPERTIES: *If c is any constant and g(X) is any function for which the moment generating function exists,*

(i)
$$M_{cg(X)}(\theta) = M_{g(X)}(c\theta)$$

(ii)
$$M_{g(X)+c}(\theta) = e^{c\theta} M_{g(X)}(\theta) .$$

These two properties enable one to dispose of a bothersome constant c which multiplies, or is added to, a function $g(X)$. By replacing integrals by sums it is easily shown that these formulas apply to discrete variables also. It is assumed that $g(x)$ and $f(x)$ are such that the integral in (24), or the corresponding sum, is finite. This implies that all the moments of $g(X)$ are finite. Applications of the preceding formulas are made in the following sections.

3.3 Rectangular Distribution

Perhaps the simplest continuous random variable is the one whose distribution is constant over some interval (a, b) and is zero elsewhere. This defines what is known as the rectangular, or uniform, distribution; hence

(26) RECTANGULAR DISTRIBUTION: $f(x) = \begin{cases} 1/(b-a), & a \leq x \leq b \\ 0, & elsewhere \end{cases}$.

The graph of a typical rectangular distribution is given in Fig. 8.

The rectangular distribution arises, for example, in the study of rounding errors when measurements are recorded to a certain accuracy. Thus, if measurements of daily temperatures are recorded to the nearest degree, it would be assumed that the difference in degrees between the true temperature and the recorded temperature is some number between $-.5$ and $.5$ and that the error is uniformly distributed throughout this interval.

The kth moment of the rectangular distribution is easy to compute. For example, if $a = 0$ and $b = 1$, application of (21) to (26) gives

$$\mu'_k = \int_0^1 x^k \, dx = \frac{1}{k+1} .$$

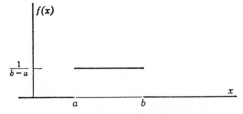

Fig. 8.   A rectangular distribution.

The moment generating function is also easy to compute. Application of (23) to (26) for this problem gives

$$M_X(\theta) = \int_0^1 e^{\theta x}\, dx = \frac{e^{\theta} - 1}{\theta}.$$

If one wished to obtain the kth moment from $M_X(\theta)$, it would be necessary to expand e^{θ} and simplify as follows:

$$M_X(\theta) = \frac{1}{\theta}\left(1 + \theta + \frac{\theta^2}{2!} + \frac{\theta^3}{3!} + \cdots - 1\right)$$

$$= 1 + \frac{\theta}{2!} + \frac{\theta^2}{3!} + \cdots + \frac{\theta^k}{(k+1)!} + \cdots.$$

Since μ_k' is the coefficient of $\theta^k/k!$, it will be seen from this expansion that $\mu_k' = k!/(k+1)! = 1/(k+1)$, which agrees with the preceding result. This computation was made for the purpose of becoming familiar with the moment generating function and not as a suggested method for computing the moments. The direct computation is obviously much simpler here.

The rectangular distribution is of somewhat limited use as a model for real-life distributions; however, it is of considerable theoretical value and is the simplest distribution of a continuous random variable on which to illustrate general formulas.

3.4 Normal Distribution

Without question, the model that has proved the most useful of all distributions for continuous random variables is a distribution called the *normal* or *Gaussian* distribution. This distribution not only serves as a model distribution for many real-life continuous random variables but it also arises naturally in many theoretical investigations. It is defined as follows:

(27)   NORMAL DISTRIBUTION: $f(x) = ce^{-\frac{1}{2}\left(\frac{x-a}{b}\right)^2}.$

Here a, b, and c are parameters that make $f(x)$ a probability density function; hence these parameters must be such that the integral of $f(x)$ from $-\infty$ to $+\infty$ is equal to 1. As a result, c can be expressed as a function of a and b. Thus there are actually only two independent parameters, which may be chosen to be a and b, determining this density.

3.4.1 Moments. The graph of a typical normal curve is given in Fig. 9. From (27) it is clear that the curve is symmetrical about the line $x = a$; hence by symmetry the mean must be given by $\mu = a$.

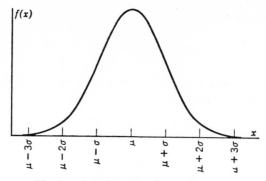

Fig. 9. Typical normal distribution.

Instead of finding the moments directly, they may be found indirectly by means of the moment generating function. Furthermore, since it is easier here to find moments about the mean than about the origin, consider the evaluation of $M_{X-\mu}(\theta)$. From definition (24), with $g(x)$ chosen equal to $x - \mu$,

$$M_{X-\mu}(\theta) = c \int_{-\infty}^{\infty} e^{\theta(x-\mu)} \cdot e^{-\frac{1}{2}\left(\frac{x-\mu}{b}\right)^2} \, dx \, .$$

Let $z = (x - \mu)/b$; then $dx = b \, dz$ and

$$M_{X-\mu}(\theta) = bc \int_{-\infty}^{\infty} e^{\theta b z - \frac{z^2}{2}} \, dz \, .$$

Complete the square in the exponent as follows:

$$\theta b z - \frac{z^2}{2} = -\tfrac{1}{2}(z - \theta b)^2 + \tfrac{1}{2}\theta^2 b^2$$

Then

$$M_{X-\mu}(\theta) = bc e^{\frac{1}{2}\theta^2 b^2} \int_{-\infty}^{\infty} e^{-\frac{1}{2}(z-\theta b)^2} \, dz \, .$$

If $t = z - \theta b$, then $dz = dt$ and

$$M_{X-\mu}(\theta) = bc e^{\frac{1}{2}\theta^2 b^2} \int_{-\infty}^{\infty} e^{-\frac{t^2}{2}} \, dt \, .$$

The value of this integral can be found in any standard table of integrals. Or it may be evaluated directly by the following device. Let

$$I = \int_{0}^{\infty} e^{-\frac{t^2}{2}} \, dt \, .$$

Then

$$I^2 = \int_0^\infty e^{-\frac{x^2}{2}} dx \int_0^\infty e^{-\frac{y^2}{2}} dy$$

$$= \int_0^\infty \int_0^\infty e^{-\frac{x^2+y^2}{2}} dx\, dy \,.$$

In polar coordinates this double integral assumes the form

$$I^2 = \int_0^{\frac{\pi}{2}} \int_0^\infty e^{-\frac{r^2}{2}} r\, dr\, d\theta$$

$$= \int_0^{\frac{\pi}{2}} -e^{-\frac{r^2}{2}} \Big|_0^\infty d\theta$$

$$= \int_0^{\frac{\pi}{2}} d\theta = \frac{\pi}{2} \,.$$

Hence,

$$\int_{-\infty}^\infty e^{-\frac{t^2}{2}} dt = \sqrt{2\pi}$$

and

(28) $$M_{X-\mu}(\theta) = \sqrt{2\pi}\, bc e^{\frac{1}{2}\theta^2 b^2} \,.$$

From (8) it follows that for any moment generating function $M(0) = 1$; hence from (28) it follows that $\sqrt{2\pi}\, bc = 1$ and that

(29) $$M_{X-\mu}(\theta) = e^{\frac{1}{2}\theta^2 b^2} \,.$$

If this exponential is expanded in a power series,

$$M_{X-\mu}(\theta) = 1 + b^2 \frac{\theta^2}{2} + b^4 \frac{\theta^4}{8} + \cdots \,.$$

Since the odd powers of θ are missing, the odd moments of X about its mean μ must be 0, which of course is true for any symmetrical distribution possessing such moments. The coefficient of $\theta^2/2!$ is the second moment of X about its mean; therefore, $b^2 = \mu_2 = \sigma^2$, or $b = \sigma$. Since $\sqrt{2\pi}\, bc = 1$, $c = 1/\sigma\sqrt{2\pi}$; consequently, (27) can be written in the form

(30) $$f(x) = \frac{1}{\sigma\sqrt{2\pi}} e^{-\frac{1}{2}\left(\frac{x-\mu}{\sigma}\right)^2} \,.$$

This result shows that a normal distribution is completely determined by specifying its mean and standard deviation. It should be noted that the only difference between (27) and (30) is that the parameters in (27) have now been reduced to two independent parameters which have been given statistical meaning.

A formula for $M_X(\theta)$, expressed in terms of statistical parameters, will be needed in subsequent sections. It can be obtained from (29) by replacing b^2 with σ^2 and using the second of the two properties in (25) with $g(X) = X$ and $c = -\mu$. Thus,

$$M_{X-\mu}(\theta) = e^{-\mu\theta}M_X(\theta) .$$

Using formula (29) on the left side and solving for $M_X(\theta)$ will yield the desired formula, namely

(31) $$M_X(\theta) = e^{\mu\theta + \frac{1}{2}\sigma^2\theta^2} .$$

For the purpose of interpreting the standard deviation geometrically, consider the points of inflection of a normal curve. When (30) is differentiated twice,

$$f' = -\frac{1}{\sigma^2}(x - \mu)f$$

$$f'' = -\frac{1}{\sigma^2}\left[1 - \left(\frac{x - \mu}{\sigma}\right)^2\right]f .$$

From these derivatives it is clear that there is but one maximum point, which occurs at $x = \mu$. From the second derivative it follows that points of inflection occur at $x = \mu \pm \sigma$. Geometrically, then, for a normal distribution the standard deviation is the distance from the axis of symmetry to a point of inflection. This implies that the normal curve of Fig. 9 is concave down between $x = \mu - \sigma$ and $x = \mu + \sigma$ and concave up outside that interval.

Additional insight into the role that the parameter σ plays in determining a normal density is obtained by computing the area under $f(x)$ within the symmetrical intervals shown in Fig. 9. Thus, from (30) the probability that X will fall inside the interval $(\mu - \sigma, \mu + \sigma)$ is given by

$$\int_{\mu-\sigma}^{\mu+\sigma} \frac{1}{\sigma\sqrt{2\pi}} e^{-\frac{1}{2}\left(\frac{x-\mu}{\sigma}\right)^2} dx .$$

Let $z = (x - \mu)/\sigma$; then $dx = \sigma\,dz$ and the integral reduces to

$$\int_{-1}^{1} \frac{1}{\sqrt{2\pi}} e^{-\frac{z^2}{2}} dz = 2\int_{0}^{1} \frac{e^{-\frac{z^2}{2}}}{\sqrt{2\pi}} dz .$$

The value of the last integral can be found in Table II in the appendix and is .3413. Hence, multiplying by 2, the value of the first integral is .68, correct to two digits. For the interval $(\mu - 2\sigma, \mu + 2\sigma)$ one may verify that $z = \pm 2$ and that the area between is .95. Similarly, the area between $\mu - 3\sigma$ and $\mu + 3\sigma$ is .997. In summary, about 68 percent of the probability lies within one standard deviation of the mean, about 95 percent within two standard deviations of the mean, and practically all the probability is found within three standard deviations of the mean. The unit of measurement given by the transformation $z = (x - \mu)/\sigma$ is called a *standard unit*. Table II is therefore a table for the normal distribution with zero mean and unit standard deviation, that is, for the normal distribution in standard units.

3.4.2 Applications. The interesting and important applications of normal distributions are considered in later chapters after further essential theory has been developed. Here, only one simple illustration of its direct applicability is given.

Many college instructors of large classes assign letter grades on examinations by means of the normal distribution. The procedure followed is to ignore that part of the distribution lying outside the interval $\mu \pm 2.5\sigma$, or $\mu \pm 3\sigma$, and then divide this interval into five equal parts corresponding to the letter grades F, D, C, B, and A. If $\mu \pm 2.5\sigma$ is used, each interval will be σ units in length; consequently, the six values of x determining these five intervals will be $\mu - 2.5\sigma$, $\mu - 1.5\sigma$, $\mu - .5\sigma$, $\mu + .5\sigma$, $\mu + 1.5\sigma$, and $\mu + 2.5\sigma$. The corresponding values of $z = (x - \mu)/\sigma$ will be -2.5, -1.5, $-.5$, $.5$, 1.5, and 2.5. From Table II it will be found that the areas within these five intervals are .06, .24, .38, .24, and .06, respectively. Since these percentages do not total 100 percent, it is customary to allow the two end intervals to extend to infinity. Then the percentages of students who will be assigned the corresponding letter grades are 7 percent F, 24 percent D, 38 percent C, 24 percent B, and 7 percent A.

3.4.3 Normal Approximation to Binomial. In 2.5.1 the Poisson distribution was introduced as an approximation to the binomial distribution when n is large and p is small. It was stated there that another distribution gives a good approximation for large n when p is not small. The normal distribution is the distribution with this property. Before investigating the nature of this approximation in general, consider a numerical example.

Let $n = 12$ and $p = \frac{1}{3}$ and construct the graph of the corresponding binomial distribution. This is hardly a large value of n, so that a good normal approximation is not to be expected here. Since $f(x)$ is to be computed for all values of x from 0 to 12, it is easier to compute each value, after the first, from the preceding one rather than to compute each value by itself. Here, by (11),

$$f(x) = \frac{12!}{x!\,(12 - x)!} \left(\frac{1}{3}\right)^{x} \left(\frac{2}{3}\right)^{12-x}.$$

It is easily verified that for this function

$$f(x + 1) = \frac{12 - x}{x + 1} \frac{1}{2} f(x) .$$

After $f(0)$ was computed, this relationship was used to obtain the remaining values.

$$
\begin{array}{llll}
f(0) = & .007707 & f(7) = \frac{3}{7}f(6) & = .047687 \\
f(1) = 6f(0) = .046242 & f(8) = \frac{5}{16}f(7) & = .014902 \\
f(2) = \frac{11}{4}f(1) = .127166 & f(9) = \frac{2}{9}f(8) & = .003312 \\
f(3) = \frac{5}{3}f(2) = .211943 & f(10) = \frac{3}{20}f(9) & = .000497 \\
f(4) = \frac{9}{8}f(3) = .238436 & f(11) = \frac{1}{11}f(10) & = .000045 \\
f(5) = \frac{4}{5}f(4) = .190749 & f(12) = \frac{1}{24}f(11) & = .000002 \\
f(6) = \frac{7}{12}f(5) = .111270
\end{array}
$$

(32)

Since $f(0)$ was computed correct to four digits only, the remaining values would not be expected to be correct to more than four digits, even though they have been recorded to six decimals for the sake of appearances. The graph of this binomial distribution is shown in Fig. 10. It appears that this histogram could be fitted fairly well by the proper normal curve.

Since a normal curve is completely determined by its mean and standard deviation, the natural normal curve to use here is the one with the same mean and standard deviation as the binomial distribution. Hence, because of (15), choose

$$\mu = 12 \cdot \tfrac{1}{3} = 4$$

and

$$\sigma = \sqrt{12 \cdot \tfrac{1}{3} \cdot \tfrac{2}{3}} = 1.63 .$$

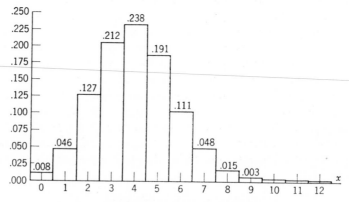

Fig. 10. Binomial distribution, $p = \tfrac{1}{3}$, $n = 12$.

As a test of the accuracy of the normal curve approximation here and as an illustration of the use of normal curve methods for approximating binomial probabilities, consider a few problems related to Fig. 10.

If the probability that a marksman will hit a target is $\frac{1}{3}$ and if he takes 12 shots, what is the probability that he will score at least six hits? The exact answer is obtained by adding the values of $f(x)$ from $x = 6$ to $x = 12$, which, by using (32), is .178, correct to three decimal places. Geometrically, this answer is the area of that part of the histogram in Fig. 10 lying to the right of $x = 5.5$. Therefore, to approximate this probability by normal curve methods, it is merely necessary to find the area under that part of the fitted normal curve which lies to the right of 5.5. Since the fitted curve has $\mu = 4$ and $\sigma = 1.63$, it follows that

$$z = \frac{x - \mu}{\sigma} = \frac{5.5 - 4}{1.63} = .92 .$$

Hence, with these values of μ and σ the change of variable $z = (x - \mu)/\sigma$ will give

$$\int_{5.5}^{\infty} \frac{e^{-\frac{1}{2}\left(\frac{x-\mu}{\sigma}\right)^2}}{\sqrt{2\pi}\,\sigma} \, dx = \int_{.92}^{\infty} \frac{e^{-\frac{z^2}{2}}}{\sqrt{2\pi}} \, dz .$$

But, from Table II, the area to the right of $z = .92$ is .179, which, compared to the correct value of .178, is in error by only about $\frac{1}{2}$ percent.

To test the accuracy of normal curve methods over a shorter interval, calculate the probability that a marksman will score precisely six hits in the 12 shots. From (32) the answer correct to three decimals is $f(6) = .111$. To approximate this answer, it is merely necessary to find the area under the fitted normal curve between $x = 5.5$ and $x = 6.5$. Thus

$$z_2 = \frac{6.5 - 4}{1.63} = 1.53 , \qquad A_2 = .4370$$

$$z_1 = \frac{5.5 - 4}{1.63} = .92 , \qquad A_1 = .3212 .$$

Therefore the required area is .116, which is in error by about 5 percent. From these two examples it appears that normal curve methods are quite accurate, even for some situations such as that considered here in which n is not very large.

The preceding examples were given to make plausible a famous limit theorem in probability that guarantees a good normal curve approximation to the binomial distribution if n is sufficiently large. This theorem, which will be stated here without proof, is a special case of a more general limit theorem that will be studied in Chapter 5.

THEOREM 2: *If X represents the number of successes in n independent trials of an event for which p is the probability of success in a single trial, then the variable $(X - np)/\sqrt{npq}$ has a distribution that approaches the normal distribution with mean 0 and standard deviation 1 as the number of trials $n \to \infty$.*

This theorem justifies the use of normal curve methods for approximating probabilities related to successive trials of an event when n is large. Experience indicates that the approximation is fairly good as long as $np > 5$ when $p \leq \frac{1}{2}$, and $nq > 5$ when $p > \frac{1}{2}$. A very small value of p, together with a moderately large value of n, would yield a small mean and thus produce a skewed distribution. Similarly, if p is very close to one and n is only moderately large, most of the distribution will be piled up close to $x = n$, thus preventing a normal curve from fitting well. If the mean is at least five units away from either extremity, the distribution has sufficient room to become fairly symmetrical. Figures 11 and 12 indicate how rapidly the distribution of the variable $(X - np)/\sqrt{npq}$ approaches normality when $p = \frac{1}{3}$, and $n = 24$ and 48, respectively. The common y scale for these two graphs is approximately 17 times that for the x axis.

There are numerous occasions when it is more convenient to work with the proportion of successes in n trials than with the actual number of successes. Since

$$\frac{X - np}{\sqrt{npq}} = \frac{\dfrac{X}{n} - p}{\sqrt{pq/n}}$$

the following useful corollary to Theorem 2 may be obtained.

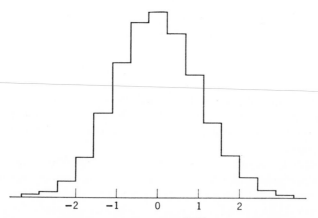

Fig. 11. Distribution of $(X - np)/\sqrt{npq}$ for $p = \frac{1}{3}$ and $n = 24$.

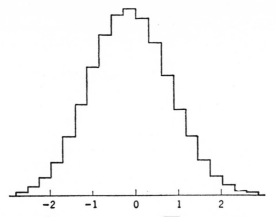

Fig. 12. Distribution of $(X - np)/\sqrt{npq}$ for $p = \frac{1}{3}$ and $n = 48$.

(33) COROLLARY: *The proportion of successes X/n will be approximately normally distributed with mean p and standard deviation $\sqrt{pq/n}$ if n is sufficiently large.*

The two approximations that have been considered for the binomial distribution, namely the Poisson and normal distributions, are sufficient to permit one to solve all the simpler problems that require the computation of binomial probabilities. If n is small, one uses formula (11) directly because the computations are then quite easy. Tables of factorials and logarithms are helpful here. If n is large and p is small or large, the Poisson approximation may be used. If n is large and p is not small or large, the normal approximation may be used. Thus all possibilities have been covered.

3.4.4 Applications. Certain types of practical problems dealing with percentages can be solved by means of the normal approximation to the binomial distribution. As a first illustration, consider the following problem related to sampling inspection. A manufacturer of machine parts claims that at most 10 percent of his parts are defective. A purchaser needs 120 such parts and to be sure of getting that many good ones he places an order for 140. If the manufacturer's claim is valid, what is the probability that the purchaser will receive at least 120 good parts? Let X represent the number of good parts that will be received. Then X may be treated as a binomial variable with $n = 140$ and $p = .9$. These values certainly justify a normal approximation here. The problem is to calculate $P\{X \geq 120\}$. The mean and standard deviation are given by

$$\mu = np = 126 \quad \text{and} \quad \sigma = \sqrt{npq} = 3.55.$$

Hence,

$$P\{X \geq 120\} = P\left\{\frac{X - \mu}{\sigma} \geq \frac{120 - \mu}{\sigma}\right\}$$

$$= P\left\{\frac{X - 126}{3.55} \geq \frac{120 - 126}{3.55}\right\}$$

$$= P\{Z \geq -1.69\} .$$

Since $Z = (X - \mu)/\sigma = (X - 126)/3.55$ may be treated as an approximate standard normal variable, this probability can be found in Table II and is .95. The purchaser is therefore fairly certain of obtaining enough good parts if the manufacturer's claim is valid. A slightly more accurate result can be obtained if the geometrical approach introduced in the preceding section is employed. The approximation is made by approximating the area of the binomial histogram to the right of 119.5, which would lead to the Z value of

$$Z = \frac{119.5 - 126}{3.55} = -1.83 .$$

From Table II the corresponding probability is .97.

As a second illustration, consider the following problem that arises in taking political polls or market surveys. In a tight political race the managers of candidate A wish to take a poll to estimate their candidate's popularity. How large a poll will be necessary if they wish to be certain with a probability of at least .95 that their poll estimate will not differ from the true proportion of voters favoring A by more than 2 percentage points?

Since the size sample to be taken will be small compared to the size of the population, it may be assumed that p will remain constant and therefore that this may be treated as a binomial problem. Furthermore, since p will be fairly close to $\frac{1}{2}$ in a tight race and n will necessarily be fairly large in order to estimate p with such high accuracy, the normal approximation to the binomial distribution may be used to solve the problem. Then, because of (33), the variable X/n, where X denotes the number of voters in the poll of size n that will favor candidate A, may be treated as a normal variable with mean p and variance pq/n. Now, for a normal variable the variable will fall within .02 unit of the mean with a probability of .95 if .02 unit is equal to 2 standard deviations of the variable; hence n should be chosen to satisfy

$$2\sigma_{\frac{X}{n}} = .02 .$$

But from (33) this implies that

$$2\sqrt{\frac{pq}{n}} = .02 .$$

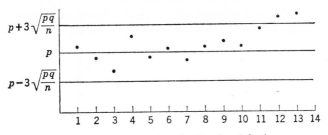

Fig. 13. Control chart for fraction defective.

Solving for n gives $n = 10,000\,pq$. It is easy to show by calculus methods that the function $pq = p(1 - p)$ assumes its maximum value for $p = \frac{1}{2}$; hence if this value of p is used, the maximum size poll needed will be obtained. Thus, if $n = 2500$ the poll will certainly be large enough. Since p will be fairly close to $\frac{1}{2}$ in a tight race, this conservative value is also very realistic; therefore the managers should expect to take a poll of 2500 voters.

As a third illustration, consider the problem of determining whether daily percentages of defectives may be treated as independent trials of an experiment for which p is constant from trial to trial. Industrial experience has shown that most production processes do not behave in this idealized manner and that much valuable information is obtained concerning the process if the order in which data are obtained is preserved. A simple graphical method, called a *quality-control chart*, has been found highly useful in the solution of this problem. Such a chart for the proportion of defectives is illustrated in Fig. 13. The middle line is thought of as corresponding to the process proportion defective, although it is usually merely the mean of past daily proportions. The other two lines serve as control limits for daily proportions of defectives. From (33) it will be observed that these two control lines are spaced three standard deviations from the mean line. The time units for successive samples are recorded along the x axis. If now the production process behaves in the idealized manner and if the normal approximation to the binomial distribution may be used, the probability that a daily proportion when plotted on this chart will fall outside the control band is approximately equal to the probability that a normal variable will assume a value more than three standard deviations away from its mean, which, from Table II, is .003. Because of this small probability, it is reasonable to assume that the production process is no longer behaving properly when a point falls outside the control band; consequently, the production engineer checks over the various steps in the process when this event occurs. From an inspection of Fig. 13 it will be observed that the process in question went out of control on the twelfth day.

Industrial experience shows that only rarely does a production process behave in this idealized manner when the control-chart technique is first applied. Nevertheless, the technique is highly useful because it enables one to discover causes of a lack of control and thus to improve on the production process until gradually statistical control has been obtained.

This illustration and discussion of a quality-control chart gives an incomplete picture of how quality-control methods operate. Such methods constitute an extensive field of applied statistics, and numerous articles and books concerning them are available.

3.5   Gamma Distribution

A probability distribution that arises naturally in the study of the length of life of industrial equipment, and which occurs frequently in various other statistical problems as well, is a distribution called the gamma distribution. The name came from the relationship of the distribution to the gamma function of advanced calculus. It involves two parameters, $\alpha > 0$ and $\beta > 0$, and is defined as follows:

$$(34) \qquad \text{GAMMA DISTRIBUTION:} \quad f(x) = \frac{x^{\alpha-1}e^{-\frac{x}{\beta}}}{\beta^{\alpha}\Gamma(\alpha)}, \qquad x > 0$$

$$= 0, \qquad x \leq 0.$$

The quantity $\Gamma(\alpha)$ is a symbol representing the value of the gamma function at the point α. This function is defined by the integral

$$(35) \qquad \Gamma(\alpha) = \int_0^{\infty} x^{\alpha-1}e^{-x}\,dx.$$

It is easily shown by integrating by parts that $\Gamma(\alpha + 1) = \alpha\Gamma(\alpha)$. If α is a positive integer, this recurrence relation gives the factorial result that $\Gamma(\alpha + 1) = \alpha!$. As a consequence of this property the gamma function is sometimes called the factorial function.

3.5.1   Moments. The moments of the gamma distribution are easily computed by means of (35). From (34)

$$E[X^k] = \frac{1}{\beta^{\alpha}\Gamma(\alpha)}\int_0^{\infty} x^{k+\alpha-1}e^{-\frac{x}{\beta}}\,dx.$$

Letting $t = x/\beta$ gives

$$E[X^k] = \frac{\beta^{k+\alpha}}{\beta^{\alpha}\Gamma(\alpha)}\int_0^{\infty} t^{k+\alpha-1}e^{-t}\,dt.$$

The use of (35) then gives

$$E[X^k] = \beta^k \frac{\Gamma(k + \alpha)}{\Gamma(\alpha)} .$$

Since k is a positive integer, it follows from repeated application of the recurrence relation $\Gamma(\alpha + 1) = \alpha\Gamma(\alpha)$ that

$$\Gamma(k + \alpha) = (k + \alpha - 1) \cdots (\alpha)\Gamma(\alpha) ,$$

and hence that

(36) $$E[X^k] = \beta^k(k + \alpha - 1)(k + \alpha - 2) \cdots (\alpha) .$$

From this formula it follows that the mean and variance of a gamma distribution are given by

(37) $$\mu = \beta\alpha \quad \text{and} \quad \sigma^2 = \beta^2\alpha .$$

3.6 Exponential Distribution

The special case of the gamma distribution that occurs when $\alpha = 1$ is called the exponential distribution. It is used sufficiently often to justify listing it separately. Thus,

(38) EXPONENTIAL DISTRIBUTION: $$f(x) = \frac{e^{-\frac{x}{\beta}}}{\beta} , \qquad x > 0$$

$$= 0 , \qquad x \leq 0 .$$

This distribution arises, for example, in studying the length of life of a radioactive material. If it is assumed that the rate at which a mass y of radioactive material is decaying is proportional to the amount of material remaining at any time t, then y will satisfy the differential equation

$$\frac{dy}{dt} = -\lambda y$$

where λ is a constant whose value depends upon the kind of radioactive material being studied. The solution of this equation is $y = y_0 e^{-\lambda t}$, where y_0 denotes the amount of material at time $t = 0$. Since $(y_0 - y)/y_0$ gives the proportion of the original material that has decayed in t units of time, this quantity may be taken as the probability that an atom selected at random from this material will decay in t units of time. Thus, if X represents the length of life of such an atom,

$$F(t) = P\{X \leq t\} = \frac{y_0 - y}{y_0} = 1 - e^{-\lambda t} .$$

Since this gives the distribution function of the random variable X at $X = t$, the density function at $X = t$ is obtained by differentiating $F(t)$ with respect to t; hence

$$f(t) = \lambda e^{-\lambda t}.$$

A comparison of this result with (38) shows that the density of the random variable X, where X is the length of life of a radioactive atom, possesses an exponential distribution with parameter $\beta = 1/\lambda$. In view of this relationship, the gamma distribution is often defined in terms of the parameters α and λ rather than α and β.

The exponential distribution has been found to be an appropriate model for calculating the probability that a piece of equipment will last for a total of t time units before it fails. The appropriateness of the model depends upon the nature of the equipment and what causes it to fail. If failure is due principally to external causes rather than internal wear, then the model is likely to be realistic. Under such circumstances the length of time to the next failure after the equipment has been repaired and placed back into use will also follow an exponential distribution. Thus, the equipment behaves as though it were new equipment after each repair.

As a simple illustration of how the exponential distribution arises in practical problems, consider the following one. A manufacturer of electronic equipment has found by experience that his equipment lasts on the average 2 years without repairs and that the time before the first breakdown follows an exponential distribution. If he guarantees his equipment to last 1 year, what proportion of his customers will be eligible for some adjustment because of failure before 1 year? Since β is the mean of the distribution given by (38), the density that applies here is $f(x) = e^{-x/2}/2$. The problem is to calculate $P\{X < 1\}$; hence letting $t = x/2$

$$P\{X < 1\} = \int_0^1 \frac{e^{-\frac{x}{2}}}{2}\, dx = \int_0^{\frac{1}{2}} e^{-t}\, dt = .39 .$$

Thus, even though the average length of life is twice the guaranteed length of life, there is a high probability of the equipment failing before the guarantee expires.

3.7 Chi-Square Distribution

Another special case of the gamma distribution that has numerous statistical applications is obtained by choosing $\beta = 2$ and writing $\alpha = \nu/2$.

It is called the chi-square distribution and its density is the following one.

$$(39) \qquad \text{CHI-SQUARE DISTRIBUTION:} \quad f(x) = \frac{x^{\frac{\nu}{2}-1} e^{-\frac{x}{2}}}{2^{\frac{\nu}{2}}\Gamma(\nu/2)}, \qquad x > 0$$

$$= 0, \qquad x \leq 0.$$

The reason for changing from the parameter α to the parameter $\nu/2$ is that the parameter ν possesses a natural intuitive meaning, called the degrees of freedom, when the chi-square distribution is applied to certain statistical problems that will be studied in later chapters. It is interesting in this connection to calculate the mean and variance of a chi-square variable in terms of this modified parameter. Setting $\beta = 2$ and $\alpha = \nu/2$ in formulas (37) yields the formulas

$$\mu = \nu \quad \text{and} \quad \sigma^2 = 2\nu.$$

Thus, the mean of a chi-square variable is equal to ν and its variance is equal to 2ν, two easily remembered facts.

Although the moments of a chi-square distribution are readily calculated by means of formula (37), other properties of the distribution are most easily demonstrated by means of its moment generating function; therefore it will be calculated now. From (39) and some algebra it follows that

$$M_X(\theta) = E[e^{\theta X}] = \frac{1}{2^{\frac{\nu}{2}}\Gamma(\nu/2)} \int_0^\infty x^{\frac{\nu}{2}-1} e^{-\frac{x(1-2\theta)}{2}} \, dx.$$

Letting $z = x(1 - 2\theta)/2$ gives

$$M_X(\theta) = \frac{2^{\frac{\nu}{2}}(1 - 2\theta)^{-\frac{\nu}{2}}}{2^{\frac{\nu}{2}}\Gamma(\nu/2)} \int_0^\infty z^{\frac{\nu}{2}-1} e^{-z} \, dz.$$

But from (35) it will be observed that the preceding integral defines $\Gamma(\nu/2)$; hence canceling common terms gives

$$(40) \qquad\qquad M_X(\theta) = (1 - 2\theta)^{-\frac{\nu}{2}}.$$

Since the gamma distribution depends upon two parameters it possesses a great deal of flexibility as a model for actual distributions. For the purpose of displaying this versatility, the graphs of several gamma densities are shown in Fig. 14. These are all for β possessing the value 2; therefore they are all chi-square densities. Since β acts as a scale parameter, changing the value of β will merely stretch or compress the corresponding curve of Fig. 14 but keeping the area under it equal to one.

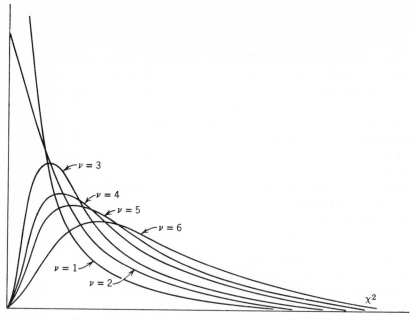

Fig. 14.   Distribution of χ^2 for various degrees of freedom.

Applications of the chi-square distribution will be made in later chapters; therefore no illustrations of its use will be given at this time.

EXERCISES

Sec. 2.1

1. A box contains 6 tickets. Two of the tickets are worth $5 each and the other four are worth $1 each. (a) If one ticket is drawn, what is the expected value of the game? (b) If two tickets are drawn, what is the expected value of the game?

2. A player tosses two coins. If two heads show he wins $4. If one head shows he wins $2, but if two tails show he must pay a $3 penalty. Calculate the expected value of the game to him.

3. If an urn contains 6 white and 4 black balls and 3 balls are drawn without replacement, what is the mean number of black balls that will be obtained? Calculate the mean by using definition (1) of the text, where X denotes the number of black balls obtained.

4. A coin is tossed until either a head or four tails occurs. Let X denote the number of tosses required and calculate $E[X]$.

Sec. 2.2

5. Calculate the mean and variance for X, the face number that comes up when rolling an honest die.

6. A die is loaded so that the probability of a given face turning up is proportional to the number on that face. Calculate the mean and variance for X, the face number showing.

7. Calculate the mean for X, the sum of the face numbers that come up when rolling two honest dice.

8. Calculate the mean of the distribution given in Table 2, Chapter 2, for the two altered dice.

9. A random variable can assume only the values 1 and 3. If its mean is $\frac{8}{3}$, find the probabilities for those two points.

10. A and B match pennies. Calculate the mean and variance of X where X is the amount won by A after two matchings.

11. In problem 10 calculate the mean and variance if A quits after the first matching, provided he wins it, but B does not employ this strategy. What does the result say concerning A's strategy?

12. A coin is tossed until a head appears. If it appears on the first toss, the player receives \$2 from the bank. If it appears for the first time on the second toss, he receives \$4. In general, if it appears for the first time on toss number k, he receives 2^k dollars. If his payment exceeds \$1,000,000, he receives only \$1,000,000. Calculate the mean amount to be won by the player. What effect would placing no limit on the amount to be won have on the mean?

13. Differentiate both sides of the equation

$$\sum_{x=1}^{\infty} p(1 - p)^{x-1} = 1$$

with respect to p and use the resulting equation to find the mean of the geometric density $f(x) = p(1 - p)^{x-1}$, $x = 1, 2, \cdots$.

Sec. 2.3

14. Given $f(x) = (\frac{1}{2})^x$, $x = 1, 2, \cdots$, and zero elsewhere, find its moment generating function. Use it to calculate the mean and variance of X.

Sec. 2.4

15. Six dice are rolled. Calling a 5 or 6 a success, find the probability of getting (a) 3 successes, (b) at most 3 successes.

16. Suppose a sample of 12 is taken from a day's output of a machine that normally produces 5 percent defective parts. If the day's production is inspected 100 percent whenever a sample of 12 gives 3 or more defectives, what is the probability that a day's production will be inspected 100 percent?

17. Experience shows that only $\frac{1}{3}$ of the patients having a certain disease will recover from it under standard treatment. A new promising drug is to be administered to a group of 12 patients having the disease. If the clinic requires that at least 7 of the patients should recover before it will accept the new drug as superior, what is the probability that the drug will be discredited even if it increases the recovery rate to $\frac{1}{2}$?

18. An electrical device consists of five separate units connected in such a manner that it will work only if all five parts operate successfully. If the probability of successful operation for each part is .9, what is the probability the device will work? What will this probability become if the device will work provided only four of the five parts operate successfully?

19. Suppose that weather records show that on the average 5 of the 30 days in November are rainy days. (a) Assuming a binomial distribution with each day as an independent trial, find the probability that next November will have at most 3 rainy days. (b) Give reasons why you may not be justified in using the binomial distribution in solving (a).

20. In calculating binomial probabilities, it is convenient to calculate $f(x + 1)$ from $f(x)$ by the formula $f(x + 1) = k(x)f(x)$, where $k(x) = [(n - x)/(x + 1)]/(p/q)$. Show that this formula is correct.

21. If X has the density $f(x) = x/2$, $0 < x < 2$, calculate the probability that (a) both of two sample values will exceed 1, (b) exactly 2 of 4 sample values will exceed 1.

22. If X has the density $f(x) = \frac{1}{2}$, $0 < x < 2$, (a) what is the probability that at least 2 of 3 sample values will exceed 1? (b) What value of x is such that the probability is $\frac{1}{2}$ that at least 2 of 3 sample values will exceed it?

23. Given that a binomial variable has mean 12 and variance 10, find the values of p and n.

24. Experience shows that 10 percent of the individuals reserving tables at a night club will not appear. If the club has 40 tables and takes 43 reservations, what is the probability that it will be able to accommodate everyone appearing?

25. In the world series for baseball, the series is concluded when one team has won 4 games. Let p be the probability of team A winning a single game and assume that this probability remains constant in the series. Show that the probabilities of the series ending in 4, 5, 6, or 7 games are .125, .25, .3125, .3125, respectively, when $p = \frac{1}{2}$ and .21, .30, .27, .22, respectively, when $p = \frac{2}{3}$.

26. Suppose that a binomial distribution is to be truncated by agreeing to discard the value $x = 0$ whenever it occurs. Find the resulting density of X, that is, find the conditional density of binomial X when $1 \leq X \leq n$.

27. Fit a binomial function to the following data on the number of seeds germinating among 10 seeds on damp filter paper for 80 sets of seeds.

x	0	1	2	3	4	5	6	7	8	9	10
f	6	20	28	12	8	6	0	0	0	0	0

Sec. 2.5

28. Use the Poisson approximation to calculate the probability that at most 2 persons in 500 will have a birthday on Christmas. Assume 365 days in the year.

29. Use the Poisson approximation to calculate the probability of getting 10 successes in 1000 trials of an experiment for which $p = .01$.

30. Metal plates are inspected regularly for flaws and average 2 flaws per square yard. Assume a Poisson process is applicable and calculate the probability of getting (a) no flaws in 4 square yards of material, (b) at most 5 flaws in 4 square yards of material.

31. Suppose the number of telephone calls an operator receives from 9:00 to 9:05 follows a Poisson distribution with $\mu = 4$. (a) Find the probability that the operator will receive no calls in that time interval tomorrow. (b) Find the probability that in the next 2 days the operator will receive a total of 3 calls in that time interval.

32. Assume that the number of particles emitted from a radioactive source follows a Poisson distribution with an average emission of 2 particles per second. (a) Find the probability that at most 1 particle will be emitted in 3 seconds. (b) How low an emission rate would be necessary before the probability of getting at most 1 emission in 3 seconds would be at least .90?

33. Assume that customers enter a store at the rate of 60 persons per hour. (a) What is the probability that during a 5-minute interval no one will enter the store? (b) What time interval is such that the probability is $\frac{1}{2}$ that no one will enter the store during that interval?

34. Assume that the number of items of a certain kind purchased in a store during a week's time follows a Poisson distribution with $\mu = 50$. How large a stock should the merchant have on hand to yield a probability of .98 that he will be able to supply the demand? Use a normal approximation.

35. Solve problem 19 using the Poisson approximation to the binomial distribution and compare answers to see how good the approximation is.

36. Customers arrive at the complaint department of a store at the rate of 5 per hour for male customers and 10 per hour for female customers. If arrivals in each case follow a Poisson process, calculate the probabilities that at most 4 customers, without regard to sex, will arrive in a 30-minute period.

37. A Geiger counter reaches 30 counts per minute in the vicinity of some radioactive material. Assuming a Poisson process is operating here, calculate the probability that there will be exactly (a) 5 counts in a 10-second period (b) x counts in a period of n seconds.

38. A source of liquid is known to contain bacteria with the mean number of bacteria per cubic centimeter equal to 4. Ten one-cubic-centimeter test tubes are filled with the liquid. Assuming the Poisson distribution is applicable, calculate the probability (a) that all 10 test tubes will show growth, that is, contain at least one bacterium each, (b) that exactly 8 test tubes will show growth.

39. Fit a Poisson function to the following famous data on the number of deaths from the kick of a horse per army corps per year, for 10 Prussian Army Corps for 20 years. The total number of units here, an army corps year, is 200.

x	0	1	2	3	4
f	109	65	22	3	1

40. (a) Given that X possesses a Poisson distribution with mean μ, show that the moment generating function of X is given by $M_X(\theta) = \exp(\mu e^\theta - \mu)$. (b) By differentiating $M_X(\theta)$, verify that the mean is μ and show that the variance is also equal to μ.

41. Show that the Poisson probabilities increase and then decrease unless $\mu \leq 1$. Determine what value of x (function of μ) has maximum probability. Consider the ratio of neighboring probabilities.

42. Assume that the mean number of persons per minute buying ferry tickets is 10. Find an expression for the probability that at least t minutes will elapse before 50 tickets will be sold. Use a Poisson model.

Sec. 2.6

43. A sample of 3 is taken from a box of 12 articles. If 4 of the articles are defective, what is the probability of getting no defectives in the sample?

44. A bag contains 2 red, 3 green, and 4 black balls. If 5 balls are drawn in succession with replacement each time, what is the probability of getting 2 red, 2 green, and 1 black ball?

45. Work problem 44 if there are no replacements of the drawn balls.

46. A box contains 100 items of which 4 are defective. Let X denote the number of defectives found in a sample of 9. (a) Calculate the probability that $X = 2$. (b) Use the binomial approximation to make the calculation. (c) Use the Poisson approximation to make the calculation.

Sec. 3.1

47. Assume $f(x) = 1/x^2, 1 < x$. Let $A_1 = \{x \mid 1 < x < 4\}, A_2 = \{x \mid 3 < x < 6\}$. Calculate (a) $P\{A_1 \cup A_2\}$, (b) $P\{A_1 \cap A_2\}$.

48. A random variable has the density $f(x) = a + bx^2, 0 < x < 1$. Determine a and b so that its mean will be $\frac{2}{3}$.

49. Calculate $P\{\mu - 2\sigma < X < \mu + 2\sigma\}$ for X possessing the density $f(x) = 6x(1 - x), 0 < x < 1$.

50. If $f(x) = cxe^{-x^2/2}, x > 0$, find (a) c, (b) the mean of X, (c) the variance of X.

51. If $f(x) = 1, 0 < x < 1$, find (a) the mean and variance of X, (b) the mean and variance of X^2.

52. Show that the Cauchy distribution, whose density is $f(x) = 1/[\pi(1 + x^2)]$, does not possess finite moments.

Sec. 3.2

53. Given that $f(x) = cx, 0 < x < 2$, find (a) c, (b) μ'_k by integration, (c) $M_X(\theta)$, (d) μ'_k from $M_X(\theta)$.

54. Given that $f(x) = ce^{-x}, x > 0$, find (a) c, (b) $M_X(\theta)$, (c) μ'_k from $M_X(\theta)$.

55. Given that $f(x) = cx^\alpha e^{-x}, x > 0$, α a positive integer, find (a) c using the fact that $\int_0^\infty x^\alpha e^{-x}\, dx = \alpha!$ for α a positive integer, (b) μ'_k from definition, (c) $M_X(\theta)$, (d) μ'_k from $M_X(\theta)$.

56. Find the moment generating function for the triangular distribution whose density is given by $f(x) = x, 0 < x < 1, f(x) = 2 - x, 1 \leq x < 2$.

57. Let $\psi_X(t) = \log M_X(t)$ where $M_X(t)$ is the moment generating function of X. Show that $\psi'(0) = \mu$ and $\psi''(0) = \sigma^2$.

58. Let $\psi_X(t) = E[t^X] = E[e^{X \log t}] = M_X(\log t)$. Show that

$$\psi_X^{(k)}(1) = E[X(X-1) \cdots (X-k+1)]$$

and hence that $\psi_X(t)$ generates factorial moments. Assume X is continuous.

Sec. 3.3

59. Two students agree to meet at a restaurant between 6 and 7 P.M. Find the probability that they will meet if each agrees to wait 15 minutes for the other and they arrive independently at random between 6 and 7.

60. Three points are chosen by chance on the circumference of a circle. What is the probability that they will lie on a semicircle?

Sec. 3.4

61. If X is normally distributed with $\mu = 2$ and $\sigma = \frac{1}{3}$, find (a) $P\{X > 3\}$, (b) $P\{2 < X < 3\}$.

62. If X is normally distributed with $\mu = 2$ and $\sigma = 2$, find a number x_0 such that (a) $P\{X > x_0\} = .10$, (b) $P\{X > -x_0\} = .20$.

63. Assume that the life in hours of a radio tube is normally distributed with mean 100 hours. If a purchaser requires at least 90 percent of them to have lives exceeding 80 hours, what is the largest value that σ can have and still have the purchaser satisfied?

64. A coin is tossed 10 times. Find the probability, both exactly and by the normal approximation, of getting (a) 5 heads, (b) at most 5 heads.

65. A die is tossed 15 times. Counting a 5 or 6 as a success, what is the probability, using the normal curve approximation, of getting (a) 4 successes, (b) at most 4 successes?

66. A die is tossed 60 times. Find the probability of getting 10 aces (a) using the binomial formula and tables of factorials, (b) using the normal curve approximation.

67. Find a number x_0 such that the probability of getting a number of heads between $500 - x_0$ and $500 + x_0$, inclusive, in 1000 tosses of a coin is .80.

68. A manufacturer has found from experience that 4 percent of his product is rejected because of flaws. A new lot of 800 units comes up for inspection. What is the approximate probability that less than 35 units will be rejected?

69. A manufacturer of cotter pins knows from experience that 6 percent of his product is defective. If he sells pins in boxes of 100 and guarantees that at most 10 pins will be defective, what is the approximate probability that a box will fail to meet the guaranteed quality?

70. Suppose you wish to construct a control chart for the proportion of words incorrectly typed by a typist per hour. If she typed 1200 words an hour, on the average, for 6 hours a day, for 10 days, and she mistyped 360 words in that total period of time, what two numbers would you use for boundaries for the control chart?

71. In the manufacturing of parts, the following data were obtained for the daily percentage defective for a production averaging 1000 parts a day. Construct

a control chart and indicate times when production was out of control. Read the
data a row at a time.

2.2	2.3	2.1	1.7	3.8	2.5	2.0	1.6	1.4	2.6
1.5	2.8	2.9	2.6	2.5	2.6	3.2	4.6	3.3	3.0
3.1	4.3	1.8	2.6	2.1	2.2	1.8	2.4	2.4	1.6
1.7	1.6	2.8	3.2	1.8	2.6	3.6	4.2.		

72. If you wished to estimate the proportion of Republicans in a certain district
and wanted your estimate to be correct within .03 unit with a probability of .90,
how large a sample should you take (a) if you know the true proportion is near .4?
(b) if you have no idea what the true proportion is?

73. Assume that telephone calls coming into a switchboard follow a Poisson
distribution at the rate of 10 calls per minute. If the switchboard can handle at
most 20 calls per minute, what is the probability that in a one-minute period the
switchboard will be overloaded? Use a normal approximation.

74. Assuming that the number of white blood cells per unit of volume of diluted
blood counted under a microscope follows a Poisson distribution with $\mu = 100$,
what is the probability, using a normal approximation, that a count of 90 or less
will be observed?

75. For $n = 12$ and $p = \frac{1}{4}$, plot on the same piece of graph paper (a) the binomial
histogram, (b) the Poisson histogram, (c) the fitted normal curve by ordinates.
Note the extent to which (b) and (c) approximate (a).

76. In firing at a target assume that the horizontal distance that a shot hits from
the center line is normally distributed with $\sigma = 4$ feet. (a) In 100 shots how many
would be expected to miss the target if it is 12 feet wide and sufficiently high?
(b) How many shots would need to be fired to be certain with a probability of .95
of getting 40 or more shots within 4 feet of the centerline?

77. Assume that the height of adult males is normally distributed with $\mu = 69$
inches and $\sigma = 3$ inches. What is the conditional probability that an individual will
be taller than 72 inches if it is known that he is taller than 70 inches?

78. Find μ_k for the normal distribution by using the integral definition and
repeated integration by parts.

Sec. 3.6

79. Let T denote the life of a radio tube in months with density $f(t) = ae^{-at}$,
$t > 0$. If $a = .02$, for how many months of life should the manufacturer guarantee
his tubes if he wants the probability to be .80 that a tube will satisfy the guarantee?

80. A mechanical system will operate only if both of two components operate.
If each of the components possesses a negative exponential distribution of time to
failure with a mean of 2 hours, what is the expected time before failure for the
system?

81. Assume that the length of a telephone conversation X has the density
$f(x) = ae^{-ax}$. Show that the probability of a conversation lasting more than
$t_1 + t_2$ minutes, given that it has already lasted at least t_1 minutes, is equal to the
unconditional probability that it will last more than t_2 minutes.

CHAPTER 4

Nature of Statistical Methods

1 INTRODUCTION

The preceding chapters have been concerned with developing the basic principles of probability and with presenting a few of the probability distributions that have been found by experience to be particularly useful in solving certain classes of problems. The motivation has been on constructing models for experiments of the repetitive type, whether real or conceptual. The advantage of such models is that they enable one to study properties of the experiment and to make predictions about the outcomes of future trials of the experiment, both of which would be difficult or impossible to do without such a model.

The process of constructing a model on the basis of experimental data and drawing conclusions from it is an example of *inductive inference*. When it is applied to statistical problems it is usually called *statistical inference*. The principal occupation of statisticians is making statistical inferences by means of experimental data.

Most often the statistician is interested in constructing a mathematical model for a random variable associated with an experiment rather than for the experiment itself. For example, if X represents the number of defective parts that will be found in a lot of 100 parts submitted for inspection, he would prefer to have a model that predicts the frequency with which the various values of X will be obtained rather than one that predicts the frequency with which the various possible experimental outcomes will occur when 100 parts are selected from the production process. As a consequence, most of the models chosen by statisticians are density functions of random variables. Statistical inferences are therefore usually inferences about density functions.

As an illustration of the preceding ideas, suppose a biologist has observed that 44 out of 200 insects of a given type possess markings that are different from those of the rest. Suppose, further, that the biologist suspects that the markings are inherited according to a law which implies that 25 percent of

97

such insects would be expected to possess the less common markings. If he assumes that the inheritance law is operating here and lets X represent the number of insects out of 200 that will possess the less common markings, then the model that he would naturally select is the binomial density

$$(1) \qquad f(x) = \frac{200!}{x!\,(200-x)!}\left(\frac{1}{4}\right)^x\left(\frac{3}{4}\right)^{200-x}.$$

If there had been no theory to suggest that $\frac{1}{4}$ of such insects should possess the unusual markings, the biologist might have chosen this same density function with the probability $\frac{1}{4}$ replaced by the observed relative frequency .22. By means of (1) it should be possible for the biologist to make predictions about future sets of 200 observations and thus detect any disagreements with his theory.

The purpose of this chapter is to introduce a few of the simpler methods for making statistical inferences so that they will be available for solving problems of this type in the following chapters. The approach here will be very informal. Precise definitions and a systematic treatment of the theory of statistical inference will be studied in Chapter 8.

2 DATA

It is convenient in discussing statistical methods to call the totality of possible experimental outcomes the *population* of such outcomes. Then a set of data obtained from performing the experiment a number of times is called a *sample* from the population. In this language statistical inference consists in drawing conclusions about a population by means of a sample extracted from the population. A basic problem, therefore, is how to extract information from samples for use in studying the populations from which the samples were drawn.

The type of information that should be extracted from a set of data depends upon the nature of the data and upon the model that is likely to be selected. In some problems one knows from theoretical considerations or from experience with similar problems what model should be used. For example, the binomial density that was introduced in (1) is such a model. All that is really needed from experimental data for such models is information that will give good estimates of the parameters involved. In other problems neither theory nor experience is available to assist one in selecting a model. Then it is necessary to use experimental data to decide on a reasonable type of model before one can proceed further.

In considering the nature of the data it is particularly important to distinguish between those sets of data for which the order in which the observations were obtained yields useful information and those sets for which it does not. For example, if one were interested in studying weather phenomena or the stock market from day to day, the order would be very important. Industrial experience indicates that the information obtained from considering the order in which articles are manufactured is indispensable for efficient production. However, if one were interested in studying certain characteristics of college students and had selected a set of students by choosing every twentieth name in a college directory, he would hardly expect the order in which the names were obtained to be of any value in the study. Methods for dealing with data for which order is important are considered in later chapters. In this chapter the emphasis is on techniques that do not use order information. The material in the later chapters will enable the investigator to decide whether he is justified in assuming that he may ignore the order information present in his data.

3   CLASSIFICATION OF DATA

Suppose one is given the weights of 200 college men and he wishes to use them to study the weight distribution of such men. Now it is very difficult to look at 200 measurements and obtain any reasonably accurate idea of how those measurements are distributed. For the purpose of obtaining a better idea of the distribution of weights it is therefore convenient to condense the data somewhat by classifying the measurements into groups. It will then be possible to graph the modified distribution and learn more about how weights are distributed. This condensation will also be useful for simplifying the computations of various averages that need to be evaluated, particularly if fast computing facilities are not available. These averages will supply additional information about the distribution. Thus the purpose of classifying data is to assist in the extraction of certain kinds of useful information concerning the underlying distribution.

In classifying data it is usually convenient to use from 10 to 20 classes. Thus, in classifying the weights of a sample of college men it would be convenient to choose a class interval of 10 pounds because such weights might be expected to range between, say, 110 and 270 pounds. Table 1 shows how a set of 200 steel rod diameters whose values ranged from .431 inch to .503 inch were classified. Since the diameters were measured to the nearest thousandth of an inch the boundaries of the class intervals were chosen to $\frac{1}{2}$ unit beyond this measurement accuracy to insure that no measurement would fall on a boundary. It is assumed in such a classification that all measurements falling in a given class interval are assigned the value at the midpoint of the interval. This midpoint value is called the class mark. After each measurement has been recorded in its proper class by means of a vertical bar, as shown in

TABLE 1

Class boundaries	Frequencies	Class marks: x	Frequencies: f
.4305–.4355	//	.433	2
.4355–.4405	/////	.438	5
.4405–.4455	///// //	.443	7
.4455–.4505	///// ///// ///	.448	13
.4505–.4555	///// ///// ///// ////	.453	19
.4555–.4605	///// ///// ///// ///// ///// //	.458	27
.4605–.4655	///// ///// ///// ///// ///// ////	.463	29
.4655–.4705	///// ///// ///// ///// /////	.468	25
.4705–.4755	///// ///// ///// ///// ///	.473	23
.4755–.4805	///// ///// ////	.478	14
.4805–.4855	///// ///// /////	.483	15
.4855–.4905	///// ////	.488	9
.4905–.4955	///// /	.493	6
.4955–.5005	////	.498	4
.5005–.5055	//	.503	2

Table 1, the results of the classification are recorded in the form of a frequency table as shown in the second half of Table 1.

4 GRAPHICAL REPRESENTATION OF EMPIRICAL DISTRIBUTIONS

A rough idea of how the values of a random variable are distributed can be obtained from inspecting the histogram. The histogram for the data of Table 1 for absolute frequencies is given in Fig. 1. It should be noted that the class marks are at the midpoints of the bases of the rectangles making up the histogram.

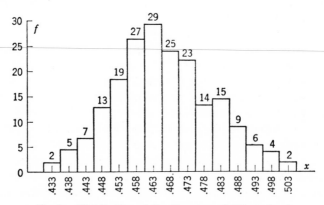

Fig. 1. Distribution of the diameters of 200 steel rods.

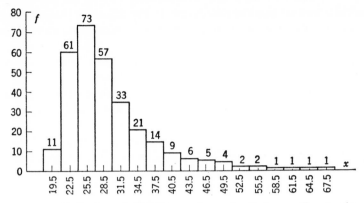

Fig. 2. Distribution of 302,000 marriages classified according to the age of the bride-groom. Frequencies are in units of 1000.

Fortunately, many important distributions to be found in nature and industry are of a relatively simple form. They usually range from a bell-shaped distribution, like that in Fig. 1, to something resembling the right half of a bell-shaped distribution. A distribution of the latter type is said to be skewed, skewness meaning a lack of symmetry with respect to a vertical axis. It will be found, for example, that the following variables have distributions that possess such forms in increasing degrees of skewness: stature, various industrial measurements, weight, age at marriage, mortality age for certain diseases, and wealth. Figures 1, 2, and 3 represent three typical distributions with increasing degrees of skewness.

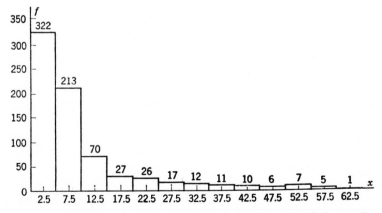

Fig. 3. Distribution of 727 deaths from scarlet fever classified according to age.

It will be observed that the distribution in Fig. 1 has the shape of a normal curve, and that therefore the normal distribution might serve as a satisfactory model for the distribution of diameters of steel rods. Many linear industrial measurements possess distributions of this type and are successfully treated by using a normal curve model.

The distribution given in Fig. 3 possesses a shape that suggests the possibility of using an exponential distribution as a model, whereas Fig. 2 requires a more sophisticated model. Since a gamma distribution with its two parameters possesses a great deal of flexibility, it might well serve as a model for the distribution of Fig. 2. An inspection of the curves shown in Fig. 13, Chapter 3, should fortify this optimism.

5 EMPIRICAL MOMENTS

Although a histogram such as those shown in Figs. 1, 2, and 3 yields a considerable amount of general information concerning the distribution of a set of sample measurements, more precise and useful information for studying a distribution can be obtained from an arithmetical description of the distribution. For example, if the histogram of weights for a sample of 200 men from one college were available for comparison with the histogram of a similar sample from another college, it might be difficult to state, except in very general terms, how the two distributions differ. Rather than compare the two weight distributions in their entirety, it might suffice to compare the average weights and the variation in weights of the two groups.

The nature of a statistical problem largely determines whether a few simple arithmetical properties of the distribution will be enough to describe it satisfactorily. Most of the problems that are encountered in this book are the type that require only a few basic properties of the distribution for their solution. For simple frequency distributions, such as those whose graphs are given in Figs. 1, 2, and 3, this description is accomplished satisfactorily by means of the low-order moments of the distribution. In many problems the statistician is concerned only with the first and second moments. In a few problems he uses the first four moments, but seldom does he use more than four. One reason for this is that the higher moments are so unstable in repeated sampling experiments that little additional reliable information can be obtained from them.

Let $x_1, x_2, \cdots, x_n$ denote the observed values of a sample of size n of the random variable X. Then, by analogy with theoretical moments, empirical moments are defined as follows:

(2) DEFINITION: *The kth moment about the origin of an empirical distribution is given by*

$$m'_k = \frac{1}{n}\sum_{i=1}^{n} x_i^k .$$

Empirical moments are also called sample moments because they are based on sample values.

The first moment, m'_1, is traditionally denoted by the symbol $\bar{x}$. It gives the center of gravity of an empirical distribution just as μ does for a theoretical distribution and it serves to measure where the empirical distribution is centered. It is called the *sample mean*.

By analogy with the definition for probability distributions, empirical moments about the mean are defined as follows:

(3) DEFINITION: *The kth moment about the mean of an empirical distribution is given by*

$$m_k = \frac{1}{n}\sum_{i=1}^{n} (x_i - \bar{x})^k .$$

Since the second moment about the mean, m_2, is used so often it is assigned the special symbol s^2 and is called the *sample variance*. Correspondingly, s is called the *sample standard deviation*. For computing s^2 it is often convenient to use the following formula which is the analogue of formula (6), Chapter 3,

$$s^2 = m'_2 - \bar{x}^2.$$

If the observational values $x_1, x_2, \cdots, x_n$ have been classified into a frequency table with x_i representing the ith class mark, f_i representing the number of observations in the ith interval, and h denoting the number of intervals, then the preceding definitions will assume the following forms:

(4) $$m'_k = \frac{1}{n}\sum_{i=1}^{h} x_i^k f_i$$

and

(5) $$m_k = \frac{1}{n}\sum_{i=1}^{h} (x_i - \bar{x})^k f_i .$$

The value of $\bar{x}$ in (5) is assumed to be the value of m_1 obtained from (4) and not from (2). Strictly speaking, formulas (4) and (5) define moments only for the classified empirical distribution and are only approximations to the values given by (2) and (3); however, the approximations are usually so good that in practice no distinction is made between these two sets of values. For example, $\bar{x}$ and s^2 are called the sample mean and sample variance whether they are obtained from formulas (2) and (3) or from formulas (4) and (5).

TABLE 2

x_i	49.5	149.5	249.5	349.5	449.5	549.5	649.5	749.5	849.5	949.5
f_i	6	28	88	180	247	260	133	42	11	5

There is no point in classifying data if only the moments of a distribution are desired. The classification is for the purpose of observing geometrically the nature of the empirical distribution. If the data have been classified for this purpose and if the values of, say, $\bar{x}$ and s^2 are desired, then it may be easier to calculate them by means of formulas (4) and (5) than by (2) and (3). As remarked before, the differences will usually be insignificant.

For the purpose of becoming a little more familiar with the standard deviation as a measure of concentration of a distribution about its mean, consider the following empirical distribution of telephone conversations in seconds shown in Table 2. The histogram for this distribution is shown in Fig. 4. From Fig. 4 it appears that a normal distribution model might be appropriate for the length of telephone conversations. If so, since the intervals $(\mu - \sigma, \mu + \sigma)$ and $(\mu - 2\sigma, \mu + 2\sigma)$ contain 68 and 95 percent, respectively, of the area under a normal curve, the intervals $(\bar{x} - s, \bar{x} + s)$ and $(\bar{x} - 2s, \bar{x} + 2s)$ would be expected to yield approximately the same percentages with respect to the empirical distribution.

Calculations for the data of Table 2 using formulas (4) and (5) gave the values $\bar{x} = 475$ and $s = 151$, correct to the nearest integer; consequently

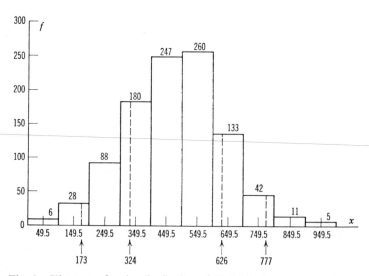

Fig. 4. Histogram for the distribution of 1000 telephone conversations.

the intervals $(\bar{x} - s, \bar{x} + s)$ and $(\bar{x} - 2s, \bar{x} + 2s)$ become the intervals (324, 626) and (173, 777). The end points of these intervals are shown on Fig. 4 by means of arrows. The number of observations lying within these intervals may be found approximately by interpolating as though the observations in a given interval were dispersed uniformly throughout the interval. This assumption implies that on the histogram any fractional part of a class interval will include the same fractional part of the frequencies in that interval. If interpolation is carried to the nearest unit, it will be found that the interval (324, 626) includes $136 + 247 + 260 + 35$ measurements, which is 67.8 percent of them. The interval (173, 777) excludes $6 + 21 + 9 + 11 + 5$ measurements, which is 5.2 percent. For a histogram as irregular as this, these results are unusually close to the theoretical percentages of 68 and 5 for a normal distribution.

6 STATISTICAL INFERENCE

In a general sense, statistical inference is a type of decision making based on probability. In a more limited sense, however, it will be found that a large share of the inferences made by statisticians fall into one of two categories. Either they involve the estimation of parameters, or other characteristics, of the density function that has been selected as a model for a random variable, or they involve the testing of some hypothesis about the model. These two types of statistical inference will be discussed only briefly here but will be studied thoroughly in Chapter 8.

6.1 Estimation

Most of the problems of estimation in statistics are those of estimating the parameters of a probability density function. For example, a telephone company interested in studying problems related to the length of telephone conversations would want to estimate the parameters μ and σ of the normal density assumed to be the appropriate model for telephone conversations.

Two kinds of estimates of parameters are in common use. One is called a point estimate and the other is called an interval estimate. A point estimate is the familiar kind of estimate; that is, it is a number obtained from computations on the observed values of the random variable which serves as an approximation to the parameter. For example, the observed proportion of defective parts in 50 consecutive parts turned out by a machine is a point estimate of the true proportion p for that machine. An interval estimate is an interval determined by two numbers obtained from computations on the observed values of the random variable that is expected to contain the true value of the parameter in its interior. Interval estimates are considered briefly in Chapter 5 and more fully in Chapter 8; therefore, only point estimates are discussed here.

The method of estimation that will be used here is known as the *method of moments*. If a parameter of a density function, such as the parameter μ for the Poisson density or the parameter β for the exponential density, is a moment of the distribution, then its estimate will be the corresponding sample moment. Since μ and β are the means of their respective densities they would both be estimated by the sample mean $\bar{x}$. The two parameters μ and σ^2 of a normal density are moments of the distribution; therefore they would be estimated by the sample mean $\bar{x}$ and sample variance s^2.

If a distribution has only one unknown parameter but that parameter is not a moment of the distribution, the parameter may still be estimated by the method of moments by calculating the first moment of the distribution, which will be a function of the parameter, and equating it to $\bar{x}$. The solution of the resulting equation for the unknown parameter value will be the desired estimate. If the distribution had, say, two unknown parameters that were not moments, the same procedure would be followed with respect to the first two moments of the distribution.

As an illustration for which the parameters are not moments, suppose both parameters of the gamma density are to be estimated by means of the first and second sample moments. From formula (37), Chapter 3, it is known that $\mu = \beta\alpha$ and $\sigma^2 = \beta^2\alpha$. Estimates of α and β are therefore obtained by solving the equations

$$\beta\alpha = \bar{x} \quad \text{and} \quad \beta^2\alpha = s^2 .$$

The solution of these equations is $\alpha = \bar{x}^2/s^2$ and $\beta = s^2/\bar{x}$.

6.2 Testing Hypotheses

In its general form a statistical hypothesis is an assertion about the density function of a random variable. Thus, the assertion that a random variable possesses a normal distribution is an example of a statistical hypothesis. The statement that the mean of a Poisson random variable is 10 is also a statistical hypothesis. Most of the hypotheses that will be considered in the next few chapters are of the latter type. That is, it will be assumed that the type of density function is given and the hypothesis will consist of an assertion about the value of a parameter of the density function.

As an illustration for a discrete variable, consider again the problem of the biologist which was discussed in section 1. If p denotes the proportion of all insects possessing the unusual markings, then the assertion that $p = \frac{1}{4}$ is a statistical hypothesis. It is assumed here that the binomial distribution is the appropriate model. As an illustration for a continuous variable, consider a problem related to the exponential distribution and radioactive decay, which was discussed in section 3.6. Chapter 3. If X denotes the time that will elapse between successive trippings of a Geiger counter and if it is assumed that the density function of X is of the form

$$(6) \qquad\qquad f(x) = \theta e^{-\theta x}$$

then the assertion that the value of θ is 2 is a statistical hypothesis.

Now consider what is meant by testing a statistical hypothesis. In its most general form a test of a statistical hypothesis is a procedure for deciding whether to accept or reject the hypothesis.

This definition permits the statistician unlimited freedom in designing a test; however, he will obviously be guided by its desirable properties. Thus a simple but ordinarily useless test is one in which a coin is tossed and it is agreed to accept the hypothesis in question if, and only if, the coin turns up a head.

In order to illustrate how the statistician proceeds in attempting to design a test that possesses desirable properties, consider a problem related to the density function (6). Suppose a physicist is certain, from theoretical or experimental considerations, that the time that elapses between two successive trippings on a counter possesses the density function (6). Suppose further that he is quite certain that for the material with which he is working the value of the parameter is either 2 or 1, with his intuition favoring the value 2. To assist him in making a choice, the statistician might proceed in the following manner.

Assume that the density (6) applies. Assume temporarily that the parameter θ has the value 2. This assumption is the statistical hypothesis to be tested. Denote this hypothesis by H_0. Let H_1 denote the alternative hypothesis that $\theta = 1$. Thus the problem is one of testing the hypothesis H_0 against the alternative H_1.

To test H_0, a single observation will be made on the random variable X; that is, a single time interval between two successive trippings of the counter will be measured. In real-life problems one usually takes several observations, but to avoid complicating the discussion at this stage only one observation is taken here. On the basis of the value of X obtained, a decision will be made either to accept H_0 or to reject it. The latter decision, of course, is equivalent to accepting H_1. The problem then is to determine what values of X should be selected for accepting H_0 and what values for rejecting H_0. If a choice has been made of the values of X that will correspond to rejection, then the remaining values of X will necessarily correspond to acceptance. It is customary to call the rejection values the *critical region* of the test. For this problem, the sample space may be considered as the positive half of the

x axis. Every possible outcome can be represented by a point on this line with its x coordinate giving the value of the associated random variable X. Since only one observation is being made here, the sample space is one dimensional. If n observations were to be taken, the corresponding sample space would be n dimensional, with one coordinate axis for each observation. In terms of the foregoing language, the problem of constructing a test for H_0 for the problem under discussion is therefore the problem of choosing a critical region on the positive x axis.

6.3 Two Types of Error

Now suppose that the statistician arbitrarily selects the part of the x axis to the right of $x = 1$ as the critical region. To decide whether this was a wise choice, consider its consequences. If H_0 is actually true and the observed value of X exceeds 1, H_0 will be rejected because it has been agreed to reject H_0 whenever the sample point falls in the critical region. This, of course, is an incorrect decision. This kind of error is called the type I error. On the other hand, if H_1 is actually true and the observed value of X does not exceed 1, H_0 will be accepted. This also is an incorrect decision. This kind of error is called the type II error. These two incorrect decisions, as well as the two correct decisions that are possible here, are displayed in Table 3.

TABLE 3

	H_0 True	H_1 True
$x > 1$ (reject H_0)	Type I error	Correct decision
$x \leq 1$ (accept H_0)	Correct decision	Type II error

It is necessary to measure in some way the seriousness of making either one of these errors before one can judge whether the choice of a critical region was wise. This can be accomplished by using what is known as the size of an error as the measure of its seriousness. The size of the type I error is the probability of making a type I error, which is the probability that the sample point will fall in the critical region when H_0 is true. The size of the type II error is the probability of making a type II error, which is the probability that the sample point will fall in the noncritical region when H_1 is true.

Now, in terms of the sizes of the two types of error, it is possible to introduce a simple principle to follow in determining good tests of hypotheses. It may be expressed as follows. Among all tests possessing the same size type I error, choose one for which the size of the type II error is as small as possible.

Other principles can easily be suggested: for example, minimizing the sum of the sizes of the two types of error. However the preceding principle has proved to be very useful in constructing tests. A statistician often determines in advance what size type I error he will tolerate. Then if the number of runs of his experiment is fixed, he will attempt to construct his test to minimize the size of the type II error. For a fixed number of runs of an experiment, the size of the type II error will usually increase if the size of the type I error is decreased; hence one cannot make the type I error as small as desired without paying for an increasingly large type II error. In real-life experiments it is often necessary to adjust the type I error until a satisfactory balance has been reached between the sizes of the two errors. The type I and type II error sizes are usually denoted by the letters α and β, respectively. For the sake of avoiding lengthy discussions regarding the practical consequences of possible choices for the sizes of these two errors, a convention of almost always choosing the size of the type I error as .05 is adopted. This means that approximately 5 percent of the time true hypotheses being tested will be rejected. The value of $\alpha = .05$ is quite arbitrary here and some other value could have been agreed on; however, it is the value of α most commonly used by applied statisticians. In any applied problem one can calculate the value of β and then adjust the value of α if the value of β is unsatisfactory when $\alpha = .05$. This works both ways, of course. For a very large experiment, with α fixed at .05, it might turn out that β would be considerably smaller than .05. If the type I error were considered more serious than a type II error, one would need to adjust the test to make α smaller than β, which would then make α smaller than .05.

Now consider the problem under discussion from the point of view of this principle. If the sizes of the two types of error for the selected critical region, namely $x \geq 1$, are denoted by α and β, respectively, then, because the two competing hypotheses here are $H_0 : \theta = 2$ and $H_1 : \theta = 1$, it follows from (6) that

$$\alpha = \int_1^\infty 2e^{-2x}\, dx = .135$$

and

$$\beta = \int_0^1 e^{-x}\, dx = .632 \, .$$

Since probabilities correspond to areas under graphs of density functions, these values may be represented geometrically as indicated in Fig. 5.

In order to decide whether the preceding test, that is, the choice of a critical region, was a good one, it is necessary to compare this test with other

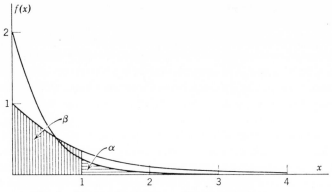

Fig. 5.   Graphs showing sizes of two types of error.

tests for which $\alpha = .135$. Here only one other test is considered as a competitor, namely, the test that uses the left "tail" rather than the right "tail" of the graph of the density function under H_0 as the critical region. Thus the critical region for the competing test consists of the part of the axis to the left of the point x_0 where x_0 is such that

$$\int_0^{x_0} 2e^{-2x}\, dx = .135 \, .$$

If the integration is performed and tables of exponentials are consulted, it will readily be found that $x_0 = .07$. From (6), it then follows that

$$\beta = \int_{.07}^{\infty} e^{-x}\, dx = .932 \, .$$

Graphs showing the sizes of the two types of error for the competing test are given in Fig. 6.

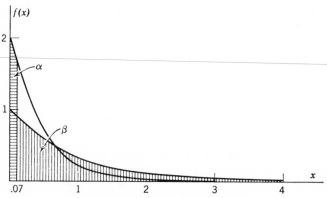

Fig. 6.   Graphs for a competing test.

It is clear from comparing the two values of β that the first test is superior to the second. The second test would incorrectly reject H_1 93 percent of the time, whereas the first test would do so only 63 percent of the time. Both tests have very large type II errors, but this is to be expected when only one observation is taken. By using methods that are developed in Chapter 8, it can be shown that the first test selected is the best test that can be constructed for this problem according to the principle of test construction that has been adopted here.

This principle of test construction applies to discrete variable problems also. As a simple illustration of how to make discrete variable computations, consider the following academic problem. A coin is known to be either an honest coin or one that yields twice as many heads as tails. A decision is to be made as to which type of coin it is by tossing it three times and observing the number of heads, X, that result.

The problem may be formalized by assuming that X is a binomial variable and choosing

$$H_0 : p = \tfrac{1}{2} \quad \text{and} \quad H_1 : p = \tfrac{2}{3} \, .$$

Here p denotes the probability that a head will be obtained when the coin is tossed once. Since the coin is to be tossed three times, the random variable X can assume only the values 0, 1, 2, or 3. The four points on the x axis corresponding to these values may be chosen as the sample space for X. When H_0 is true, the probabilities that should be assigned to the four sample points in this space are those displayed in Fig. 7. These values were calculated using the binomial distribution with $p = \tfrac{1}{2}$.

Now consider two different choices for the critical region of this test, namely, the point $x = 0$ and the point $x = 3$. Except for convenience, these two parts of the sample space were chosen quite arbitrarily. They should serve to illustrate techniques and principles in test construction for discrete variables. From Fig. 7 it will be seen that both of these critical regions yield a type I error of size $\alpha = \tfrac{1}{8}$; hence they are equally good as far as making type I errors is concerned.

When H_1 is true, calculations using the binomial density with $p = \tfrac{2}{3}$ will show that the probabilities that should be assigned to the four sample points are those listed in Fig. 8. Now when $x = 0$ is chosen as the critical region for the test, the size of the type II error is equal to $\beta = \tfrac{26}{27}$ because that is the probability that X will not assume the value 0. On the other hand, when $x = 3$ is chosen as the critical region, the value of β is $\tfrac{19}{27}$ because that is

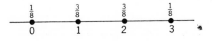

Fig. 7. H_0 probabilities.

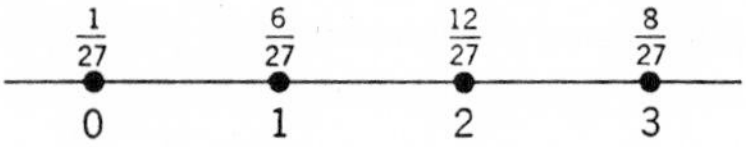

$$\frac{1}{27} \qquad \frac{6}{27} \qquad \frac{12}{27} \qquad \frac{8}{27}$$

0   1   2   3

Fig. 8. H_1 probabilities.

the probability that X will not assume the value 3. Thus it is clear that $x = 3$ is a better critical region than $x = 0$ for testing H_0 against H_1.

In discussing critical regions, it is customary to call them *critical regions of size* α if the magnitude of the type I error is α. Thus in the preceding illustration the two competing critical regions were of size $\frac{1}{8}$.

One difficulty in applying these methods to discrete variable problems is that critical regions of specified sizes cannot always be chosen without resorting to other devices. In the preceding illustration, for example, one cannot directly choose a critical region of size $\alpha = \frac{1}{10}$. This difficulty is seldom much of a problem in real-life applications because then experiments are usually sufficiently large to permit a wide choice of sizes for the type I error. Moreover, there are techniques available, which are discussed in a later chapter, that enable one to construct critical regions of any desired size, even for problems such as the preceding one.

In the following chapters tests of hypotheses are made without being concerned whether the critical region selected is the best possible; however, after Chapter 8 has been studied it will be found that the tests in the earlier chapters were well chosen from this point of view. For the simpler problems, the critical region that the experimenter carefully selects on an intuitive basis is likely to be a good one from the point of view of the basic principle.

EXERCISES

Sec. 4

1. Given the following frequency table of the heights in centimeters of 1000 students, draw its histogram, indicating the class marks.

x	155–157	158–160											
f	4	8	26	53	89	146	188	125	92	60	22	4	1	1

Sec. 5

2. Given the following frequency table of the diameters in feet of 56 shrubs from a common species, (a) draw its histogram and (b) guess by merely inspecting the histogram the values of $\bar{x}$ and s.

x	1	2	3	4	5	6	7	8	9	10	11	12
f	1	7	11	16	8	4	5	2	1	0	0	1

3. For the data of problem 2 calculate $\bar{x}$.

4. For the data of problem 2 calculate s.

5. For the histogram of problem 2, using the results of problems 3 and 4, calculate the approximate percentages of the data that lie within the intervals $\bar{x} \pm s$ and $\bar{x} \pm 2s$.

6. Show that $\sum_{i=1}^{h} (x_i - \bar{x})f_i = 0$.

7. If stature of adult males may be assumed to possess a normal distribution, what would you guess the standard deviation of stature to be if you estimate a two-standard deviation interval about the mean through your knowledge of male stature?

8. If the scores on a set of examination papers are changed by (a) adding 10 points to all scores, (b) increasing all scores by 10 percent, what effects will these changes have on the mean and standard deviation?

9. What would you judge a distribution to be like if the variable can assume only positive values and the mean and standard deviation have the same value?

10. By expanding the binomial in formula (5) in the text and summing term by term, derive a formula for calculating the kth moment about the mean in terms of the kth and lower order moments about the origin.

11. Suppose only the two means $\bar{x}_1$ and $\bar{x}_2$ are available from two sets of observations of sizes n_1 and n_2 made on the variable X. Find a formula for $\bar{x}$ in terms of $\bar{x}_1$ and $\bar{x}_2$.

12. If the two standard deviations s_1 and s_2 are also available in problem 11, show that the variance s^2 of the combined set can be obtained from the formula

$$s^2 = (n_1 s_1{}^2 + n_2 s_2{}^2)/(n_1 + n_2) + n_1 n_2 (\bar{x}_1 - \bar{x}_2)^2/(n_1 + n_2)^2 .$$

Sec. 6

13. Suppose you wish to test a hypothesis H_0 against an alternative H_1 by tossing a coin once and agreeing to accept H_0 if a head shows and to accept H_1 otherwise. (a) What are the values of α and β for this test? (b) What would α and β become if you tossed the coin twice and agreed to accept H_0 if 2 heads showed and to accept H_1 otherwise?

14. Given the density $f(x; \theta) = 1/\theta$, $0 < x < \theta$, and 0 elsewhere, if you are testing the hypothesis $H_0: \theta = 3$ against $H_1: \theta = 2$ by means of a single observed value of X, (a) what would the sizes of the type I and type II errors be if you chose the interval $x < 1$ as the critical region? (b) What would the sizes of those errors be if you chose the interval $1 < x < 2$ as the critical region?

15. Given that X has the density $f(x; \theta) = \frac{1}{4}$, $\theta - 2 < x < \theta + 2$, and 0 elsewhere, if $H_0: \theta = 4$ and $H_1: \theta = 5$ and the critical region is to be of size $\alpha = .25$ and to consist of a single interval, show by a sketch what critical region you would choose and determine the value of β for that choice, assuming the test is based on a single observed value of X.

16. What critical region consisting of a single interval with $\alpha = .5$ would you choose in problem 14 if you wanted a critical region that minimizes β?

17. Given $f(x; \theta) = (1 + \theta)x^\theta$, $\theta > 0$, $0 \leq x \leq 1$, and 0 elsewhere, if the hypothesis $H_0: \theta = 2$ is to be tested by taking a single observation on X and using the interval $x < .25$ as the critical region, (a) calculate the value of α, and (b) calculate the probability of determining that H_0 is false if the true value of θ is 3.

18. Let X be a random variable whose density values under H_0 and H_1 are as follows.

x	1	2	3	4	5	6	7
$f(x \mid H_0)$	.01	.02	.03	.05	.05	.07	.77
$f(x \mid H_1)$	.03	.09	.10	.10	.20	.18	.30

(a) List all critical regions whose size is equal to .10.
(b) List all critical regions whose size does not exceed .10.
(c) Among the regions in (a) which has the smallest β?
(d) Are there any in (b) that have a still smaller β?

19. A box is known to contain either 3 red and 4 black balls or 4 red and 3 black balls. Three balls are to be drawn and on the basis of their colors a decision as to the contents of the box will be made. If H_0 corresponds to 3 red and 4 black balls and if H_0 will be accepted unless 3 red balls are obtained, what are the values of α and β here?

20. A bag is known to contain 8 balls, of which either 1 or 2 are white and the rest black. To test the hypothesis that there is only 1 white ball, balls are drawn until a white ball appears. Let X equal the number of balls drawn and find $f(x)$ under both hypotheses. Choose a good critical region for the test and find its value of α and β.

21. If you rolled a die 240 times and obtained 50 sixes would you decide the die was biased in favor of sixes? (b) If you repeated the experiment and obtained 48 sixes, would you conclude that the second experiment justified your decision in (a) or would you conclude differently?

CHAPTER 5

Sampling Theory

1 RANDOM SAMPLING

In the applications of the binomial distribution it was pointed out that the binomial model is strictly valid only if the trials of the experiment are independent and p is constant from trial to trial. In the language of sampling, this means that samples must be obtained by a method that possesses these two properties.

The theory that is about to be developed for continuous variables is based on assumptions very similar to those used to derive the binomial distribution. The first assumption is that the successive trials of the experiment are independent and the second is that the density function of the random variable remains the same from trial to trial. If the theory is to be applicable to real experimental data, it is necessary that the data be obtained by a sampling method that possesses these two properties. In order to express these properties in a mathematical form, consider the following notation and procedure.

Let $f(x)$ be the density function of the continuous random variable X and let a sample of size n be drawn. The resulting sample values are denoted by x_1', x_2', $\cdots$, x_n'. If a second sample of size n were drawn, the resulting sample values would be denoted by x_1'', x_2'', $\cdots$, x_n'', and similarly for additional samples. These values are conveniently arranged as follows:

$$x_1', \quad x_2', \quad \cdots, \quad x_n'$$
$$x_1'', \quad x_2'', \quad \cdots, \quad x_n''$$
$$x_1''', \quad x_2''', \quad \cdots, \quad x_n'''$$
$$\cdot \qquad \cdot \qquad \qquad \cdot$$
$$\cdot \qquad \cdot \qquad \qquad \cdot$$
$$\cdot \qquad \cdot \qquad \qquad \cdot$$

Now consider the values in the first column. These values may be treated as the values of a random variable X_1 with a density function $f_1(x_1)$. In the same manner the values in the second column may be treated as the values of a random variable X_2 with density function $f_2(x_2)$, and similarly for the remaining columns.

In this notation the requirement that the density function of the random variable X shall remain constant from trial to trial means that the random variables $X_1, X_2, \cdots, X_n$ must possess the original density function, that is, that

$$f_1(x) = f_2(x) = \cdots = f_n(x) = f(x) \,.$$

In this same notation the requirement that the trials shall be independent means that the variables $X_1, X_2, \cdots, X_n$ must be independent. A method of sampling that possesses these two properties is called *random sampling*. In view of formula (24), Chapter 2, and the preceding discussion, random sampling may be defined mathematically in the following manner.

(1) DEFINITION: *Random sampling is a method of sampling for which*

$$f(x_1, x_2, \cdots, x_n) = f(x_1)f(x_2) \cdots f(x_n)$$

where $f(x)$ is the density function of the random variable X for the population being sampled and where $X_1, X_2, \cdots, X_n$ are random variables corresponding to the n trials of the sample.

Although the variable X in the preceding discussion was treated as a continuous variable, definition (1) applies to both continuous and discrete variables.

As an illustration of a continuous random variable for which the sampling method approximates random sampling, let X be the distance the end of a spinning pointer is from the 0 point, as measured along the circumference, after it comes to rest. Figure 1 indicates the nature of this variable. If a sample of size 5 were desired, the pointer would be spun five times and the distances recorded. Now, if a pointer is spun repeatedly and the resulting values of X are marked off into consecutive sets of five, it will usually be found that the empirical distributions of the variables $X_1, \cdots, X_5$ will approach the rectangular distribution $f(x) = 1/c$, where c is the circumference.

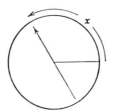

Fig. 1. A game of chance.

It will also be found that tests of independence, which will be studied later, usually substantiate independence of trials here.

It should be noted that definition (1) defines a method of sampling and says nothing about particular samples. It is legitimate to call a sample a random sample only if it has been obtained by a random sampling method.

It is frequently not feasible to check many real-life sampling methods for randomness because of the expense or difficulty of obtaining enough data to test whether the properties in definition (1) are reasonably satisfied. Then one must rely on judgment and experience to determine whether the method is sufficiently random to permit the use of models derived on the basis of random sampling.

2 MOMENTS OF MULTIVARIATE DISTRIBUTIONS

Since random sampling involves the multivariate density $f(x_1, \cdots, x_n)$, it is necessary to study properties of this function. In particular, it is necessary to define moments and the moment generating function for multivariate functions. The moment notation that was introduced in Chapter 3 becomes quite cumbersome when it is applied to multivariate situations. Furthermore, it lacks flexibility in deriving formulas; consequently all formulas will be expressed in terms of the expected value operator E.

Consider a generalization of definition (22), Chapter 3, to multivariate functions of a set of continuous random variables $X_1, \cdots, X_n$. In this connection let $h(X_1, \cdots, X_n)$ be any function of the random variables $X_1, \cdots, X_n$, whose density function is $f(x_1, \cdots, x_n)$. Then the expected value of $h(X_1, \cdots, X_n)$ is defined by

$$(2) \qquad E[h] = \int_{-\infty}^{\infty} \cdots \int_{-\infty}^{\infty} h(x_1, \cdots, x_r) f(x_1, \cdots, x_n) \, dx_1 \cdots dx_n \,.$$

The random variables of which h is a function have been omitted on the left side for notational convenience. It is assumed here that this integral exists. Infinite limits are used in this definition even though f may be positive over a finite domain only. Since f will be zero outside such a domain, this integral will then reduce to one with finite limits. Just as in the case of a function of one random variable, it can be shown that this definition yields the same value as that obtained by finding the density of the random variable h and calculating the expected value of h directly by using its density. The advantage of (2) lies in the fact that (2) does not require one to find the density of h.

The particular quantities that are needed in this chapter are the kth moment of a function $g(X_1, \cdots, X_n)$ and the moment generating function of

$g(X_1, \cdots, X_n)$. By choosing $h(X_1, \cdots, X_n) = g^k(X_1, \cdots, X_n)$ in (2), the kth moment of $g(X_1, \cdots, X_n)$ is defined by

$$(3) \qquad E[g^k] = \int_{-\infty}^{\infty} \cdots \int_{-\infty}^{\infty} g^k(x_1, \cdots, x_n) f(x_1, \cdots, x_n)\, dx_1 \cdots dx_n \,.$$

Corresponding to this definition, the moment generating function of $g(X_1, \cdots, X_n)$ is defined by

$$(4) \quad M_g(\theta) = E[e^{\theta g}] = \int_{-\infty}^{\infty} \cdots \int_{-\infty}^{\infty} e^{\theta g(x_1, \cdots, x_n)} f(x_1, \cdots, x_n)\, dx_1 \cdots dx_n \,.$$

That (4) generates moments in the same manner as does (24), Chapter 3, is easily verified by expanding $e^{\theta g}$ and integrating term by term.

Since expected value methods are used to assist in the development of the theory in this chapter, several of the most useful properties of the expected value operator are derived next.

3 PROPERTIES OF E

If c is any constant, it follows directly from (2) by choosing $h = cg$ and factoring out c from the integral on the right side that

$$E[cg] = cE[g] \,.$$

Next, since the integral of a sum is equal to the sum of the integrals, it follows from (2) by choosing $h = g_1 + g_2$ that

$$(5) \qquad\qquad E[g_1 + g_2] = E[g_1] + E[g_2] \,.$$

Here g_1 and g_2 are any two functions of the same set of random variables such that the indicated expectations exist.

Finally, since g_1 and g_2 are random variables that are functions of the same set of random variables $X_1, \cdots, X_n$, it is theoretically possible to derive the joint density function of g_1 and g_2 from the density function of $X_1, \cdots, X_n$. Consequently, if g_1 and g_2 are two functions that are independently distributed, which means that their joint density function can be factored into the product of the two individual density functions, it will follow that

$$(6) \qquad\qquad E[g_1 g_2] = \int_{-\infty}^{\infty} \int_{-\infty}^{\infty} g_1 g_2 f_1(g_1) f_2(g_2)\, dg_1\, dg_2 \,.$$

Here f_1 and f_2 are the density functions of the individual random variables g_1 and g_2, respectively. The formulation in (6) is in terms of the random variables g_1 and g_2 themselves and not in terms of the basic random variables $X_1, \cdots, X_n$ as in (2). But the double integral in (6) can be written in the product form

$$\int_{-\infty}^{\infty} g_1 f_1(g_1)\, dg_1 \int_{-\infty}^{\infty} g_2 f_2(g_2)\, dg_2 .$$

Since this is the product of the individual expected values, it follows that when g_1 and g_2 are independently distributed

(7) $$E[g_1 g_2] = E[g_1]E[g_2] .$$

It should be noted from (5) that the expected value of a sum of random variables is equal to the sum of their expected values, whether or not the variables are independently distributed, whereas the expected value of a product need not be equal to the product of the expected values unless the variables are independently distributed.

As an illustration of expected value techniques, which at the same time will illustrate how useful such methods are, consider the problem of finding the mean and variance of a sum of independent variables.

Let $X_1, \cdots, X_n$ be a set of independent variables with means $\mu_1, \cdots, \mu_n$ and variances $\sigma_1^2, \cdots, \sigma_n^2$ and let

$$W = X_1 + \cdots + X_n .$$

From formula (5) it follows that

(8) $$\mu_W = E[W] = \sum_{i=1}^{n} E[X_i] = \sum_{i=1}^{n} \mu_i .$$

Using this result, write

$$W - \mu_W = (X_1 - \mu_1) + \cdots + (X_n - \mu_n) .$$

Then

$$(W - \mu_W)^2 = \sum_{i=1}^{n} \sum_{j=1}^{n} (X_i - \mu_i)(X_j - \mu_j) .$$

Application of formula (5) then gives

(9) $$E(W - \mu_W)^2 = \sum_{i=1}^{n} \sum_{j=1}^{n} E(X_i - \mu_i)(X_j - \mu_j) .$$

The bracket notation for expected values is usually omitted, as it is here, when it becomes cumbersome and no confusion results from doing so. Now, since the variables X_i and X_j are independent when $i \neq j$, formula (7) may be applied to give

$$E(X_i - \mu_i)(X_j - \mu_j) = E(X_i - \mu_i)E(X_j - \mu_j), \qquad i \neq j .$$

But $E(X_i - \mu_i) = 0$; therefore (9) reduces to

$$E(W - \mu_W)^2 = \sum_{i=1}^{n} E(X_i - \mu_i)^2 .$$

Since $E(X_i - \mu_i)^2 = \sigma_i^2$, this result can be expressed as

(10)
$$\sigma_W^2 = \sum_{i=1}^{n} \sigma_i^2 \,.$$

Formula (8) states that the mean of a sum of random variables is equal to the sum of the means of the variables. The derivation of that formula did not require the independence of the variables; therefore, as stated before, the formula holds regardless of this assumption. Formula (10) states that the variance of a sum of independent random variables is equal to the sum of the variances of the variables. Here the independence assumption is vital.

As a particular application of these two formulas, consider the problem of finding the mean and variance of a binomial variable. This problem was solved in Chapter 3 by a direct application of moment definitions.

Let $X_1, \cdots, X_n$ be binomial variables corresponding to the n independent trials of an experiment for which p is the probability of success in a single trial. Thus $X_i = 1$ if a success occurs and $X_i = 0$ if a failure occurs on the ith trial. A random variable of this type is usually called a Bernoulli random variable, whereas the name binomial variable is used when a sequence of n Bernoulli trials occurs. Now let

$$W = X_1 + \cdots + X_n \,.$$

In view of the definition of X_i, it follows that W is equal to the total number of successes that will occur in the n trials of the experiment; consequently, the problem is to find the mean and variance of the binomial variable W.

For a discrete random variable that can assume the values 0 and 1 only, the expected value, as given by definition (1), Chapter 3, assumes the form

$$E[X] = \sum_{x=0}^{1} x f(x) \,.$$

Since each of the variables X_i assumes the values 0 and 1 with the probabilities q and p, respectively, it follows from this last formula that

$$\mu_i = E[X_i] = 0 \cdot q + 1 \cdot p = p \,.$$

This result when applied to (8) gives

$$\mu_W = np \,.$$

The technique used to calculate $E[X_i]$ may be applied to obtain the value of $E(X_i - \mu_i)^2 = E(X_i - p)^2$. Thus

$$\sigma_i^2 = E(X_i - p)^2 = (0 - p)^2 q + (1 - p)^2 p = pq \,.$$

As a result, formula (10) gives

$$\sigma_W{}^2 = npq \, .$$

These two results, of course, agree with formulas (15), Chapter 3.

4 SUM OF INDEPENDENT VARIABLES

A very useful formula for developing theory about sample means can be obtained when the variables $X_1, \cdots, X_n$ are independent and when $g(X_1, \cdots, X_n)$ is the special function

$$g(X_1, \cdots, X_n) = X_1 + \cdots + X_n \, .$$

The moment generating function of this sum is

$$M_{X_1 + \cdots + X_n}(\theta) = E[e^{\theta(X_1 + \cdots + X_n)}]$$

$$= E[e^{\theta X_1} \cdots e^{\theta X_n}] \, .$$

But because of the independence of these exponential functions, it follows from application of (7) that

$$M_{X_1 + \cdots + X_n}(\theta) = E[e^{\theta X_1}] \cdots E[e^{\theta X_n}]$$

$$= M_{X_1}(\theta) \cdots M_{X_n}(\theta) \, .$$

Since this result is used so often, it is stated in the form of a theorem.

THEOREM 1: *The moment generating function of the sum of n independent variables is equal to the product of the moment generating functions of the individual variables, that is*

$$M_{X_1 + \cdots + X_n}(\theta) = M_{X_1}(\theta) \cdots M_{X_n}(\theta) \, .$$

5 DISTRIBUTION OF $\bar{X}$ FROM A NORMAL DISTRIBUTION

In this section the distribution of a sample mean based on a random sample of size n from a normal population is derived.

Let X be normally distributed with mean μ and variance σ^2. Consider a random sample of size n from this normal population. The mean of such a sample

$$\bar{X} = \frac{1}{n}(X_1 + \cdots + X_n)$$

will be a random variable because $X_1, \cdots, X_n$ corresponding to the n trials of the sample are random variables. After a particular random sample has been taken $\bar{X}$ will be a number, but before it has been drawn it will be a random variable capable of assuming any value that the original variable X can assume. For the purpose of finding the density function of $\bar{X}$, consider its moment generating function.

If the first formula given in (25), Chapter 3, is used, it will follow that

$$M_{\bar{X}}(\theta) = M_{\frac{1}{n}(X_1+\cdots+X_n)}(\theta) = M_{X_1+\cdots+X_n}\left(\frac{\theta}{n}\right).$$

Since the sampling is random, the variables $X_1, \cdots, X_n$ are independent, and therefore Theorem 1 may be applied to give

$$M_{\bar{X}}(\theta) = M_{X_1}\left(\frac{\theta}{n}\right) \cdots M_{X_n}\left(\frac{\theta}{n}\right).$$

But random sampling as given by definition (1) also implies that all the variables $X_1, \cdots, X_n$ have the same density function, namely that of X, and hence the same moment generating function. Consequently, all the M's on the right are the same function, namely the moment generating function of the variable X. Thus

(11) $$M_{\bar{X}}(\theta) = M_X{}^n\left(\frac{\theta}{n}\right).$$

Now, from formula (31), Chapter 3, it is known that if X is normally distributed

(12) $$M_X(\theta) = e^{\mu\theta+\frac{1}{2}\sigma^2\theta^2}.$$

If this result, with θ replaced by θ/n, is used in (11), that formula will assume the form

$$M_{\bar{X}}(\theta) = \left[e^{\mu\frac{\theta}{n}+\frac{1}{2}\sigma^2\left(\frac{\theta}{n}\right)^2}\right]^n = e^{\mu\theta+\frac{1}{2}\frac{\sigma^2}{n}\theta^2}$$

The function on the right, when compared with (12), is seen to be the moment generating function of a normal variable with mean μ and variance σ^2/n. This is sufficient to ensure that $\bar{X}$ is such a variable because there exists a theorem in probability theory which states that the moment generating function of a variable uniquely determines the distribution of the variable. The preceding calculations together with this uniqueness theorem yield the following result.

THEOREM 2: *If X is normally distributed with mean μ and variance σ^2 and a random sample of size n is taken, then the sample mean $\bar{X}$ will be normally distributed with mean μ and variance σ^2/n.*

This theorem shows how the precision of a sample mean for estimating the population mean increases as the sample size is increased. If two normal

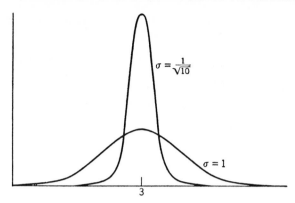

Fig. 2. Normal distribution of X and $\bar{X}$ for $n = 10$.

distributions have the same mean but one has half as large a standard deviation as the other, then there will be as much probability mass inside a central interval of length $L/2$ for the first distribution as inside a central interval of length L for the second distribution. In this sense the standard deviation of a normal variable does measure the degree of concentration of the random variable about its mean. Since the standard deviation of a normal $\bar{X}$ measures the concentration of sample $\bar{X}$'s about μ and therefore may be treated as a measure of the precision of estimating μ by means of $\bar{X}$, it is clear from this theorem that it is necessary to take four times as large a sample if one wishes to double the precision of an estimate at hand. Figure 2 shows the graph of a normal distribution with $\mu = 3$ and $\sigma = 1$, together with the graph of the distribution of $\bar{X}$ for samples of size 10 drawn from it. Here $\bar{X}$ possesses $\sqrt{10}$ times as much precision in estimating μ as does a single measurement.

This property of $\bar{X}$ possessing the mean μ and variance σ^2/n is true not only for a normal variable but for any variable X that possesses a second moment. The property is easily demonstrated by means of formulas (8) and (10) as follows.

From (8) it follows that

$$E[\bar{X}] = E\left(\frac{W}{n}\right) = \frac{1}{n} E[W] = \frac{1}{n}\sum_{i=1}^{n}\mu = \mu .$$

From (10) and the fact that $\sigma_{aX}^2 = a^2\sigma_X^2$ it follows that

$$\sigma_{\bar{X}}^2 = \sigma_{\frac{W}{n}}^2 = \frac{1}{n^2}\sigma_W^2 = \frac{1}{n^2}\sum_{i=1}^{n}\sigma^2 = \frac{\sigma^2}{n} .$$

5.1   Applications

As an illustration of the application of Theorem 2, consider the following problem. A manufacturer of a certain type of synthetic fishing line has found from long experience of testing that the breaking strength of his product possesses an approximate normal distribution with a mean of 30 pounds and a standard deviation of 4 pounds. A time and money saving change in the manufacturing process of the product is tried. A sample of 25 testing-length pieces of the new process line is taken and tested with a resulting sample mean of 28 pounds. What is the probability of obtaining a mean as low as 28 if the new process has had no harmful effect on breaking strength?

If X denotes the breaking strength of a randomly selected piece of line and if X is assumed to be a normal variable with $\mu = 30$ and $\sigma = 4$, then according to Theorem 2 the sample mean $\overline{X}$ based on a sample of size 25 will be normally distributed with mean $\mu = 30$ and standard deviation

$$\sigma_{\overline{X}} = \frac{\sigma}{\sqrt{n}} = \frac{4}{\sqrt{25}} = .8 \ .$$

As a result

$$P\{\overline{X} \le 28\} = P\left\{ \frac{\overline{X} - \mu}{\sigma_{\overline{X}}} \le \frac{28 - \mu_{\overline{X}}}{\sigma_{\overline{X}}} \right\}$$

$$= P\left\{ \frac{\overline{X} - 30}{.8} \le \frac{28 - 30}{.8} \right\}$$

$$= P\{Z \le -2.5\} \ .$$

Since Z is a standard normal variable, this probability may be obtained from Table II and will be found to be .006. Thus, there is a very small chance of obtaining a sample mean as low as 28 if there had been no change in the quality of the line due to the new process.

As a second illustration, consider the following problem. Since the mean in the preceding illustration was undoubtedly lowered by the change in the production process, it would be desirable to have an accurate estimate of the new process mean. Suppose the manufacturer wishes to estimate the new mean with an error of at most $\frac{1}{2}$ pound. How large a sample should he take if he wants to be assured with a probability of .95 that the error in his estimate will not exceed $\frac{1}{2}$ pound?

It will be assumed here that X is normally distributed with an unknown mean μ but with the same standard deviation $\sigma = 4$ as before. This implies that the mean may have changed but that the variability of the product has not been affected by the new production process. With these assumptions the problem can be solved in the same manner as the problem of determining how large a sample would be required to estimate a politician's popularity to a certain accuracy and which was discussed in section 3.4.4, Chapter 3. If the desired maximum error in $\bar{X}$ as an estimate of μ is to be $\frac{1}{2}$ pound and this maximum is to be exceeded only 5 percent of the time, then it is necessary that the maximum error of $\frac{1}{2}$ correspond to 2 standard deviations of $\bar{X}$. Therefore n must satisfy the equation

$$2\sigma_{\bar{X}} = \frac{1}{2}.$$

This is equivalent to the equation

$$2\frac{4}{\sqrt{n}} = \frac{1}{2}$$

whose solution is $n = 256$. Since a sample of size 25 is already available, an additional sample of 231 should suffice.

6 DISTRIBUTION OF $\bar{X}$ FROM A NON-NORMAL DISTRIBUTION

Since many variables of interest possess distributions that are not even approximately normal, it is important to know whether the preceding properties of $\bar{X}$ are approximately satisfied when the sampling is from a non-normal distribution. If it were true that $\bar{X}$ does possess an approximate normal distribution even though the distribution of X is far from being normal, the methods of the preceding section would be of very wide applicability. Now the remarkable fact is that no matter what type of distribution X has, provided that its mean and variance exist, the distribution of $\bar{X}$ will tend toward a normal distribution as $n \to \infty$. The proof of this fact is quite sophisticated; however a statement of the mathematical theorem that expresses it is quite simple. This theorem, which is called a *central limit theorem*, is one of the famous theorems of mathematics. An outline of a proof based on two advanced theorems of probability theory is given in the appendix.

(13) CENTRAL LIMIT THEOREM: *Let X be a random variable with mean μ and variance σ^2, then the random variable $Z = (\bar{X} - \mu)\sqrt{n}/\sigma$ has a distribution that approaches the standard normal distribution as $n \to \infty$.*

From a practical point of view, this theorem is exceedingly important because it permits the use of normal curve methods on problems related to sample means of the type illustrated in the preceding sections even when the basic variable X has a distribution that differs considerably from normality.

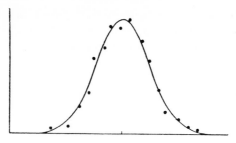

Fig. 3.   Distribution of $\bar{X}$ from a rectangular
distribution.

Of course the more the distribution of X differs from normality the larger n
must become to guarantee approximate normality for $\bar{X}$. Sampling experiments have shown that for $n > 50$ the form of $f(x)$ has little influence on the
form of the distribution of $\bar{X}$ for ordinary types of $f(x)$. Figure 3, for
example, shows the empirical distribution of $\bar{X}$ for 100 samples of size $n = 10$
each from the rectangular distribution given by $f(x) = 1$, $0 \leq x \leq 1$,
together with the corresponding limiting normal curve. The convergence
toward normality appears to be quite rapid here because $n = 10$ is hardly a
large sample.

   In Chapter 3 various binomial distribution problems were solved by using
a normal approximation with mean $\mu = np$ and variance $\sigma^2 = npq$. This
approximation was justified by comparing the graphs of these two distributions for a few values of n and p and by stating that a theoretical justification
would be given later. The central limit theorem is the promised theoretical
justification for the following reasons. If $X_i = 1$ or 0, according as a success
or failure occurs at the ith trial and if X denotes the total number of successes
in n trials, then

$$\bar{X} = \frac{1}{n}(X_1 + \cdots + X_n) = \frac{X}{n}.$$

Since, as was observed in section 3, the mean and variance of X_i are given by
p and pq, respectively, it follows that the variable Z of the central limit
theorem becomes

$$Z = \frac{(\bar{X} - \mu)\sqrt{n}}{\sigma} = \frac{\left(\dfrac{X}{n} - p\right)\sqrt{n}}{\sqrt{pq}} = \frac{X - np}{\sqrt{npq}}.$$

As a result, the claim made in Theorem 2, Chapter 3, that $(X - np)/\sqrt{npq}$
possesses a distribution approaching that of a standard normal variable as
$n \to \infty$ is now seen to be justified.

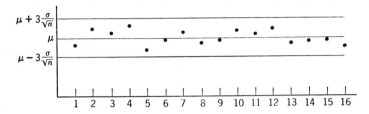

Fig. 4.   Control chart for the mean.

6.1 Applications

In view of the central limit theorem, the problems that were solved in section 5.1 could have been solved in the same manner without requiring that the basic variable X be normally distributed because n was sufficiently large in all those problems to justify treating $\bar{X}$ as a normal variable, provided it is assumed that X possesses a second moment.

As a slightly different type of illustration, consider the problem of constructing a control chart for sample means similar to the chart for sample proportions that was introduced in section 3.4.4, Chapter 3. Because of the central limit theorem it is not essential that the basic variable be normally distributed for such charts; consequently, they are of wide applicability. Such a chart is shown in Fig. 4. It will be observed that the process appears to be under control. It should be noted that the control band is a three-standard-deviation band about the mean, and therefore for a normal variable the probability should be only .003 that a point will fall outside this band. Since many industrial variables are not normally distributed, and since the sample means used in control charts are often based on only four or five measurements each, one could hardly expect the probability of .003 to be very realistic. Three-standard-deviation control limits are chosen because industrial experience has found them to be especially useful rather than because they correspond to a desired probability.

7 DISTRIBUTION OF LINEAR FUNCTIONS

An important generalization of the mean of a set of variables is the weighted mean. If $X_1, \cdots, X_n$ represent a set of random variables and $a_1, \cdots, a_n$ denote a set of weights, then

$$W = a_1 X_1 + \cdots + a_n X_n$$

is called a weighted mean, provided the a's are non-negative and sum to one.

Linear combinations such as this are useful even though such restrictions are not placed on the weights.

In particular, suppose that $X_1, \cdots, X_n$ are a set of independent normal variables with means $\mu_1, \cdots, \mu_n$ and variances $\sigma_1^2, \cdots, \sigma_n^2$. Then to show that W is also a normal variable, consider its moment generating function. Because of the assumptions made on the X's, it follows that

$$
\begin{aligned}
M_W(\theta) &= E[e^{\theta(a_1 X_1 + \cdots + a_n X_n)}] \\
&= E[e^{\theta a_1 X_1}] \cdots E[e^{\theta a_n X_n}] \\
&= M_{X_1}(a_1 \theta) \cdots M_{X_n}(a_n \theta) \\
&= e^{\mu_1 a_1 \theta + \frac{1}{2}\sigma_1^2 a_1^2 \theta^2} \cdots e^{\mu_n a_n \theta + \frac{1}{2}\sigma_n^2 a_n^2 \theta^2} \\
&= e^{(a_1 \mu_1 + \cdots + a_n \mu_n)\theta + \frac{1}{2}(a_1^2 \sigma_1^2 + \cdots + a_n^2 \sigma_n^2)\theta^2} .
\end{aligned}
$$

It will be observed by inspecting (12) that this is the moment generating function of a normal variable with its mean given by the coefficient of θ and its variance given by the coefficient of $\frac{1}{2}\theta^2$. From the uniqueness theorem, it therefore follows that W is a normal variable. This result may be expressed as follows:

THEOREM 3: *The variable $W = a_1 X_1 + \cdots + a_n X_n$, where $X_1, \cdots, X_n$ are independent normal variables with means $\mu_1, \cdots, \mu_n$ and variances $\sigma_1^2, \cdots, \sigma_n^2$, possesses a normal distribution with mean $\mu_W = a_1 \mu_1 + \cdots + a_n \mu_n$ and variance $\sigma_W^2 = a_1^2 \sigma_1^2 + \cdots + a_n^2 \sigma_n^2$.*

This result is useful, for example, in determining what set of weights, summing to 1, to attach to a set of measurements if a linear combination of those measurements is to be used to estimate the quantity being measured and the measurements are obtained by means of instruments of differing accuracy. If X_i denotes the measurement that will be obtained when using the ith instrument and if all the instruments measure without bias, then the means of the random variables $X_1, \cdots, X_n$ will all be equal to the quantity being measured. It is assumed here that this is what is meant by the phrase "without bias." Since the variance is a valid basis for determining measurement precision for a normal variable in estimating its mean and since measurement errors tend to be normally distributed, the a's should be chosen to minimize σ_W^2. Thus, the problem is to choose the a's satisfying $\sum_{i=1}^{n} a_i = 1$ to minimize

$$
\sigma_W^2 = \sum_{i=1}^{n} a_i^2 \sigma_i^2 .
$$

Letting $a_n = 1 - \sum_{i=1}^{n-1} a_i$ to eliminate the restriction on the a's gives

$$
\sigma_W^2 = \sum_{i=1}^{n-1} a_i^2 \sigma_i^2 + \left(1 - \sum_{i=1}^{n-1} a_i\right)^2 \sigma_n^2 .
$$

Then, by standard calculus techniques,

$$\frac{\partial \sigma_W{}^2}{\partial a_j} = 2a_j\sigma_j{}^2 - 2\left(1 - \sum_{i=1}^{n-1} a_i\right)\sigma_n{}^2$$

$$= 2a_j\sigma_j{}^2 - 2a_n\sigma_n{}^2 .$$

For a minimum, all such partial derivatives must vanish; hence setting these derivatives equal to zero and solving gives

$$a_j\sigma_j{}^2 = a_n\sigma_n{}^2, \qquad j = 1, \cdots, n .$$

This says that $a_j\sigma_j{}^2$ is the same for all j; therefore it follows that

$$a_j = \frac{c}{\sigma_j{}^2}$$

where c is some constant. The value of c, if desired, can be obtained by summing both sides of this equation and setting the left side equal to 1. It is easily seen that these necessary conditions for a minimum do produce a minimum. This shows that the weights should be chosen inversely proportional to the variances of the variables. Thus, if one measuring device has twice as large a variance as another, the measurement obtained from it should be weighted only half as heavily as the measurement obtained from the more accurate device. The preceding result is valid for non-normal variables also because its derivation did not use the normality assumption. That assumption is needed only if probability statements concerning W are to be made. The formulas given in Theorem 3 for the mean and variance of W can be obtained without assuming normality by using the techniques employed in section 3.

8   DISTRIBUTION OF THE SAMPLE VARIANCE

The techniques that have been employed in this chapter can be used to solve certain problems related to the variance of a normal distribution. In this connection let X be normally distributed with mean μ and variance σ^2. It will be assumed that the value of μ is known but that the value of σ^2 is not known. The problem is to obtain information concerning σ^2 by means of a random sample of size n of X.

Let $Z_i = (X_i - \mu)/\sigma$. Since the X_i represent random samples, the Z_i are independent standard normal variables. Let

$$V = \sum_{i=1}^{n} Z_i{}^2 .$$

From Theorem 1 it follows that

$$M_V(\theta) = M_{Z_1{}^2}(\theta) \cdots M_{Z_n{}^2}(\theta)$$

(14) $$= M_{Z_1{}^2}^n(\theta) .$$

Since Z_1 is a standard normal variable,

$$M_{Z_1{}^2}(\theta) = \int_{-\infty}^{\infty} e^{\theta z^2} \frac{e^{-\frac{z^2}{2}}}{\sqrt{2\pi}} \, dz$$

$$= \frac{1}{\sqrt{2\pi}} \int_{-\infty}^{\infty} e^{-\frac{z^2}{2}(1-2\theta)} \, dz .$$

Let $y = z\sqrt{1 - 2\theta}$; then this integral reduces to

$$M_{Z_1{}^2}(\theta) = (1 - 2\theta)^{-\frac{1}{2}} \int_{-\infty}^{\infty} \frac{e^{-\frac{y^2}{2}}}{\sqrt{2\pi}} \, dy$$

$$= (1 - 2\theta)^{-\frac{1}{2}} .$$

From this result and (14) it therefore follows that

$$M_V(\theta) = (1 - 2\theta)^{-\frac{n}{2}} .$$

From formula (40), Chapter 3, it will be observed that this is the moment generating function of a chi-square variable with n degrees of freedom. In view of the uniqueness theorem relating moment generating functions and distributions it follows that V must possess a chi-square distribution with n degrees of freedom. This result is expressed in the form of a theorem.

THEOREM 4: *If X is normally distributed with mean μ and variance σ^2 and $X_1, \cdots, X_n$ is a random sample of size n of X, then the random variable $V = \sum_{i=1}^{n} (X_i - \mu)^2/\sigma^2$ will possess a chi-square distribution with n degrees of freedom.*

It should be noted that the quantity $\sum_{i=1}^{n} (X_i - \mu)^2/n$ looks very much like the definition of σ^2 for a discrete random variable and therefore it should come as no surprise that it may serve as a sample estimate of σ^2 and be used to give information concerning the value of σ^2 for a normal variable. This is accomplished by means of the variable V and its chi-square distribution, as shown in the next section.

8.1 Applications

Theorem 4 can be used to solve problems concerning the parameter σ of a normal distribution similar to those solved by means of Theorem 2 for the parameter μ. As an illustration, assume that X is normally distributed with mean $\mu = 10$ and unknown variance σ^2 and consider the problem of estimating σ^2 by means of a random sample of size 30. The quantity $\sum_{i=1}^{30} (X_i - 10)^2/30$ will be chosen to estimate σ^2. What is the probability that this estimate will not be more than 20 percent off, that is, will not be more than 20 percent larger or smaller than σ^2?

This requirement can be expressed by means of the inequality

$$.8\sigma^2 < \frac{\sum\limits_{i=1}^{30}(X_i - 10)^2}{30} < 1.2\sigma^2 .$$

The following equivalent inequality is obtained by multiplying through by $30/\sigma^2$

$$24 < \frac{\sum\limits_{i=1}^{30}(X_i - 10)^2}{\sigma^2} < 36 .$$

In terms of the notation of Theorem 4, this can be written

$$24 < V < 36 .$$

Here V is a chi-square variable with 30 degrees of freedom. From Table III it will be found that the probability of a chi-square variable with 30 degrees of freedom exceeding 36 is approximately .20 and of being smaller than 24 is approximately .25; therefore the probability of its lying inside this interval is approximately .55. From this result it appears that σ^2 is not estimated with much accuracy by means of a sample as small as 30.

As a second illustration that is somewhat different, consider the problem of calculating the probability that an antiaircraft shell designed to burst at a specified point in space will have a radial error of at most 100 feet if it is assumed that the x, y, and z errors with respect to a coordinate system with the origin at the specified point are independently normally distributed with a common standard deviation of 50 feet.

For convenience of notation let $X_1 = X/50$, $X_2 = Y/50$, and $X_3 = Z/50$, where X, Y, and Z represent the coordinates of the bursting shell. Then X_1, X_2, X_3 will be independent standard normal variables. Since the variables $(X_i - \mu)/\sigma$, $i = 1, \cdots, n$, of Theorem 4 are independent standard normal variables, the variable V of that theorem is merely the sum of the squares of n independent standard normal variables; consequently for the standard normal variables X_1, X_2, X_3 defined above it follows from Theorem 4 that

$$V = \sum_{i=1}^{3} X_i^2$$

will possess a chi-square distribution with 3 degrees of freedom. Thus, the desired probability is given by

$$P\{\sqrt{X^2 + Y^2 + Z^2} < 100\} = P\{X^2 + Y^2 + Z^2 < (100)^2\}$$
$$= P\{X_1{}^2 + X_2{}^2 + X_3{}^2 < 4\}$$
$$= P\{V < 4\}.$$

From Table III it will be found that this probability is approximately .73.

The chi-square distribution has many other interesting applications, several of which will be discussed in later chapters.

9   HYPOTHESIS TESTING APPLICATIONS

In this section the theory that has been developed in this chapter will be applied to several important hypothesis testing problems. In every case an appropriate random variable will be selected and a critical region chosen. Although the choice of a critical region will be made on an intuitive basis, criteria developed in Chapter 8 will show that the choice was an excellent one.

9.1   Testing a Mean

Consider a slight modification of the first illustration of section 5.1. There X was assumed to be a normal variable with mean $\mu = 30$ and standard deviation $\sigma = 4$ and a random sample of size $n = 25$ gave a sample mean $\bar{x} = 28$. Suppose in addition that the sample data gave $s = 5$. As before, the problem is to determine whether the new manufacturing process had a harmful effect on the mean breaking strength. This problem is a typical hypothesis testing problem with the choices

$$H_0 : \mu = 30 \quad \text{and} \quad H_1 : \mu < 30.$$

The statistic $\bar{X}$ will be chosen as the random variable with which to construct the test. Since $\bar{X}$ is a normal variable with mean $\mu_{\bar{X}} = \mu = 30$ and standard deviation $\sigma_{\bar{X}} = \sigma/\sqrt{n} = 4/\sqrt{25} = .8$, the variable $Z = (\bar{X} - 30)/.8$ will be a standard normal variable. Because of the nature of H_1 the natural critical region to choose is the left tail of the $\bar{X}$ distribution. Since the experiment produced $\bar{X} = 28$, $Z = (28 - 30)/.8 = -2.5$. The critical value of Z for $\alpha = .05$ is $z = -1.64$; therefore H_0 is rejected.

Suppose now that the value of σ were not known or there was reason to believe that both μ and σ had been affected by the new process. Then it would have been necessary to replace σ in the above calculations by its sample estimate $s = 5$. With this substitution, it follows that $\sigma_{\bar{X}} \doteq 5/\sqrt{25} = 1$ and $Z \doteq 28 - 30 = -2$. This value is also in the critical region. Approximating σ by s makes this test only an approximate test here; however, a method for eliminating this approximation is available and will be studied in Chapter 10.

9.2 Testing the Difference of Two Means

A frequently occurring problem in the various sciences is that of comparing an experimental and a control group of animals, individuals, or objects, for the purpose of determining whether the experiment produced a meaningful result. One method of treating the problem is to test whether the means of the two populations from which the data were obtained are essentially equal. The random variables corresponding to these two populations will be denoted by X and Y, respectively.

Let $\bar{X}$ and $\bar{Y}$ represent sample means of the two sets of data based on random samples of sizes n_X and n_Y. Since the experiments are independent, $\bar{X}$ and $\bar{Y}$ will be independently distributed. If X and Y are normally distributed, or if n_X and n_Y are sufficiently large to justify application of the central limit theorem, $\bar{X}$ and $\bar{Y}$ will be normally distributed, or at least approximately so. It is assumed therefore that $\bar{X}$ and $\bar{Y}$ are normally distributed.

Let $W = \bar{X} - \bar{Y}$ and choose $n = 2$, $X_1 = \bar{X}$, $X_2 = \bar{Y}$, $a_1 = 1$, and $a_2 = -1$ in Theorem 3. Application of that theorem will then yield the following useful result.

THEOREM 5: *If $\bar{X}$ and $\bar{Y}$ are normally and independently distributed, then $\bar{X} - \bar{Y}$ will be normally distributed with mean $\mu_{\bar{X}-\bar{Y}} = \mu_X - \mu_Y$ and variance $\sigma^2_{\bar{X}-\bar{Y}} = \sigma_X{}^2/n_X + \sigma_Y{}^2/n_Y$.*

As an illustration of how this theorem may be applied to solve the comparison problem, consider the following particular problem. A potential buyer of light bulbs bought 50 bulbs of each of two brands. On testing these bulbs, he found that brand A had a mean life of 1282 hours with a standard deviation of 80 hours, whereas brand B had a mean life of 1208 hours with a standard deviation of 94 hours. Can a buyer be quite certain that the two brands differ in quality? Since mean life may be taken to measure quality, to answer this question it will suffice to test the hypothesis

$$H_0 : \mu_X = \mu_Y$$

against the alternative

$$H_1 : \mu_X \neq \mu_Y .$$

Here X and Y represent the length of life of a randomly selected light bulb from brand A and brand B, respectively.

Experience with industrial variables such as this one indicates that light bulb life has an approximate normal distribution and therefore that $\bar{X}$ and $\bar{Y}$ may be assumed to be normally distributed. The samples are obviously independent; hence Theorem 5 may be applied to this problem. Under the assumption that H_0 is true, $\mu_{\bar{X}-\bar{Y}} = 0$; therefore it follows from Theorem 5 that $\bar{X} - \bar{Y}$ may be treated as a normal variable with

$$\mu_{\bar{X}-\bar{Y}} = 0 \qquad \text{and} \qquad \sigma_{\bar{X}-\bar{Y}} = \sqrt{\frac{\sigma_X{}^2}{50} + \frac{\sigma_Y{}^2}{50}}\,.$$

Since σ_X and σ_Y are unknown it is necessary to estimate them by means of their sample values. Such approximations introduce an error, but for samples as large as 50 this error is not serious. It can be shown that the error in $\sigma_{\bar{X}-\bar{Y}}$ almost certainly does not exceed 10 percent here. With these approximations

$$\mu_{\bar{X}-\bar{Y}} = 0 \qquad \text{and} \qquad \sigma_{\bar{X}-\bar{Y}} \doteq \sqrt{\frac{(80)^2}{20} + \frac{(94)^2}{50}} = 17.5\,.$$

Hence

$$Z = \frac{\bar{X} - \bar{Y} - \mu_{\bar{X}-\bar{Y}}}{\sigma_{\bar{X}-\bar{Y}}} \doteq \frac{74}{17.5} = 4.23\,.$$

Because of the choice of H_1, the critical region for this test will be chosen to consist of the two equal tails of the distribution of $\bar{X} - \bar{Y}$. This choice is made because the larger the value of $|\bar{X} - \bar{Y}|$ the less faith one should have in the truth of H_0. Since the mean of $\bar{X} - \bar{Y}$ is 0 when H_0 is true, values of $\bar{X} - \bar{Y}$ would be expected to concentrate about 0, and therefore large values of $|\bar{X} - \bar{Y}|$ would tend to favor H_1. If a critical region of size .05 is selected, the critical region will therefore consist of those values of $\bar{X} - \bar{Y}$ that are more than two standard deviations away from 0. In terms of the corresponding standard normal variable Z the critical region is the region where $|z| > 2$, or more accurately $|z| > 1.96$. The value $Z = 4.23$ for this problem lies in the critical region, and therefore H_0 is rejected. It seems quite certain that the two brands differ in quality as far as mean burning time is concerned and that brand A is to be preferred.

9.3   Testing the Difference of Two Proportions

If two sets of data drawn from two binomial populations are to be compared, it is necessary to work with the proportion of successes rather than with the number of successes, unless the number of trials in each set is the same. For example, 40 heads in 100 tosses of a coin would not be compared with 30 heads in 50 tosses unless they were both placed on a percentage basis. Now from the central limit theorem it follows that the proportion of successes $\hat{p} = X/n$ may be assumed to be normally distributed with mean p and variance pq/n, provided that n is large.

Let $\hat{p}_1$ and $\hat{p}_2$ represent two independent sample proportions based on n_1 and n_2 trials, respectively, from two binomial populations with probabilities p_1 and p_2 respectively, and assume that n_1 and n_2 are large enough to treat $\hat{p}_1$ and $\hat{p}_2$ as normal variables. Then Theorem 3 may be applied to the variable $W = \hat{p}_1 - \hat{p}_2$ in precisely the same manner as it was applied to the variable $W = \bar{X} - \bar{Y}$ in the preceding section. The result may be expressed in the following manner.

THEOREM 6: *When the number of trials n_1 and n_2 are sufficiently large, the difference of the sample proportions $\hat{p}_1 - \hat{p}_2$ will be approximately normally distributed with mean $\mu_{\hat{p}_1 - \hat{p}_2} = p_1 - p_2$ and variance $\sigma^2_{\hat{p}_1 - \hat{p}_2} = p_1 q_1/n_1 + p_2 q_2/n_2$.*

Just as for the simple binomial distribution, the normal approximation will usually be satisfactory in applications if each $n_i p_i$ exceeds 5 when $p_i < \frac{1}{2}$ and $n_i q_i$ exceeds 5 when $p_i > \frac{1}{2}$.

As an illustration of how this theorem may be applied, consider the following problem. A railroad company installed two sets of 50 red oak ties each. The two sets were treated with creosote by two different processes. After a number of years of service, it was found that 22 ties of the first set and 18 ties of the second set were still in good condition. Is one justified in claiming that there is no real difference between the preserving properties of the two processes? To answer this question, let p_1 and p_2 denote the respective probabilities that a railroad tie treated by the corresponding process will be in good condition after this number of years of service. Then set up the hypothesis

$$H_0 : p_1 = p_2$$

against the alternative

$$H_1 : p_1 \neq p_2 .$$

If the common value of p_1 and p_2 under H_0 is denoted by p, then by Theorem 6 it follows that

$$\mu_{\hat{p}_1 - \hat{p}_2} = 0 \quad \text{and} \quad \sigma_{\hat{p}_1 - \hat{p}_2} = \sqrt{\frac{pq}{50} + \frac{pq}{50}} = \frac{\sqrt{pq}}{5} .$$

The value of p is unknown, and so its value must be estimated from sample values. Since the hypothesis H_0 treats the two samples as though they were drawn from populations with the same p, the samples may be combined into one sample of 100 for which there were 40 successes. Hence a good estimate

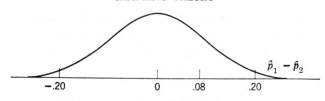

Fig. 5.   Distribution of $\hat{p}_1 - \hat{p}_2$.

of p here is .40. With this estimate,

$$\mu_{\hat{p}_1-\hat{p}_2} = 0 \qquad \text{and} \qquad \sigma_{\hat{p}_1-\hat{p}_2} \doteq .10 .$$

The situation is described geometrically in Fig. 5.

As in the case of testing the difference of two means, the critical region for this test will be chosen to consist of the two equal tails of the distribution of $\hat{p}_1 - \hat{p}_2$ because the larger $|\hat{p}_1 - \hat{p}_2|$ becomes the less faith one would have in the truth of H_0. Since $\hat{p}_1 - \hat{p}_2 = .44 - .36 = .08$ lies well within a two-standard-deviation interval of the mean, the hypothesis H_0 is accepted. In terms of standard normal variables, one could calculate

$$Z = \frac{\hat{p}_1 - \hat{p}_2}{\sigma_{\hat{p}_1-\hat{p}_2}} = \frac{.08}{.10} = .8 .$$

Just as in the problem of testing the difference of two means, the critical region would correspond to $|z| > 2$; therefore since $|.8| < 2$ the hypothesis H_0 would be accepted. The fact that the value of p must be estimated from the sample values and that $\hat{p}_1 - \hat{p}_2$ is only approximately normally distributed makes this test somewhat inaccurate. Both samples are large enough in this illustration, however, to insure a reliable test.

As a second illustration, consider a problem that arises frequently in the construction of tests. A civil-service examination is given to a group of 200 candidates. On the basis of their total scores, the 200 candidates are divided into two groups, the upper 30 percent and the remaining 70 percent. Consider the first question on this examination. In the first group, 40 had the correct answer; in the second group, 80 had the correct answer. On the basis of these results, can one conclude that the first question is no good at discriminating ability of the type being examined here? To solve this problem, set up the hypothesis

$$H_0 : p_1 = p_2$$

where p_1 and p_2 denote the respective probabilities of an individual from each of the two groups getting the correct answer on the first question. The natural alternative hypothesis here is

$$H_1 : p_1 > p_2$$

because the better candidates would be expected to do at least as well as the weaker candidates on all questions. As before, it follows that

$$\mu_{\hat{p}_1 - \hat{p}_2} = 0 \quad \text{and} \quad \sigma_{\hat{p}_1 - \hat{p}_2} = \sqrt{\frac{pq}{60} + \frac{pq}{140}}$$

where p is the common value of $p_1 = p_2$ under H_0. To estimate p, combine the two groups to give 120 successes in 200 trials, or an estimate of .60. Using this estimate for p,

$$\sigma_{\hat{p}_1 - \hat{p}_2} \doteq .076 .$$

Now, $\hat{p}_1 - \hat{p}_2 = 40/60 - 80/140 = .10$; therefore,

$$Z = \frac{\hat{p}_1 - \hat{p}_2}{\sigma_{\hat{p}_1 - \hat{p}_2}} \doteq 1.32 .$$

Since $Z < 1.64$, the hypothesis H_0 will be accepted. This implies that the first question is not satisfactory for distinguishing between the stronger and the weaker candidates and therefore should be deleted from the examination. It might happen, however, that quite a few of the questions will fail to show discriminating ability as judged by individual tests such as this but when taken together will show such ability. Hence, from a practical point of view, one does not always reject a question merely because it does not reject the hypothesis H_0.

EXERCISES

Sec. 1

1. Explain which features of random sampling are satisfied and which are not satisfied if you wish to estimate the distribution of students' grade point averages and do so by taking a sample of 100 students from the registration files by consulting a table of random numbers but always ignoring any grade point average less than .8. Assume the student enrollment is (a) large, (b) small.

Sec. 3

2. Give an example of two random variables for which $E[XY] \neq E[X]E[Y]$.

3. Given $f(x, y) = e^{-(x+y)}$, $x > 0$, $y > 0$, (a) calculate $E[Z]$ where $Z = X + Y$, (b) calculate $E[Z^2]$, (c) find $M_Z(\theta)$.

4. Let X have the distribution given by $f(x) = pq^x$, $x = 0, 1, 2, \cdots$. (a) Calculate $E[X]$ by using the formula $1/(1 - q) = 1 + q + q^2 + \cdots$ and its derivative. (b) Calculate the variance of X by first calculating $E[X(X - 1)]$ by means of similar techniques.

5. Cards numbered 1 through n are shuffled and laid out in a line. Let $X_k = 1$ if the number on the kth card is smaller than the number on the $(k + 1)$th card and

let $X_k = 0$ otherwise. Let $Z = \sum_{k=1}^{n-1} X_k$, which means that Z represents the total number of increases in the sequence. Calculate the mean and variance of Z.

Sec. 4

6. By using moment generating function methods, show that the sum of two independent binomial variables with the same parameter p is also a binomial variable. How could you argue this result directly?

7. By using moment generating function methods show that the sum of two independent Poisson variables with means μ_1 and μ_2 is also a Poisson variable with mean $\mu_1 + \mu_2$.

8. Using the methods of problem 7, explain why the difference of two independent Poisson variables is not a Poisson variable. How could you argue this directly?

Sec. 5

9. If X is normally distributed with $\mu = 20$ and $\sigma = 4$, calculate the probability that (a) $X > 21$, (b) $\bar{X} > 21$ if $\bar{X}$ is based on a random sample of size 16.

10. If you wish to estimate the mean of a normal population whose variance is 9, how large a sample should you take so that the probability is .80 that your estimate will not be in error by more than $\frac{1}{2}$ unit?

Sec. 6

11. Past experience indicates that wire rods purchased from a certain company have a mean breaking strength of 400 pounds and a standard deviation of 20 pounds. How many rods should you select so that you would be certain with a probability of .95 that your sample mean would not be in error by more than 3 pounds?

12. A research worker wishes to estimate the mean of a population by using a sample large enough that the probability will be .90 that the sample mean will not differ from the true mean by more than 20 percent of the standard deviation. How large a sample should he take?

13. Have each member of the class perform the following experiment 10 times. Select 10 one-digit random numbers and calculate their mean. Bring these 10 sample means to class, where the total set of such means may be classified, histogram drawn, and mean and standard deviation calculated. The results should be compared with those to be expected under theorem (13). The population here has $\mu = 4.5$ and $\sigma = 2.87$.

14. (a) Construct a control chart for $\bar{X}$ for the following data on the blowing time of fuses, samples of 5 being taken every day. Each set of 5 has been arranged in order of magnitude. Estimate $\sigma_{\bar{X}}$ by first estimating σ by means of s calculated from all 60 values. (b) Comment on whether production seems to be under control.

42	42	19	36	42	51	60	18	15	69	64	61
65	45	24	54	51	74	60	20	30	109	91	78
75	68	80	69	57	75	72	27	39	113	93	94
78	72	81	77	59	78	95	42	62	118	109	109
87	90	81	84	78	132	138	60	84	153	112	136

15. Given $f(x) = e^{-x/\beta}/\beta$, $x > 0$, and a sample of size n, use the central limit theorem to find an interval for β based on $\bar{X}$ which will hold with high probability.

16. Let X have a Poisson distribution with mean μ. Give arguments why the limiting distribution of the random variable $Z = (X - \mu)/\sqrt{\mu}$ as $\mu \to \infty$ is the standard normal distribution.

Sec. 7

17. Give an example of two dependent random variables for which the variance of their sum is (a) larger than the sum of their variances, (b) smaller than the sum of their variances.

18. Given independent $X \sim N(0, 4)$ and $Y \sim N(1, 9)$, calculate $P(\bar{X} < \bar{Y})$ if $\bar{X}$ and $\bar{Y}$ are based on samples of size 10 each. This symbol means $N(\mu, \sigma^2)$.

19. Prove that if X and Y are independent variables having the same rectangular distribution over $(0, 1)$ then $Z = X + Y$ will not have a rectangular distribution.

20. Find the density of $n\bar{X}$ for a random sample of size n from $f(x) = e^{-x}$, $x > 0$. Use moment generating function techniques.

21. If $X_1, X_2, \cdots, X_n$ is a random sample from the distribution with continuous density $f(x)$, calculate $P\{X_1 < t, \cdots, X_n < t\}$ and use it to find the density of the variable $Z = \max \{X_1, \cdots, X_n\}$. Express the result in terms of $f(x)$ and its distribution function $F(x)$. First find the distribution function of Z.

22. Use the result in problem 21 to find the distribution of the lifetime of a piece of electronic equipment that has n vital parts with lifetimes $X_1, \cdots, X_n$ which are independently and identically distributed with density $f(x) = ae^{-ax}$, $x > 0$, if it is assumed that one functioning vital part is sufficient to operate the equipment.

Sec. 8

23. Use the moment generating function of a chi-square variable with v degrees of freedom to find its mean and variance.

24. Use expected value operator methods on $W = \sum_{i=1}^{n} X_i^2$ to find the mean and variance of a chi-square variable with n degrees of freedom. Assume that the X_i are independent standard normal variables.

25. Given $X \sim N(\mu, \sigma^2)$ and the following interval which is satisfied with a certain probability,

$$\frac{\sum (X_i - \mu)^2}{b} < \sigma^2 < \frac{\sum (X_i - \mu)^2}{a},$$

calculate the expected value of the length of this interval.

Sec. 9

26. Suppose one mixes the ingredients for concrete to attain a mean breaking test of 2000 pounds and believes he has such a mixture. How many breaking tests will be needed in order that $\alpha = .05$ and $\beta = .10$ in testing the hypothesis $H_0: \mu = 2000$ against $H_1: \mu = 1900$ if $\sigma = 200$ and one assumes normality?

27. The same test was given to two classes. The first class of 20 students averaged 120 points with a standard deviation of 20 points. The second class of 30 students averaged 130 points with a standard deviation of 18 points. Is it safe to conclude that the second class is superior?

28. Two sets of 100 students each were taught to read by two different methods. After instruction was over a reading test gave the following results: $\bar{x} = 73.4$, $\bar{y} = 70.3$, $s_x = 8$, $s_y = 10$. (a) Test the hypothesis $\mu_X = \mu_Y$. (b) Determine how large an equal size sample from each group should have been used if it were desired to estimate $\mu_X - \mu_Y$ to within 1 unit with a probability of .95.

29. Suppose you wish to test whether there is a tendency for an individual's right foot to be longer than his left foot. (a) Explain why it would be incorrect to take a random sample of, say, 100 individuals and apply the usual technique for testing $\mu_X = \mu_Y$, where X and Y are the right and left foot measurements. (b) Explain how you could sample differently or handle the data differently to overcome the difficulty here.

30. Suppose $\bar{X}$ and $\bar{Y}$ are the means of two samples of size n each from a normal population with variance σ^2. Determine n so that the probability will be about .95 that the two sample means will differ by less than σ.

31. Two different samplers were sent into the same forest to select trees at random. Each took a sample of 100 trees and measured their diameters with the following results: $\bar{x} = 19.2$, $\bar{y} = 20.3$, $s_x = 3.2$, $s_y = 2.6$. (a) Does the smaller standard deviation for the second sampler imply that he is more accurate than the first one? (b) What conclusions can be drawn concerning the accuracy of the two samplers?

32. Suppose $\mu_X - \mu_Y = \frac{1}{2}$ and $\sigma_X = \sigma_Y = 1$ for two independent normal variables. How large an equal sample from each population should be taken so that the probability of rejecting the false hypothesis $H_0 : \mu_X = \mu_Y$ will be .90 if the critical region is two-sided and $\alpha = .10$?

33. In a large scale experiment 2000 children were split into two groups of 1000 each. One group received a serum for the prevention of a disease; the other group did not. The number of children in each group who contracted the disease was 30 and 50, respectively. Treating these as sample values of two Poisson variables which may be considered as approximately normally distributed, test the hypothesis that $\mu_1 = \mu_2$.

34. In a poll taken among college students, 46 of 200 fraternity men favored a certain proposition, whereas 51 of 300 nonfraternity men favored it. Is there a real difference of opinion on this proposition?

35. A manufacturer of housedresses sent out advertising by mail. He sent samples of material to each of two groups of 1000 women, but for one group he used a white envelope and for the other he used a blue envelope. He received orders from 10 and 12 percent, respectively. Is it quite certain that the blue envelope will help sales?

36. A civil service examination was given to 200 people. On the basis of their total scores they were divided into the upper 30 percent, the middle 40 percent, and the lower 30 percent. On a certain question, 45 of the upper group and 30 of the lower group answered correctly. Is this question likely to be useful for discriminating ability of the type being tested?

37. A test of 100 youths and 200 adults showed that 42 of the youths and 50 of the adults were poor drivers. Use these data to test the claim that the youth percentage of poor drivers is larger than the adult percentage by 8 percentage points, against the possibility of a still larger difference.

38. Two players each play a game of chance 100 times. If one dollar is paid for every win and the probability of winning at a single trial is $\frac{1}{4}$, what is the approximate probability that the first player will finish with at least 8 dollars more than the second player?

CHAPTER 6

Probability Distributions for Correlation and Regression

Probability density functions of two variables were defined in Chapter 2 for both discrete and continuous variables. Although a number of their properties were discussed there, most of the discussion was concerned with discrete variables. This chapter will concentrate on continuous variables and probability models for them.

1 DISTRIBUTIONS FOR TWO VIARABLES

The concepts of a marginal distribution and of a conditional distribution were introduced in Chapter 2 for discrete random variables. They are equally important for continuous random variables. For the purpose of introducing them here, let $f(x, y)$ be the joint density of two continuous random variables X and Y.

The geometrical representation of $f(x, y)$ as a surface in three dimensions as displayed in Fig. 20, Chapter 2, is convenient for interpreting probability as a volume under the surface; however, in discussing marginal and conditional distributions, it is more convenient to think of $f(x, y)$ as giving the density distribution of probability mass over the x, y plane, with the total mass being equal to 1. This was easy to do in section 12, Chapter 2, for discrete variables because only a finite number of mass points was involved. Here, however, it is necessary to conceive of a continuous distribution of mass such as in a sheet of metal. The density of the metal sheet at a point (x, y) is given by $f(x, y)$ and the mass of the entire sheet is equal to 1. Figure 1 attempts to portray this density interpretation for the density function that is graphed as a surface in Fig. 20, Chapter 2.

From the density point of view, the probability that a single sample will yield a point (x, y) lying in a given rectangle is equal to the mass of the

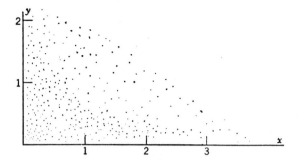

Fig. 1.   Probability density distribution.

rectangle. This interpretation of probability, as well as the volume interpreta-tion, clearly holds for regions other than rectangles in the x, y plane.

1.1   Marginal Distributions

For the purpose of obtaining a formula for continuous variables corre-sponding to formula (27), Chapter 2, let $f(x, y)$ be the joint density function of two continuous random variables X and Y and consider the following relations.

$$P\{\alpha < X < \beta\} = P\{\alpha < X < \beta, -\infty < Y < \infty\}$$
$$= \int_\alpha^\beta \int_{-\infty}^\infty f(x, y) \, dy \, dx$$
$$= \int_\alpha^\beta h(x) \, dx$$

where, as indicated,

(1)
$$h(x) = \int_{-\infty}^\infty f(x, y) \, dy .$$

Now, if X is considered independently of Y, then by definition

$$P\{\alpha < X < \beta\} = \int_\alpha^\beta f(x) \, dx$$

where $f(x)$ is the density function of X alone. If these two expressions for $P\{\alpha < X < \beta\}$ are equated,

(2)
$$\int_\alpha^\beta h(x) \, dx = \int_\alpha^\beta f(x) \, dx .$$

Since this equality is to hold for all intervals (α, β), α may be held fixed and β allowed to vary, in which event these integrals may be treated as functions of β. By the well-known calculus formula that has been used before, if

$$F(\beta) = \int_{\alpha}^{\beta} f(x)\, dx$$

then

$$\frac{dF(\beta)}{d\beta} = f(\beta)$$

provided that $f(x)$ is continuous at $x = \beta$. If both sides of (2) are differentiated with respect to β, this formula gives

$$h(\beta) = f(\beta)\,.$$

Since this is an identity in β, it follows that the function $h(x)$ defined by (1) is the density function $f(x)$. These arguments therefore show that the marginal density function $f(x)$ is given by the following formula.

(3)   MARGINAL DISTRIBUTION: $f(x) = \int_{-\infty}^{\infty} f(x, y)\, dy\,.$

This formula uniquely defines $f(x)$ provided $f(x)$ is continuous for all x. However, some other function which differs from this function at, say, a finite number of points could serve equally well as a marginal density because integrating it over any interval would yield the same probability as integrating $f(x)$ over that interval. If $f(x, y)$ is a continuous function of x and y, $f(x)$ will certainly be a continuous function of x, and hence will be the unique marginal density. The densities that will be considered in this chapter are mostly of this type.

Formula (3) is the continuous analogue of formula (27), Chapter 2, for the discrete case. In a similar manner the integration of $f(x, y)$ with respect to x from $-\infty$ to $+\infty$ will yield the Y marginal density function $g(y)$. From the density point of view $f(x)$ may be thought of as giving the probability density distribution along the x axis after the entire probability mass in the x, y plane has been projected perpendicularly onto the x axis.

As a simple illustration of how formula (3) applies, consider the joint density function

(4) $f(x, y) = \begin{cases} 2 - x - y, & 0 < x < 1, 0 < y < 1\,. \\ 0 & , \quad \text{elsewhere} \end{cases}$

Here, formula (3) gives

(5) $f(x) = \int_{0}^{1} (2 - x - y)\, dy = \tfrac{3}{2} - x\,, \qquad 0 < x < 1$

and

$$g(y) = \int_{0}^{1} (2 - x - y)\, dx = \tfrac{3}{2} - y\,, \qquad 0 < y < 1\,.$$

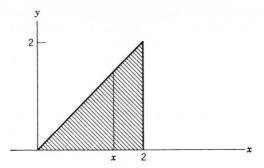

Fig. 2. A particular sample space.

As a second illustration, consider the following density function:

$$f(x, y) = \begin{cases} \frac{1}{2}xy, & 0 < x < 2, 0 < y < x. \\ 0, & \text{elsewhere.} \end{cases}$$

In this problem the sample space is the triangle bounded by the lines $x = 2$, $y = x$, and $y = 0$ as shown in Fig. 2. Although the limits in formula (3) are written $-\infty$ and ∞, this is merely for notational convenience and it is understood that when the limits are not infinite one must determine the limits from the sample space boundaries. The limits of integration in this problem certainly depend on the chosen value of x. Formula (3) and Fig. 2 give

$$f(x) = \int_0^x \tfrac{1}{2}xy \, dy = \frac{x^3}{4}, \qquad 0 < x < 2.$$

Similarly,

$$g(y) = \int_y^2 \tfrac{1}{2}xy \, dx = \tfrac{1}{4}y(4 - y^2), \qquad 0 < y < 2.$$

1.2 Conditional Distributions

Now that marginal distributions have been determined, it is possible to proceed with the problem of defining conditional distributions for continuous variables. For the purpose of obtaining a formula for continuous variables corresponding to formula (28), Chapter 2, consider the function defined by

(6)
$$\frac{f(x, y)}{f(x)}.$$

If x is held fixed and is such that $f(x) > 0$, then (6) defines a non-negative function of y for which, in view of (3),

$$\int_{-\infty}^{\infty} \frac{f(x, y)}{f(x)} \, dy = \frac{1}{f(x)} \int_{-\infty}^{\infty} f(x, y) \, dy = 1.$$

Thus, according to (30), Chapter 2, $f(x, y)/f(x)$ has properties that enable it to serve as a density function for Y when X is fixed as indicated. Because of this property, $f(x, y)/f(x)$ is called the conditional density function of Y for fixed X and is denoted by $f(y \mid x)$. This definition may be expressed as follows:

(7)   CONDITIONAL DISTRIBUTION: $f(y \mid x) = \dfrac{f(x, y)}{f(x)}$.

By going back to the definition of conditional probability for events as given by (6), Chapter 2, and working with integrals, it is possible to derive (7) directly in a natural manner; however (7) is treated here as a definition. Formula (7) is identical with the corresponding formula for the discrete case. The conditional density function of X for Y fixed is defined in a similar manner.

From a density point of view $f(y \mid x)$ may be thought of as giving the probability density distribution along the vertical line in the x, y plane corresponding to the fixed value x, the total mass of this line being equal to 1. The density function $f(x, y)$ with x fixed as it stands could not be used as a probability density function along such a line because by (3) it would not give a total probability mass of one for the entire line unless $f(x)$ happened to be equal to 1. The factor $1/f(x)$ insures that the total mass of the line will be 1.

In the surface representation of $f(x, y)$, the conditional distribution of Y for $X = x_0$, say, is represented by a modification of the curve of intersection of the surface and the plane whose equation is $x = x_0$. Since the area under the curve is ordinarily not equal to 1, the ordinates of the curve must be multiplied by the proper number to make the area equal 1 before the curve will be the graph of a density function. The proper number, of course, is $1/f(x_0)$. Figure 3 indicates this geometrical interpretation for the density function given by (4). The cross hatched curve in the x, y plane should be ignored here.

For the problem discussed in (4), the equation for the conditional density function is obtained by applying (7) to (4) and (5); hence for this problem

(8) $f(y \mid x) = \dfrac{2 - x - y}{\frac{3}{2} - x}$, $0 < x < 1, \quad 0 < y < 1$.

For a fixed x this is a linear function of y; hence the graph of $f(y \mid x)$ must be a straight line, which of course is obvious from Fig. 3 and the geometrical interpretation of $f(y \mid x)$. It will be observed that the only curve of intersection of the type being considered on the surface $z = f(x, y)$ that has unit area under it is the one for which $x = \frac{1}{2}$. All other curves of intersection

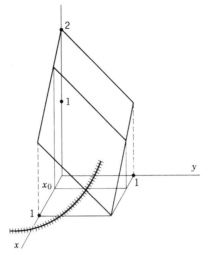

Fig. 3. Geometrical representation of
a conditional distribution.

must have their ordinates multiplied by $1/(\frac{3}{2} - x)$ before they will possess
unit area.

1.3 Curve of Regression

This section uses the preceding material on conditional distributions to
define the concept of a regression curve. From a geometrical point of view a
regression curve is the locus of the means of the conditional distributions
whose densities are given by $f(y \mid x)$. Here it is convenient to use the density
interpretation of $f(y \mid x)$. Let x have the fixed value x_0. Then along the line
$x = x_0$ the mean value of Y will determine a point whose ordinate is denoted
by $\mu_{Y \mid x_0}$. As different values of x are selected, different mean points along
the corresponding vertical lines will be obtained. Thus the ordinate $\mu_{Y \mid x}$
of the mean point for any such line is a function of the value of x selected.
The locus of such mean points, that is, the graph of $\mu_{Y \mid x}$ as a function of x,
will be a curve that is called the curve of regression of Y on X. Analytically,
the equation of the curve of regression is given by the following formula.

$$(9) \qquad \text{CURVE OF REGRESSION:} \quad \mu_{Y \mid x} = \int_{-\infty}^{\infty} y f(y \mid x)\, dy \,.$$

Because of (7), this formula may also be expressed in the form

$$(10) \qquad \mu_{Y \mid x} = \int_{-\infty}^{\infty} y\, \frac{f(x, y)}{f(x)}\, dy \,.$$

The curve of regression of X on Y is defined in an analogous manner.

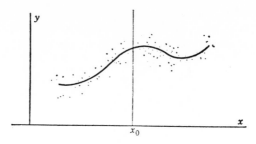

Fig. 4.   Curve of regression.

Figure 4 indicates the geometrical nature of the preceding definition of the curve of regression for a general density distribution.

The density function given in (4) will be used to illustrate the preceding definition. From the result obtained in (8), a direct application of (9) gives

$$\mu_{Y\,|\,x} = \int_0^1 y\, \frac{2 - x - y}{\frac{3}{2} - x}\, dy$$

$$= \frac{1}{\frac{3}{2} - x} \cdot \int_0^1 [(2 - x)y - y^2]\, dy$$

$$= \frac{3x - 4}{6x - 9}, \qquad 0 < x < 1.$$

This is the equation of a hyperbola. The graph of this curve of regression is shown as the crossed line lying in the x, y plane of Fig. 3.

The second illustration in section 1.1 will be used here to illustrate the technique of finding the equation of a regression curve when the limits are variable. From the results obtained there, it follows that

$$f(y \mid x) = \frac{\frac{1}{2}xy}{\frac{1}{4}x^3} = \frac{2y}{x^2}.$$

In view of the triangular nature of the sample space, as seen in Fig. 2, when x is fixed y can range over the values from 0 to x only; consequently (9) becomes

$$\mu_{Y\,|\,x} = \frac{2}{x^2} \int_0^x y^2\, dy = \tfrac{2}{3}x, \qquad 0 < x < 2.$$

The fact that the regression curve is a straight line with slope $\tfrac{2}{3}$ might have been anticipated because of the nature of the density function and the sample space.

1.4 Moments

Moments for general functions of several random variables were defined in section 2, Chapter 5. In dealing with two random variables there is a special type of moment that has been found very useful. It is known as a product moment and is defined as follows:

(11) PRODUCT MOMENT: $\mu'_{pq} = E(X^p Y^q) = \int_{-\infty}^{\infty} \int_{-\infty}^{\infty} x^p y^q f(x, y)\, dy\, dx$.

Here p and q are any non-negative integers. The corresponding product moment about the mean is defined by the formula

(12)
$$\mu_{pq} = E[(X - \mu_X)^p (Y - \mu_Y)^q] = \int_{-\infty}^{\infty} \int_{-\infty}^{\infty} (x - \mu_X)^p (y - \mu_Y)^q f(x, y)\, dy\, dx.$$

It will be observed that these definitions are special cases of the general definition for expected values given in (2), Chapter 5, in which $g(x_1, x_2, \cdots, x_n)$ is chosen as either the function $x_1^p x_2^q$ or $(x_1 - \mu_1)^p (x_2 - \mu_2)^q$ and $n = 2$.

By using formula (3) it is easily seen that the kth moment of X, for example, can be obtained from (11) by choosing $p = k$ and $q = 0$. Thus it follows from similar choices that

(13) $\quad \mu'_{00} = 1$, $\quad \mu'_{10} = \mu_X$, $\quad \mu'_{01} = \mu_Y$, $\quad \mu_{20} = \sigma_X{}^2$, $\quad \mu_{02} = \sigma_Y{}^2$.

The particular product moment μ_{11}, which is called the *covariance* of the two variables, is of special interest because the *correlation coefficient* ρ between the two variables is defined in terms of it.

(14) $\qquad\qquad$ CORRELATION COEFFICIENT: $\rho = \dfrac{\mu_{11}}{\sigma_X \sigma_Y}$.

As an illustration of how to calculate a correlation coefficient, consider the application of formula (14) to the problem first considered in (4). By symmetry and (5), it follows that

$$\mu_X = \mu_Y = \int_0^1 x(\tfrac{3}{2} - x)\, dx = \tfrac{5}{12} .$$

Formula (12) yields

$$\mu_{20} = \mu_{02} = \int_0^1 (x - \tfrac{5}{12})^2 (\tfrac{3}{2} - x)\, dx = \tfrac{11}{144} .$$

Formula (12) applied to (4) gives

$$\mu_{11} = \int_0^1 \int_0^1 \left(x - \frac{5}{12} \right) \left(y - \frac{5}{12} \right)(2 - x - y)\, dy\, dx$$

$$= \int_0^1 \left(x - \frac{5}{12} \right) \int_0^1 \left(y - \frac{5}{12} \right)(2 - x - y)\, dy\, dx$$

$$= \int_0^1 \left(x - \frac{5}{12} \right) \left(\frac{1}{24} - \frac{x}{12} \right) dx$$

$$= -\frac{1}{144}.$$

Formula (14) applied to these results gives

$$\rho = \frac{-\frac{1}{144}}{\frac{11}{144}} = -\frac{1}{11}.$$

If the variables X and Y are independent, the covariance and hence the correlation coefficient of those variables must be equal to zero because under independence

$$E(X - \mu_X)(Y - \mu_Y) = E(X - \mu_X)E(Y - \mu_Y) = 0.$$

The converse is not true. Examples are easily constructed for which $\rho = 0$ but for which X and Y are heavily dependent. Thus, consider the distribution shown in Fig. 5 in which the density is uniform over the two bands shown there. It follows from symmetry that the covariance is zero because to each point in the first quadrant band there is a symmetrical point in the second quadrant band for which the value of $(x - \mu_X)(y - \mu_Y)$ is numerically the same but has the opposite sign. The variables X and Y are, however, highly dependent in the sense that for any given X, the corresponding value of Y is determined to within a very small interval of Y values and varies considerably with X.

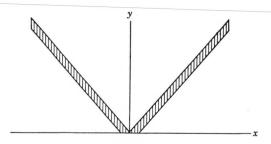

Fig. 5.   A distribution for which $\rho = 0$.

It can be demonstrated mathematically that the correlation coefficient satisfies the inequality $-1 \le \rho \le 1$ and that it will assume the value -1 or 1 if, and only if, all the probability mass is concentrated on some straight line in the x, y plane. This assumes that σ_X and σ_Y are not zero.

In view of the discussion in the two preceding paragraphs, it follows that nothing much can be said concerning the relationship between X and Y if $\rho = 0$ but that X and Y must be linearly related if $\rho = \pm 1$. One would expect that as $|\rho|$ increases the variables X and Y would tend increasingly toward a linear relationship. These ideas will be exploited further in a following section after a normal distribution of two variables has been defined and studied.

2 NORMAL DISTRIBUTION OF TWO VARIABLES

Now, since the normal density function has been shown to be a useful mathematical model for distributions of a single continuous variable, it is to be expected that a joint normal density function for two continuous variables will also prove to be a useful model. If two random variables X and Y are normally distributed but in addition are independently distributed, then their joint density function is easily written down because, from (23), Chapter 2, the joint density function is then the product of the two marginal density functions. In this case, therefore,

$$(15) \qquad f(x, y) = \frac{e^{-\frac{1}{2}\left(\frac{x-\mu_X}{\sigma_X}\right)^2}}{\sqrt{2\pi}\,\sigma_X} \cdot \frac{e^{-\frac{1}{2}\left(\frac{y-\mu_Y}{\sigma_Y}\right)^2}}{\sqrt{2\pi}\,\sigma_Y} = \frac{e^{-\frac{1}{2}\left[\left(\frac{x-\mu_X}{\sigma_X}\right)^2+\left(\frac{y-\mu_Y}{\sigma_Y}\right)^2\right]}}{2\pi\sigma_X\sigma_Y}.$$

If the variables X and Y are not independently distributed it is necessary to modify (15) to take account of the relationship between X and Y. This is done by introducing a cross-product term in the exponent of (15) which is such that its coefficient will be equal to 0 when X and Y are independent. The desired modification is accomplished by means of the following definition.

(16) DEFINITION: *The normal density function of two random variables X and Y is given by the following formula, where* $-1 < \rho < 1$,

$$f(x, y) = \frac{e^{-\frac{1}{2(1-\rho^2)}\left[\left(\frac{x-\mu_X}{\sigma_X}\right)^2-2\rho\left(\frac{x-\mu_X}{\sigma_X}\right)\left(\frac{y-\mu_Y}{\sigma_Y}\right)+\left(\frac{y-\mu_Y}{\sigma_Y}\right)^2\right]}}{2\pi\sigma_X\sigma_Y\sqrt{1-\rho^2}}.$$

If the same approach had been used here as for a normal distribution of one variable, one would have defined a joint normal density function as an exponential function of two variables in which the exponent is a quadratic function of those variables, and then one would have proceeded to show that the parameters defining the function can be expressed in terms of familiar statistical parameters. The result of such an approach is the expression given in (16). As a consequence, the function defined in (16) possesses the properties of a joint density function and its parameters are consistent with the general moment properties given in (13) and (14). This implies, for example, that the parameter ρ in (16) is actually the correlation coefficient as defined in (14). These facts can be verified by evaluating the necessary integrals.

2.1 Marginal Distribution

The marginal distributions of a joint normal distribution are obtained by applying formula (3), and its Y version, to (16). For example, the X marginal density function is given by

$$(17) \qquad f(x) = \int_{-\infty}^{\infty} f(x, y)\, dy$$

where $f(x, y)$ is given in (16). In order to simplify the integration, let $u = (x - \mu_X)/\sigma_X$ and introduce the change of variable $v = (y - \mu_Y)/\sigma_Y$. Then $dy = \sigma_Y\, dv$ and (17) reduces to

$$f(x) = \frac{1}{2\pi\sigma_X\sqrt{1 - \rho^2}} \int_{-\infty}^{\infty} e^{-\frac{1}{2(1-\rho^2)}(u^2 - 2\rho uv + v^2)}\, dv .$$

Adding and subtracting $\rho^2 u^2$ to the exponent in order to complete the square in v gives

$$f(x) = \frac{1}{2\pi\sigma_X\sqrt{1 - \rho^2}} \int_{-\infty}^{\infty} e^{-\frac{1}{2(1-\rho^2)}(v^2 - 2\rho uv + \rho^2 u^2 - \rho^2 u^2 + u^2)}\, dv$$

$$= \frac{e^{-\frac{u^2}{2}}}{2\pi\sigma_X\sqrt{1 - \rho^2}} \int_{-\infty}^{\infty} e^{-\frac{1}{2(1-\rho^2)}(v - \rho u)^2}\, dv .$$

Now make the change of variable $z = (v - \rho u)/\sqrt{1 - \rho^2}$. Then $dv = \sqrt{1 - \rho^2}\, dz$ and $f(x)$ reduces to

$$f(x) = \frac{e^{-\frac{u^2}{2}}}{2\pi\sigma_X} \int_{-\infty}^{\infty} e^{-\frac{z^2}{2}}\, dz .$$

Substituting back the value of u in terms of x and inserting the value $\sqrt{2\pi}$ for this familiar integral will yield the result

$$(18) \qquad f(x) = \frac{e^{-\frac{1}{2}\left(\frac{x - \mu_X}{\sigma_X}\right)^2}}{\sqrt{2\pi}\,\sigma_X} .$$

Since the corresponding result for y follows from symmetry, (18) shows that the marginal distributions of a joint normal distribution are normal.

This result was to be expected, because one would certainly have been unhappy with the definition of a joint normal distribution if the individual variables had not been normally distributed.

The result obtained in (18) is very convenient for demonstrating the consistency of definition (16) with several of the general moment properties given in (13). For example, in order to demonstrate that the constant in (16) has been properly chosen, it is necessary to show that the volume under the surface whose equation is given by (16) is equal to 1. Hence it is necessary to evaluate the double integral.

$$(19) \qquad \int_{-\infty}^{\infty} \int_{-\infty}^{\infty} f(x, y) \, dy \, dx$$

where $f(x, y)$ is given by (16). But from (17) the result of integrating with respect to y is given by (18); therefore, the evaluation of (19) is reduced to the integration of (18) with respect to x over all values of x. The value of this integral is, of course, 1.

If one sets $\rho = 0$ in (16), it will be observed that (16) reduces to (15), which is the density function of two independent normal variables. This shows that if two normal variables are uncorrelated, they are independently distributed. From the discussion of correlation given in section 1.4, particularly with respect to Fig. 5, it should be clear that a lack of linear correlation does not ordinarily guarantee a lack of relationship of every kind between the two variables.

2.2 Conditional Distribution

A joint normal distribution of two variables possesses conditional distributions with interesting properties. In order to study these properties, it will suffice to examine the conditional density function $f(y \mid x)$.

For ease of writing, let $u = (x - \mu_X)/\sigma_X$ and $v = (y - \mu_Y)/\sigma_Y$. Then, a direct application of definition (7) to (16) and (18), together with a few algebraic reductions, will give

$$f(y \mid x) = \frac{e^{-\frac{1}{2(1-\rho^2)}(u^2 - 2\rho uv + v^2)}}{2\pi\sigma_X\sigma_Y\sqrt{1-\rho^2}} \div \frac{e^{-\frac{u^2}{2}}}{\sqrt{2\pi}\,\sigma_X}$$

$$= \frac{e^{-\frac{1}{2(1-\rho^2)}(v^2 - 2\rho uv + \rho^2 u^2)}}{\sqrt{2\pi}\,\sigma_Y\sqrt{1-\rho^2}}$$

$$= \frac{e^{-\frac{1}{2}\left(\frac{v-\rho u}{\sqrt{1-\rho^2}}\right)^2}}{\sqrt{2\pi}\,\sigma_Y\sqrt{1-\rho^2}}.$$

If the values of u and v in terms of x and y are inserted and if the value of y is denoted by y_x to show its dependence on the selected value of x, $f(y \mid x)$ will reduce to

(20)
$$f(y \mid x) = \frac{e^{-\frac{1}{2}\left[\frac{y_x - \mu_Y - \rho\frac{\sigma_Y}{\sigma_X}(x - \mu_X)}{\sigma_Y\sqrt{1 - \rho^2}}\right]^2}}{\sqrt{2\pi}\,\sigma_Y\sqrt{1 - \rho^2}}.$$

Since x has a fixed value and Y_x is the random variable here, (20) shows that Y_x possesses a normal distribution with mean $\mu_Y + \rho(\sigma_Y/\sigma_X)(x - \mu_X)$ and standard deviation $\sigma_Y\sqrt{1 - \rho^2}$. By symmetry a similar result holds for X and Y interchanged. Thus the conditional distributions of a joint normal distribution are also normal.

Since by definition (9) a curve of regression is the locus of the means of the conditional distributions, it follows from (20) that the curve of regression of Y on X for X and Y jointly normally distributed is the straight line whose equation is

(21)
$$\mu_{Y \mid x} = \mu_Y + \rho\frac{\sigma_Y}{\sigma_X}(x - \mu_X).$$

This property of a joint normal distribution, namely, that the curve of regression of Y on X is a straight line, helps to justify the frequent use of linear regression because variables that are approximately normally distributed are encountered frequently.

2.3   Normal Surface

Instead of thinking in terms of probability density in the plane, consider now the geometry of (16), treating it as the equation of a surface in three dimensions. If (7) and the particular results (18) and (20) are applied, the equation of this surface may be written

(22)
$$z = f(x)\frac{e^{-\frac{1}{2}\left[\frac{y - \mu_Y - \rho\frac{\sigma}{\sigma_X}(x - \mu_X)}{\sigma_Y\sqrt{1 - \rho^2}}\right]^2}}{\sqrt{2\pi}\,\sigma_Y\sqrt{1 - \rho^2}}.$$

For the purpose of studying this surface consider its intersections with planes perpendicular to the x axis. The equations of the intersecting curves are obtained by replacing x with the constant values corresponding to the cutting planes. From (22) it will be observed that these curves are normal curves, although not the graphs of normal density functions because the area under any such curve is not usually equal to one, with their means lying on the regression line (21), all having the same standard deviation $\sigma_Y\sqrt{1 - \rho^2}$ and varying in maximum height according to the factor $f(x)$. The tallest

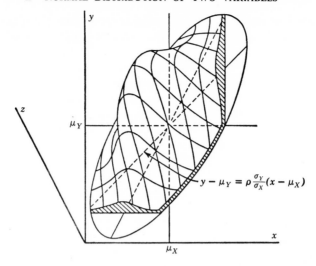

Fig. 6. Normal correlation surface.

such normal curve is the one lying in the cutting plane $x = \mu_X$, since this value makes $f(x)$ a maximum. By symmetry, planes perpendicular to the y axis will intersect the surface in normal curves with corresponding properties. A sketch of a normal correlation surface which shows these various geometrical properties is given in Fig. 6.

Further information is obtained by considering the intersection of the surface by planes perpendicular to the z axis. In this connection it is more convenient to use the original form (16) with $f(x, y)$ replaced by z. If z assumes different constant values, the quantity in brackets in the exponent will assume corresponding values that can be calculated from the constant values assigned to z. Hence the equations of such intersecting curves may be written in the form

$$\left(\frac{x - \mu_X}{\sigma_X}\right)^2 - 2\rho\left(\frac{x - \mu_X}{\sigma_X}\right)\left(\frac{y - \mu_Y}{\sigma_Y}\right) + \left(\frac{y - \mu_Y}{\sigma_Y}\right)^2 = k$$

where k corresponds to the selected value of z. Since this is a quadratic function in x and y, these curves of intersection must be conic sections. Furthermore, since the type of conic section depends only on the quadratic terms, the discriminant for testing conic sections may be applied directly to give

$$B^2 - 4AC = \left(\frac{2\rho}{\sigma_X\sigma_Y}\right)^2 - 4\frac{1}{\sigma_X^2}\frac{1}{\sigma_Y^2}$$

$$= \frac{4(\rho^2 - 1)}{\sigma_X^2\sigma_Y^2} < 0.$$

This result shows that the intersecting curves are ellipses, because by definition (16) $\rho^2 < 1$. Allowing k to assume different values will merely change the sizes of these ellipses; consequentlv, these ellipses have the same centers and the same orientation of principal axes. Part of one such ellipse and its projection onto the x, y plane are shown in Fig. 6. Since the regression line is the locus of the means of the conditional distributions it is clear from Fig. 6 that the regression line and the major axis of the ellipse are different lines, provided that ρ does not have the value 0 or ± 1.

For the purpose of observing how the value of ρ affects a distribution of two normal variables, consider a set of such distributions for which $\mu_X = \mu_Y = 0$ and $\sigma_X = \sigma_Y = 1$. The equation of an ellipse of constant probability density is then given by $x^2 - 2\rho xy + y^2 = k$, where k is a constant. The graph of a set of such curves gives a good geometrical representation of the distribution. Figure 7 gives the graphs of such ellipses for several values of ρ and the same value of $f(x, y)$. The lines shown in Fig.

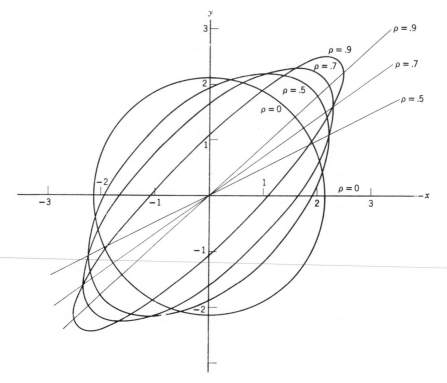

Fig. 7. Constant probability density curves for several values of ρ.

7 with corresponding values of ρ are the regression lines for those values of ρ. It should be noted that a regression line must cut its ellipse at a point where the tangent to the ellipse is a vertical line.

3 NORMAL REGRESSION

A frequently occurring problem in the various sciences is that of predicting the value of one variable from a knowledge of the value of one or more related variables. For example, it would be desirable to be able to predict the breaking strength of a steel shaft with good accuracy from a measurement of its hardness because the latter type of measurement is nondestructive. For the purpose of studying problems of this type, consider first the problem of predicting the value of a random variable Y from a knowledge of a related random variable X.

If X and Y possess a joint normal distribution, the theory derived in the preceding sections shows that the regression curve of Y on X is a straight line whose equation is given by (21). In view of this fact, it appears that linear functions might serve as satisfactory models for regression functions in many real-life situations because variables that possess approximate normal distributions are not uncommon. The preceding theory also shows that the conditional distribution of Y for X fixed is normal with its mean lying on the regression line and with its variance independent of the fixed value of X.

For two normal variables the problem of prediction is therefore quite simple. Given the value of $X = x$, one predicts the value of Y to be $\mu_Y + \rho(\sigma_Y/\sigma_X)(x - \mu_X)$ because that is the mean of the conditional normal distribution of Y for $X = x$. This is based on the principle that the best prediction for a normal random variable is its mean value. Since the variance of this conditional normal variable Y_x is known to be $\sigma_Y^2(1 - \rho^2)$, it is possible to use the normal curve techniques of Chapter 5 to make probability statements concerning the size of the error of prediction. For example, one can state that

$$P\left\{ \left| Y_x - \mu_Y - \rho \frac{\sigma_Y}{\sigma_X}(x - \mu_X) \right| < 2\sigma_Y\sqrt{1 - \rho^2} \right\} = .95 .$$

The foregoing solution to the prediction problem by means of a linear regression function is satisfactory provided the assumption that X and Y possess a joint normal distribution is satisfied. Unfortunately, most of the interesting prediction problems that arise in practice do not fit into this framework. In most industrial experiments, the experimenter decides what X values interest him and conducts his experiments to predict Y for those X values only. For the purpose of obtaining an estimate of the underlying regression function, he will usually choose equally spaced X values over the range of X values that are of interest and conduct his experiments at those selected values. Thus, the variable X is not a random variable in such situations.

Although the X values are not obtained by random selection, and therefore the preceding formulas are not valid for solving the prediction problem, there is no reason why the linearity of the regression function and the normality of the conditional distributions should be dropped. Thus, a satisfactory model for the preceding type of experiment is one that assumes a linear regression function, and assumes conditional normal distributions with means lying on the regression line and having a common variance. This type of model may be formalized as follows.

Let $x_1, \cdots, x_n$ be a set of selected X values. Let $Y_1, \cdots, Y_n$ be conditional random variables corresponding to the selected X values. Assume that $Y_1, \cdots, Y_n$ are independently normally distributed with means given by

$$E[Y_i] = \alpha + \beta(x_i - \bar{x}), \qquad i = 1, \cdots, n,$$

and with a common variance σ^2. As usual, $\bar{x} = (1/n)(x_1 + \cdots + x_n)$. In terms of this notation, the density function of the variable Y_i is given by

$$f(y_i \mid x_i) = \frac{e^{-\frac{1}{2\sigma^2}[y_i - \alpha - \beta(x_i - \bar{x})]^2}}{\sqrt{2\pi}\,\sigma}.$$

A sketch illustrating the preceding assumptions concerning the conditional distribution of the Y_i is given in Fig. 8. Because of the independence assumption concerning the Y's, their joint density is given by

$$f(y_1, \cdots, y_n) = \prod_{i=1}^{n} \frac{e^{-\frac{1}{2\sigma^2}[y_i - \alpha - \beta(x_i - \bar{x})]^2}}{\sqrt{2\pi}\,\sigma}.$$

Before these formulas can be used to make probability statements about prediction errors, it is necessary to be told the values of the parameters α, β, and σ. In practice it is necessary to estimate those parameters from experimental data. Such problems are considered in the next chapter.

If one has more than one variable X to use in making predictions on the variable Y, a normal linear regression model is easily constructed by generalizing the preceding model. Thus, suppose $X_1, \cdots, X_k$ are random variables associated with the random variable Y. Let $x_{1i}, \cdots, x_{ki}$, $i = 1, \cdots, n$, denote n fixed values of $X_1, \cdots, X_k$ and let $Y_1, \cdots, Y_n$ be conditional random variables corresponding to the selected X values. Assume that $Y_1, \cdots, Y_n$ are independently normally distributed with means given by

$$E[Y_i] = \alpha + \beta_1(x_{1i} - \bar{x}_1) + \cdots + \beta_k(x_{ki} - \bar{x}_k), \qquad i = 1, \cdots, n,$$

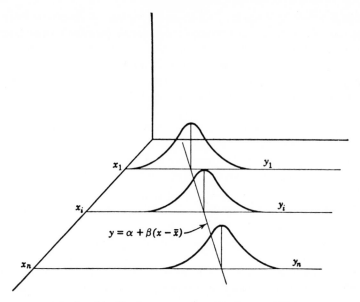

Fig. 8. Distribution assumptions for linear regression.

and with a common variance σ^2. The density of Y_i is then given by

$$f(y_i \mid x_{1i}, \cdots, x_{ki}) = \frac{e^{-\frac{1}{2\sigma^2}[y_i - \alpha - \beta_1(x_{1i} - \bar{x}_1) - \cdots - \beta_k(x_{ki} - \bar{x}_k)]^2}}{\sqrt{2\pi}\,\sigma}, \qquad i = 1, \cdots, n.$$

The joint density is the product of these individual densities.

EXERCISES

Sec. 1

1. Assume that X and Y possess a uniform distribution over the square with vertices $(1, 1)$, $(1, -1)$, $(-1, 1)$, $(-1, -1)$. Calculate the probability of the following events:
(a) $X + Y < 1$
(b) $X^2 + Y^2 < 1$
(c) $|X + Y| < 1$.

2. Assume that X and Y possess a uniform distribution over the circle $x^2 + y^2 \leq 1$. Calculate the following probabilities:
(a) $P\{Y < X\}$
(b) $P\{X^2 + Y^2 < \frac{1}{2}\}$
(c) $P\{Y < cX\}, c > 0$.

Sec. 1.2

3. If $f(x, y) = 1$, $0 < x < 1$, $0 < y < 1$, find the probability that (a) $X > .4$ and $Y > .6$, (b) $X > .4$, (c) $X > Y$, (d) $X = Y$, (e) $X > .4$ given that $Y = .6$, (f) $X > Y$ given that $Y < .4$, (g) $X + Y < 2$, (h) $X^2 + Y^2 < 1$.

4. If $f(x, y) = e^{-(x+y)}$, $x > 0$, $y > 0$, find the probability that (a) $X < 2$, (b) $X < 2$ given that $Y = 1$, (c) $X > Y$, (d) $X + Y < 2$, (e) $X > Y$ given that $Y < 2$.

5. Given $f(x, y) = c(xy^2 + e^x)$, $0 < x < 1$, $0 < y < 1$, determine (a) the value of c, (b) $f(x)$, (c) whether X and Y are independent.

6. Given $f(x, y) = 6(1 - x - y)$, $x > 0$, $0 < y < 1 - x$, and 0 elsewhere. Calculate (a) $g(y)$, (b) $h(x \mid y)$.

7. Let a random point X be chosen in the interval $(0, 1)$. If $X = x$, let a random point Y be chosen in the interval $(0, x)$. Calculate (a) $P\{X + Y > 1\}$, (b) $E[Y \mid x]$, (c) $E[X \mid y]$.

8. Assume that X and Y possess a uniform distribution over the square with vertices $(0, 0)$, $(1, 0)$, $(1, 1)$, $(0, 1)$. Calculate (a) the marginal distributions, (b) the condition distribution of Y given X, (c) the distribution function $F(z)$ of $Z = \max(X, Y)$. Use the fact that that $Z \leq z$ provided that $X \leq z$ and $Y \leq z$.

9. Given the conditional density $f(y \mid x) = x^y e^{-x}/y!$, where Y can assume the values $Y = 0, 1, 2, \cdots$, and given $f(x) = e^{-x}$, $x > 0$, show that the marginal density of Y is given by $g(y) = (\frac{1}{2})^{y+1}$, $y = 0, 1, \cdots$. Use the factorial property of the gamma function integral.

10. The length of life X of a physical particle is a random variable whose distribution depends on a parameter α. This parameter characterizes the type of particle. A population of particles is made up of various types of particles with the proportion having parameter α given by $g(\alpha) = e^{-\alpha}$, $\alpha > 0$. If the distribution of length of life for fixed α is given by $f(x \mid \alpha) = \alpha e^{-\alpha x}$, $x > 0$, find the unconditional density function $f(x)$.

Sec. 1.3

11. Given $f(x, y) = xe^{-x(y+1)}$, $x > 0$, $y > 0$, find (a) the marginal density functions, (b) the conditional density functions, (c) the curve of regression of Y on X.

12. Find the equation of the regression curve of Y on X, given that $f(x, y) = \frac{9}{2}(1 + x + y)/(1 + x)^4(1 + y)^4$, $x > 0$, $y > 0$.

13. Given $f(x, y) = x + y$, $0 < x < 1$, $0 < y < 1$, calculate (a) $f(y \mid x)$, (b) the mean of Y, given $X = x$.

14. If $f(x, y) = 2$, $0 < x < 1$, $0 < y < x$, find (a) the marginal density functions, (b) the conditional density functions, (c) the curve of regression of Y on X.

15. Assume X and Y possess a uniform distribution over the circle $x^2 + y^2 \leq 9$. Calculate the following: (a) $f(x)$, (b) $f(y \mid x)$, (c) $\mu_{Y \mid x}$.

16. Find a nontrivial joint distribution of two variables X and Y such that the regression curve of Y on X is the parabola $y = x^2$.

17. If X and Y are independent variables, show that the curve of regression of Y on X will be a horizontal straight line. What line?

18. Find a nontrivial joint distribution of two variables such that both regression curves are straight lines.

Sec. 1.4

19. Given $f(x, y) = 2,\ 0 < x < 1,\ 0 < y < x$, show that $E[XY] \neq E[X]E[Y]$.

20. Given

$$f(x, y) = \begin{cases} 1, & -y < x < y,\ \ 0 < y < 1 \\ 0, & \text{elsewhere} \end{cases}$$

show that X and Y are uncorrelated but not independent.

21. Given $f(x, y) = 8xy,\ 0 < x < 1,\ 0 < y < x$, show that X and Y are not independent variables.

22. Given $f(x, y) = \frac{2}{3}(x + 2y),\ 0 < x < 1,\ 0 < y < 1$, calculate the value of μ_{11}.

23. Given $f(x, y) = 1,\ 0 < x < 1,\ 0 < y < 1$, find (a) μ'_{pq}, (b) ρ, (c) $\mu_{Y|x}$.

24. Given $f(x, y) = 2/a^2,\ 0 < x < a,\ 0 < y < x$, find (a) μ'_{pq}, (b) ρ, (c) $\mu_{Y|x}$.

25. Given $f(x, y) = c$ in the two triangular regions bounded by the lines $x = -1,\ y = 0,\ y = -x$, and $x = 1,\ y = 0,\ y = x$, find the value of (a) c, (b) ρ, (c) $\mu_{Y|x}$.

26. If X and Y are independent variables, find expressions for the mean and variance of $Z = XY$ in terms of the means and variances of X and Y.

27. Let n independent trials be made of an experiment for which p is the probability success in a single trial. Let X equal the number of successes and let Y equal the sum of the numbers of the trials at which successes occur. Write

$$X = X_1 + \cdots + X_n$$

where $X_i = 1$ or 0, depending on success or failure, and write $Y = Y_1 + \cdots + Y_n$ where $Y_i = i$ or 0, depending on success or failure. Calculate $E[X]$, $E[Y]$, $E(XY)$, μ_{11}.

Sec. 2

28. Assume that a bomber is making a run along the positive y axis at a square target 200 feet by 200 feet whose center is at the origin. Assume the X and Y errors in repeated runs are normally distributed about 0. (a) If X and Y are independent with $\sigma_X = \sigma_Y = 300$ feet, what is the probability of a hit on one run? (b) Under (a) what is the probability of at least one hit in 10 runs? (c) Under (a) how many runs would be needed to make the probability of at least one hit greater than .8? (d) Why would (a) be difficult to solve if X and Y were correlated with $\rho = \frac{1}{2}$?

29. In problem 28 assume the X and Y errors are with respect to the point of aiming rather than the center of the target and that the coordinates of the aiming errors are independent normal variables about the center of the target with $\sigma_X = \sigma_Y = 100$ feet. Let $X = Z + U$ and $Y = W + V$, where Z and W are the aiming errors and U and V are the bombing errors. Solve part (a) of problem 28 by

using the fact that X and Y are independent normal variables because they are the sums of such variables.

30. If X and Y are independent normal variables with $\mu_X = \mu_Y = 0$ and $\sigma_X = \sigma_Y = 1$, (a) find the probability that $X^2 + Y^2 < 2$, (b) determine what size circle with the center at the origin is such that the probability is .95 that the sample point will fall inside it.

31. Verify by integration that definition (16) is consistent with the general moment properties given in (13) and (14).

Sec. 2.1

32. Construct or describe a joint non-normal distribution of two variables whose marginal distributions are both normal.

Sec. 2.2

33. If the exponent in the normal density of X and Y is

$$-\tfrac{5}{8}[4(x - 1)^2 - 9.6(x - 1)(y + 2) + 16(y + 2)^2]$$

find (a) μ_X, μ_Y, σ_X, σ_Y, ρ, (b) $f(x)$, (c) $\mu_{Y|x}$.

34. How must the density function of a normal variable X be modified if X is restricted to (a) values larger than μ, (b) positive values?

35. Suppose X and Y are jointly normally distributed with $\mu_X = 4$, $\mu_Y = 2$, $\sigma_X = \sigma_Y = 2$, and $\rho = \tfrac{1}{4}$. If you wish to estimate the value of Y for an individual whose X value is equal to 6, how will the size of the variance of the error of this estimate differ from that when nothing is known about his X value?

Sec. 2.3

36. Prove that all vertical plane sections of a normal correlation surface are normal curves.

CHAPTER 7

Empirical Methods for Correlation
and Regression

The preceding chapter introduced probability models that involve parameters not previously encountered. In particular, it was concerned with the correlation coefficient and with the coefficients of a linear regression function. This chapter treats statistical inference problems related to those parameters.

1 THE ESTIMATION OF ρ

Let X and Y possess a joint normal distribution with density function given by formula (16), Chapter 6, and let $(x_1, y_1), \cdots, (x_n, y_n)$ denote the values obtained from a random sample of size n from this distribution. This means that a set of n individuals is selected at random from the population and then the attributes X and Y are measured for each of the sampled individuals.

There are five parameters needed to specify a normal density, namely μ_X, μ_Y, σ_X, σ_Y, and ρ. From earlier work on the estimation of the mean and variance of a single random variable it follows that μ_X, μ_Y, σ_X, and σ_Y can be estimated by means of the sample values $\bar{x}$, $\bar{y}$, s_X, and s_Y. Thus, all that remains is to estimate ρ. Since the method of moments is being employed for estimation, ρ will be estimated indirectly in this manner by replacing the parameters involved in its definition, as given by formula (14), Chapter 6, by their sample estimates that are obtained by the method of moments.

The product moment μ_{11} is defined as $E(X - \mu_X)(Y - \mu_Y)$. If μ_X and μ_Y are unknown and are estimated by $\bar{x}$ and $\bar{y}$, respectively, then the corresponding sample product moment is

$$\frac{1}{n} \sum_{i=1}^{n} (x_i - \bar{x})(y_i - \bar{y}) .$$

Since σ_X and σ_Y are being estimated by s_X and s_Y, the desired sample estimate of ρ, which is denoted by r, is therefore given by the formula

(1)
$$r = \frac{\sum_{i=1}^{n}(x_i - \bar{x})(y_i - \bar{y})}{ns_X s_Y}.$$

This formula defines the *sample correlation coefficient*.

A somewhat better form for computational purposes is obtained by inserting computing formulas for s_X and s_Y and performing some algebra. Thus, omitting indices of summation,

(2)
$$r = \frac{n \sum xy - \sum x \sum y}{\sqrt{[n \sum x^2 - (\sum x)^2][n \sum y^2 - (\sum y)^2]}}.$$

This last form requires the sums of x_i, y_i, x_i^2, y_i^2, and $x_i y_i$, all of which are readily obtained with the aid of desk calculators.

As an illustration, consider the data of Table 1 consisting of the scores of 30 students on a language test X and a science test Y. The maximum possible score on each of these tests was 50 points. The choice of which variable to call X and which to call Y is arbitrary here.

TABLE 1

x	y	x	y	x	y
34	37	28	30	39	36
37	37	30	34	33	29
36	34	32	30	30	29
32	34	41	37	33	40
32	33	38	40	43	42
36	40	36	42	31	29
35	39	37	40	38	40
34	37	33	36	34	31
29	36	32	31	36	38
35	35	33	31	34	32

For the purpose of observing whether the preceding theory is applicable to these two variables, this set of 30 pairs of measurements will be plotted as points in the x, y plane. A graph of this type is called a *scatter diagram*. By means of it one can quickly discern whether there is any pronounced relationship between the variables and whether a normal distribution assumption is plausible. The scatter diagram for the data of Table 1 is shown in Fig. 1.

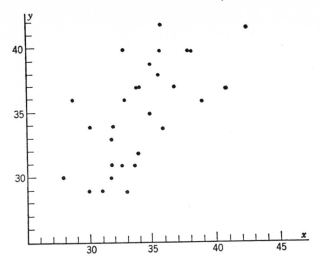

Fig. 1. Scatter diagram for language and science scores.

An inspection of this scatter diagram shows that there is a tendency for small values of X to be associated with small values of Y and for large values of X to be associated with large values of Y. Furthermore, the general trend of the scatter is that of a straight line. Experience with test scores indicates that they frequently possess individual normal distributions; therefore in view of the reasonableness of a linear regression assumption here, it appears that the assumption that X and Y possess a joint normal distribution is a reasonable one. As a result, r may be used to estimate the value of ρ in the joint normal density. Calculations based on formula (2) for the data of Table 1 gave $r = .66$.

The interpretation of this value of r as a measure of the strength of the linear relationship between X and Y is the same as that for ρ, and which was discussed in Chapter 6. There several graphs were presented of ellipses of constant probability density corresponding to various values of ρ. Here it is instructive to look at several scatter diagrams with their corresponding values of r, as shown in Fig. 2. The first four could have been obtained from sampling a two-dimensional normal distribution, whereas the fifth illustrates an empirical distribution for which r is unsatisfactory as a measure of the relationship between X and Y.

The diagrams of Fig. 2 with their associated values of r make plausible two properties of r, namely, that the value of r must satisfy the inequality $-1 \leq r \leq 1$ and that the value of r will be equal to ± 1 if, and only if, the points of the scatter lie on a straight line. A demonstration of these properties is given in the appendix. Similar properties hold for ρ.

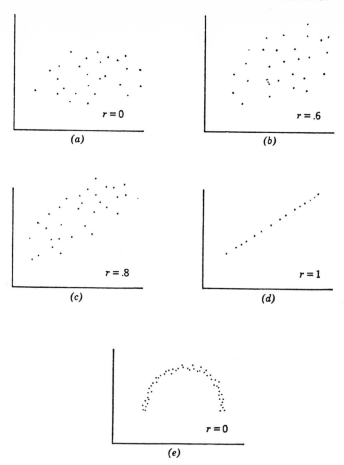

Fig. 2.   Scatter diagrams and their associated values of r.

2 THE RELIABILITY OF r

If X and Y are jointly normally distributed the pairs of random sample variables X_i, Y_i, $i = 1, \cdots, n$, will be independently distributed each with the same distribution as X and Y. In terms of these random variables the sample means and variances are denoted by $\bar{X}$, $\bar{Y}$, S_X, and S_Y, and the sample correlation coefficient by

$$r = \frac{\sum_{i=1}^{n}(X_i - \bar{X})(Y_i - \bar{Y})}{nS_X S_Y}.$$

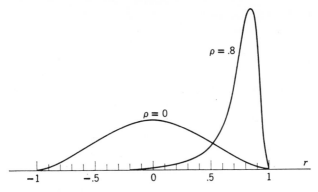

Fig. 3. Distribution for r for $\rho = 0$ and $\rho = .8$ when $n = 9$.

After a sample of size n has been taken and the observational values $(x_1, y_1), \cdots, (x_n, y_n)$ made available, r as given by formula (1) can be computed and is merely a number. However, in discussing how the function r behaves in repeated sampling experiments, r is a random variable which is a function of the n pairs of random variables X_i, Y_i, $i = 1, \cdots, n$. It is theoretically possible to derive the probability density function of r from the density function of those n pairs of variables; however, both the form and the derivation of this density are too complicated to be considered here. It turns out that the density function of r depends only on the parameters ρ and n, where n is the number of points in the scatter diagram. Graphs of the density function of r for $\rho = 0$ and for $\rho = .8$ when $n = 9$ are shown in Fig. 3.

It is clear from Fig. 3 that the distribution of r is decidedly non-normal for large values of ρ; consequently it will not suffice to obtain the standard deviation of r and use it to determine the accuracy of r as an estimate of ρ. Fortunately, there exists a simple change of variable which transforms the complicated distribution of r into an approximately normal distribution. The resulting normal distribution may then be used to determine the accuracy of r as an estimate of ρ in the same way that the normal distribution of $\bar{X}$ was used to determine the accuracy of $\bar{X}$ as an estimate of μ. This change of variable is from r to z, where

(3)
$$z = \tfrac{1}{2} \log_e \frac{1 + r}{1 - r}.$$

It can be shown that when the preceding assumptions are satisfied, the random variable z will be approximately normally distributed with mean

$$\mu_z = \tfrac{1}{2} \log_e \frac{1 + \rho}{1 - \rho}$$

and standard deviation

$$\sigma_z = \frac{1}{\sqrt{n-3}}.$$

As an illustration of how this transformation is used, consider the problem of determining an interval of values within which r could reasonably be expected to fall if $\rho = .8$ and if r is based on a sample of size 28. Let reasonably be understood to mean with a probability of .95. The construction of such an interval can be accomplished by first constructing such an interval for z and then transforming it into an interval for r. The simplest interval for z that possesses the desired property is the interval with end points $z_1 = \mu_z - 2\sigma_z$ and $z_2 = \mu_z + 2\sigma_z$. For $\rho = .8$ and $n = 28$, it follows from (3) that those end points are

$$z_1 = \tfrac{1}{2}\log 9 - \frac{2}{\sqrt{25}} = .70$$

and

$$z_2 = \tfrac{1}{2}\log 9 + \frac{2}{\sqrt{25}} = 1.50 .$$

From tables of the exponential function it will be found that values of r that correspond to these values of z are $r_1 = .60$ and $r_2 = .91$. Thus it can be stated that the probability is approximately .95 that the sample correlation coefficient will satisfy the inequality $.60 < r < .91$ when r is based on a sample of 28 and $\rho = .80$. This example illustrates how unreliable r is as an estimate of ρ unless one had a very large sample.

Although the preceding relationship simplifies the problem of determining the accuracy of r as an estimate of ρ, it has the disadvantage of being unreliable if X and Y do not have a joint normal distribution; consequently unless one is quite certain that these variables possess such a distribution, at least to a good approximation, the results should not be relied upon.

3 INTERPRETATION OF r

Given any two random variables X and Y one can ask the question whether those variables are independent. Since two variables are independent if, and only if, $f(x, y) = g(x)h(y)$ where $g(x)$ and $h(y)$ are the marginal densities of X and Y and since an extremely large sample would be required to determine whether this relationship is being satisfied, it is clear that some other method is needed to solve this problem. One approach is to introduce some measure of the relationship between two variables, whose value is zero for independent variables, and use it to determine whether the variables are independent.

In view of the definition of $f(x, y)$, it is seen that the normal variables X and Y are independent if, and only if, they are uncorrelated. Thus, the parameter ρ completely determines whether or not two normal variables are independently distributed. As a result, it suffices to determine whether $\rho = 0$ for such a pair of variables to ascertain independence. Since the sample correlation coefficient r serves as an estimate of ρ, it can be used to determine whether it is reasonable to assume that $\rho = 0$,

If X and Y cannot be assumed to be normally distributed, even approximately, then ρ can no longer be used as a basis for determining the extent to which X and Y are related. Figure 2 and Fig. 5, Chapter 6, both indicate the inefficiency of ρ and r for measuring the extent of the relationship when X and Y are not normally distributed.

Even though two variables may possess a joint normal distribution, and therefore that ρ may be used as a measure of the strength of the relationship of the two variables, it does not follow that the relationship as measured by ρ is meaningful in a practical sense. The fact that two variables tend to increase or decrease together does not imply that one has any direct or indirect effect on the other. Both may be influenced by other variables in a manner that will give rise to a strong mathematical relationship. The favorite example to illustrate this fact is the one concerned with teachers' salaries. Over a period of years the correlation coefficient between teachers' salaries and the consumption of liquor turned out to be .98. During that period of time there was a steady rise in wages and salaries of all types and a general upward trend of good times. Under such conditions teachers' salaries would also increase. Moreover, the general upward trend in wages and buying power would be reflected in increased purchases of liquor. Thus, the high correlation merely reflected the common effect of the upward trend on the two variables. This is a type of correlation that has received the name of *spurious correlation*. The preceding discussion should make it clear that success with correlation coefficients requires familiarity with the field of application as well as with their mathematical properties and that both the reliability and interpretation of r depend heavily upon the extent to which X and Y are jointly normally distributed.

4 LINEAR REGRESSION

As has been observed, empirical correlation methods are often useful in studying how two variables are related. It frequently happens, however, that one studies the relationship between the variables in the hope that any relationship that is discovered can be used to assist in making estimates or predictions of one of the variables. Thus if the two variables are the high school record X and the college record Y of college students, any relationship between X and Y would be useful for assisting in predicting a student's college success from a knowledge of how well he did in high school. The correlation coefficient is not capable of solving such prediction problems. For problems such as this the correlation coefficient serves only as an exploratory tool for determining which variables may be worth incorporating in a regression function for predicting a given variable.

Suppose the variables X and Y are jointly normally distributed. Then formula (21), Chapter 6, shows that the regression curve of Y on X is a straight line whose equation is given by

$$(4) \qquad \mu_{Y|x} = \mu_Y + \rho \frac{\sigma_Y}{\sigma_X}(x - \mu_X).$$

Since the value of Y corresponding to $X = x$ will be predicted by $\mu_{Y|x}$, the empirical problem of prediction is the problem of estimating the regression function (4). The parameters in this function are readily estimated from the sample. The desired sample estimate of Y_x is therefore given by

$$\hat{\mu}_{Y|x} = \overline{Y} + r \frac{S_Y}{S_X}(x - \overline{X}).$$

If X and Y do not possess a joint normal distribution, the preceding technique cannot be applied. Even though X and Y are not normally distributed, it may well be that the regression curve is a straight line. If so, the equation of the regression function corresponding to (4) can be expressed in the form

$$(5) \qquad \mu_{Y|x} = \alpha + \beta(x - \mu_X).$$

The empirical problem of prediction is now reduced to the problem of estimating the parameters α, β, and μ_X.

5 LEAST SQUARES

Since no assumption is being made concerning the nature of the density $f(x, y)$, except that the regression curve is a straight line, it is not possible to express the parameters α and β in terms of familiar parameters as was true for normal variables. Thus, the method of moments that has been employed thus far for the estimation of parameters cannot be used here.

For the purpose of arriving at a method for estimating α and β, let the sample values $(x_1, y_1), \cdots, (x_n, y_n)$ be plotted as points in the x, y plane. Then the problem of estimating a linear regression function can be treated as a problem of fitting a straight line to this set of points. The coefficients of the fitted line, when it is expressed in the form of (5), will serve as the estimates of α and β. Although there are numerous methods for fitting a curve to a set of points, the best known one is the *method of least squares*, which will now be described.

Since the desired curve is to be used for estimating, or predicting, purposes, it is reasonable to require that the curve be such that it makes the errors of estimation small. By an error of estimation, or prediction, is meant the difference between an observed value of Y and the corresponding fitted curve value of Y. If the value of the variable to be predicted is denoted by Y and the corresponding curve value by Y', then the error of estimation, or prediction, is given by $Y - Y'$. Since the errors may be positive or negative and might add up to a small value for a poorly fitting curve, it will not do to require merely that the sum of the errors be as small as possible. This difficulty can be avoided by requiring that the sum of the absolute values of the errors be as small as possible. However, sums of absolute values are not convenient to work with mathematically; consequently the difficulty is avoided by requiring that the sum of the squares of the errors be a minimum. The values of the parameters obtained by this minimization determine what is known as the best fitting curve in the sense of least squares.

Consider the application of this principle to the fitting of a straight line to a set of n points. Corresponding to equation (5) the equation of the line may be written in the form

$$(6) \qquad y' = a + b(x - \bar{x})$$

where b is its slope and a is the y intercept on the line $x = \bar{x}$. The y intercept on the y axis is $a - b\bar{x}$. It will be seen shortly why it is so convenient to express the equation of the line in this form rather than in the slope-intercept form, $y = a + bx$, of analytical geometry. The problem now is to determine the parameters a and b so that the sum of the squares of the errors of prediction will be a minimum. Since the coordinates of the ith point are denoted by (x_i, y_i), this sum of squares will be $\sum_{i=1}^{n} (y_i - y_i')^2$. When y_i' is replaced by its value, as given by (6), it becomes clear that this sum is a function of a and b only. If this function is denoted by $G(a, b)$, then

$$G(a, b) = \sum_{i=1}^{n} [y_i - a - b(x_i - \bar{x})]^2 .$$

For this function to have a minimum value, it is necessary that its partial derivatives vanish there; hence a and b must satisfy the equations

$$\frac{\partial G}{\partial a} = \sum 2[y - a - b(x - \bar{x})][-1] = 0$$

$$\frac{\partial G}{\partial b} = \sum 2[y - a - b(x - \bar{x})][-(x - \bar{x})] = 0$$

where the subscripts and range of summation have been omitted for convenience. When the summations are performed term by term and the sums that involve y are transposed, these equations assume the form

$$an + b \sum (x - \bar{x}) = \sum y$$
$$a \sum (x - \bar{x}) + b \sum (x - \bar{x})^2 = \sum (x - \bar{x})y .$$

Since $\sum (x - \bar{x}) = 0$, the solution of these equations is given by

$$a = \bar{y} \quad \text{and} \quad b = \frac{\sum (x - \bar{x})y}{\sum (x - \bar{x})^2} .$$

These values when inserted in (6) yield the desired least squares line because it can be shown that these values do minimize G. This line is usually called the sample regression line of y on x; hence the preceding derivation gives the following result:

(7)   REGRESSION LINE: $y' - \bar{y} = b(x - \bar{x})$, where

$$b = \frac{\sum (x_i - \bar{x})y_i}{\sum (x_i - \bar{x})^2} .$$

A pioneer in the field of applied statistics gave the least squares line this name in connection with some studies he was making on estimating the extent to which the stature of sons of tall parents reverts or regresses toward the mean stature of the population.

For computational purposes it is convenient to change the form of b slightly in the following manner.

(8) $$b = \frac{\sum xy - \bar{x} \sum y}{\sum x^2 - 2\bar{x} \sum x + \sum \bar{x}^2}$$
$$= \frac{\sum xy - n\bar{x}\bar{y}}{\sum x^2 - n\bar{x}^2} .$$

As an illustration of the least squares technique for estimating a linear regression function, consider the data of Table 2 on the amount of water

TABLE 2

Water (x)	12	18	24	30	36	42	48
Yield (y)	5.27	5.68	6.25	7.21	8.02	8.71	8.42

applied in inches and the yield of alfalfa in tons per acre on an experimental farm. The graph of these data is given in Fig. 4. From this graph it appears

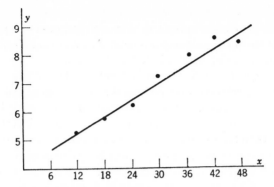

Fig. 4.   Hay yield as a function of amount of irrigation.

that X and Y are approximately linearly related for this range of X values. For the purpose of predicting Y from X, it should therefore suffice to use a linear function of X.

Table 3 illustrates the computational procedure for the data of Table 2.

TABLE 3

x	y	$x - \bar{x}$	$(x - \bar{x})y$	$(x - \bar{x})^2$
12	5.27	−18	−94.86	324
18	5.68	−12	−68.16	144
24	6.25	−6	−37.50	36
30	7.21	0	0	0
36	8.02	6	48.12	36
42	8.71	12	104.52	144
48	8.42	18	151.56	324
210	49.56		103.68	1008

Here (7) was used to calculate b instead of the suggested computing formula (8) because $\bar{x}$ is an integer.   As a result of these computations, the equation of the regression line was found to be

$$y' = 4.0 + .10x .$$

The graph of this line is shown in Fig. 4.

There is an important difference between the scatter diagrams of Fig. 1 and Fig. 4 that should be noted.   In Fig. 1 the points correspond to a random sample of 30 students;   consequently, both X and Y are random variables. In Fig. 4, however, the X values were chosen in advance, so that Y is the only random variable.   Since least squares can be applied whether the X values were fixed in advance or were obtained from random samples, the regression approach to studying the linear relationship between two variables is more flexible than the correlation approach. The interpretation of r as a measure of the strength of the linear relationship between two variables obviously does not apply if the values of X are selected as desired because the value of r will usually depend heavily on the choice of X values. In addition to being more flexible, regression methods also possess the advantage of being the natural methods to use in many experimental situations. The experimenter often wishes to change X by uniform amounts over the range of interest for that variable rather than take a random sample of X values. Thus, if he wanted to study the effect of an amino acid on growth, he would increase the amount of amino acid by a fixed amount, or factor, each time he ran the experiment.

6   MULTIPLE LINEAR REGRESSION

It happens quite often that the method of the preceding section for predicting one variable by means of a related variable yields poor results not because the relationship is far removed from the linear one assumed there but because there is no single variable related closely enough to the variable being predicted to yield good results. However, it may happen that there are several variables that, when taken jointly, will serve as a satisfactory basis for predicting the desired variable. Since linear functions are so simple to work with and experience shows that many sets of variables are approximately linearly related, it is reasonable to attempt to predict the desired variable by means of a linear function of the remaining variables. For this purpose let $Y, X_1, X_2, \cdots, X_k$ represent the available variables and consider the problem of predicting the variable Y by means of a linear function of the remaining variables. The linear predicting function may be expressed in the form

$$(9) \qquad y' = c_0 + c_1 x_1 + c_2 x_2 + \cdots + c_k x_k$$

where the c's are to be determined by means of available data.

As in the case of two variables, the unknown coefficients are estimated by the method of least squares. This implies that n sets of values of the $k + 1$ variables are available for obtaining the estimates. Geometrically, the problem is one of finding the equation of the plane which fits best, in the sense of least squares, a set of n points in $k + 1$ dimensions and in which errors are measured in the direction parallel to the y axis.

Let $x_{1i}, \cdots, x_{ki}, i = 1, \cdots, n$, denote the values of $X_1, \cdots, X_k$ that have been selected. Thus, the X's are not random variables here. Let $y_i, i = 1, \cdots, n$, denote the value of Y that was obtained when an observation was taken at the point $(x_{1i}, \cdots, x_{ki})$ in the k dimensional space of the x's. The problem now is to find the set of c's in (9) that will minimize the sum $\sum_{i=1}^{n} (y_i - y_i')^2$. In terms of the notation in (9) this sum may be written

$$(10) \qquad G(c_0, c_1, \cdots, c_k) = \sum_{i=1}^{n} [y_i - c_0 - c_1 x_{1i} - \cdots - c_k x_{ki}]^2 .$$

If this function is to have a minimum value, it is necessary that its partial derivatives vanish there; hence the c's must satisfy the equations

$$\frac{\partial G}{\partial c_0} = \frac{\partial G}{\partial c_1} = \cdots = \frac{\partial G}{\partial c_k} = 0 .$$

Differentiation of (10) produces the following equations, in which the summation indices have been deleted for simplicity of notation.

$$\sum 2[y - c_0 - c_1 x_1 - \cdots - c_k x_k][-1] = 0$$
$$\sum 2[y - c_0 - c_1 x_1 - \cdots - c_k x_k][-x_1] = 0$$
$$\cdots \cdots \cdots \cdots \cdots \cdots \cdots \cdots \cdots \cdots$$
$$\sum 2[y - c_0 - c_1 x_1 - \cdots - c_k x_k][-x_k] = 0 .$$

If these equations are multiplied by $\frac{1}{2}$, the summations performed term by term, and the first sum transferred to the right side, these equations will assume the form

$$c_0 n + c_1 \sum x_1 + \cdots + c_k \sum x_k = \sum y$$
$$(11) \qquad c_0 \sum x_1 + c_1 \sum x_1^2 + \cdots + c_k \sum x_1 x_k = \sum x_1 y$$
$$\cdots \cdots \cdots \cdots \cdots \cdots \cdots \cdots \cdots \cdots \cdots$$
$$c_0 \sum x_k + c_1 \sum x_k x_1 + \cdots + c_k \sum x_k^2 = \sum x_k y .$$

Equations such as these which are obtained by the method of least squares are commonly called the *normal equations of least squares* or more briefly the *normal equations*. These equations are easily solved, provided the number of equations is small. For large sets of equations, high speed computing facilities are needed. If such facilities are not available, there is considerable saving of time if one of the compact computing schemes designed for desk calculators, such as the Doolittle method, is used.

The derivation of (11) did not require that all the n values of X_1, nor of the remaining variables, be different. It is necessary only that there be a sufficient number of distinct values of the variables $X_1, \cdots, X_k$ to determine uniquely the least squares plane. Ordinarily this means that $k + 1$ distinct values will suffice because a plane in $k + 1$ dimensions is determined by $k + 1$ points, provided that the $k + 1$ points do not lie in a lower dimensional plane. For example, a plane in three dimensions is determined uniquely by three points, provided that the three points do not lie on a straight line.

Although satisfying equations (11) was a necessary condition imposed upon the c's to minimize G, it can be shown that a set of c's that satisfies these equations does minimize G.

As in the case of two variables, the calculations become somewhat simpler if the variables are measured from their sample means. If, for example, x_{1i} represents $x_{1i} - \bar{x}_1$ rather than x_{1i} itself, then $\sum x_{1i} = 0$; and similarly for the remaining variables. With such variables, the first equation reduces to $c_0 n = \sum y$. Since then $c_0 = \bar{y}$, there will be only k equations in k unknowns to solve.

As an illustration of the preceding methods, consider the problem of predicting the amount of hay from a knowledge of the spring rainfall and temperature based on the following data which came from an experiment conducted in England. Here Y denotes the amount of hay in units of 100 pounds per acre, X_1 the spring rainfall in inches, and X_2 the accumulated temperature above $42°F$ in the spring. After the data were acquired, calculations of the sums needed in (11) were carried out in terms of variables measured from their sample mean. A wave is placed over a variable to indicate that it is measured from its mean. The data yielded the following values

$$\bar{y} = 28.0, \qquad \bar{x}_1 = 4.91, \qquad \bar{x}_2 = 594, \qquad \frac{1}{n}\sum \tilde{x}_1 \tilde{y} = 3.872,$$

$$\frac{1}{n}\sum \tilde{x}_2 \tilde{y} = -149.6, \qquad \frac{1}{n}\sum \tilde{x}_1 \tilde{x}_2 = -52.36,$$

$$\frac{1}{n}\sum \tilde{x}_1{}^2 = 1.21, \qquad \frac{1}{n}\sum \tilde{x}_2{}^2 = 7225.$$

The normal equations (11) then become, after multiplying through by n,

$$1.21c_1 - 52.36c_2 = 3.872$$
$$52.36c_1 - 7225c_2 = 149.6.$$

The solution of these equations is $c_1 = 3.3$ and $c_2 = .004$; consequently (9) becomes

$$\tilde{y}' = 3.3\tilde{x}_1 + .004\tilde{x}_2.$$

Expressed in terms of the original variables, the equation is

$$y' = 9.4 + 3.3x_1 + .004x_2.$$

This equation indicates that if x_2 is held fixed the amount of hay will increase about 330 pounds per acre with each inch increase in spring rainfall. On the other hand, if spring rainfall is held fixed, the accumulated spring temperature would have to increase about 250 units, which will be observed to be about three standard deviations for variable x_2, in order to increase the amount of hay by 100 pounds per acre. Thus it appears that the spring temperature is relatively unimportant compared with spring rainfall. Such conclusions, of course, are only approximately true. They depend on the variables being approximately linearly related, and they express only average relationships. They also assume that the function in (9) is a satisfactory estimate of a "true" linear regression function for those variables. As in the case of linear regression for one independent variable, the problem of the accuracy of the coefficients in the least squares regression function as estimates of the coefficients in a theoretical regression function is postponed to Chapter 10.

One final remark should be made concerning the preceding problem. Strictly speaking, this problem does not fit the regression model that has been studied in this chapter because the values of the X's were not selected in advance of observing the Y's. They represent rainfall and temperature in England over a period of several years and hence are more like random variables than predetermined variables. They are not strictly random either because the experimenter did not select a random set of years to determine the X's. As remarked before, regression methods possess the flexibility of permitting the X's to be preselected, or to be random, or to be determined by some other process. Only when discussing reliability, which will be done in a later chapter, is it necessary to be concerned about these matters.

7 CURVILINEAR REGRESSION

If a scatter diagram in the x, y plane indicates that a straight line will not fit a set of points satisfactorily because of the nonlinearity of the relationship, it may be possible to find some simple curve that will yield a satisfactory fit. Since an investigator always strives to explain relationships as simply as possible, with the restriction that his explanation be consistent with previous knowledge, he will prefer to use a simple type of curve. It follows, therefore, that the type of curve to use will depend largely on the amount of theoretical information one has concerning the relationship and thereafter on convenience.

7.1 Polynomial Regression

If there are no theoretical reasons for expecting a curve of a certain type to represent the relationship, polynomials are usually selected because of their simplicity and flexibility. The lowest degree polynomial that will suffice can often be determined by an inspection of the scatter diagram. After the degree has been determined, the best-fitting polynomial of that degree may then be fitted by the method of least squares.

Let the degree of the polynomial be k and let the equation of the polynomial be written in the form

$$(12) \qquad y' = a_0 + a_1 x + a_2 x^2 + \cdots + a_k x^k .$$

The normal equations here need not be derived because they can be obtained from the normal equations (11) of multiple regression by letting $x_i = x^i$. This is permissible because the derivation of equations (11) did not place any restriction on the nature of the variables $x_1, x_2, \cdots, x_k$, and therefore they may be related in any manner desired. With this choice of the x's the normal equations for polynomial regression become

$$(13) \qquad \begin{aligned} a_0 n + a_1 \sum x + \cdots + a_k \sum x^k &= \sum y \\ a_0 \sum x + a_1 \sum x^2 + \cdots + a_k \sum x^{k+1} &= \sum xy \\ \cdots\cdots\cdots\cdots\cdots\cdots\cdots\cdots\cdots\cdots\cdots\cdots\cdots\cdots \\ a_0 \sum x^k + a_1 \sum x^{k+1} + \cdots + a_k \sum x^{2k} &= \sum x^k y. \end{aligned}$$

As in the case of multiple linear regression, if the number of equations is large and no high speed facilities are available, the equations should be solved by one of the compact computing schemes for such problems. From the discussion following (11), it follows that all n points of the scatter diagram for polynomial curve fitting need not have distinct x values. It will suffice to have $k + 1$ distinct x values since a polynomial of degree k is uniquely determined by $k + 1$ points. In evaluating sums such as $\sum x^m$ it is understood that the sum is over all the x values and not over just the distinct x values.

If the investigator is not certain what degree polynomial should be used in a given problem and wishes to compare different degree polynomials for their adequacy, he would prefer a fitting technique that requires little additional labor to increase the degree of the fitted polynomial by one unit. Such a technique is available if one uses *orthogonal polynomials*. These polynomials possess the desirable property of leaving unchanged the coefficients of the previously fitted polynomial when a higher degree term is added. If orthogonal polynomials are not used, the entire set of coefficients would have to be recomputed. Orthogonal polynomials are particularly convenient when there is but one value of y to each value of x and the x values are equally spaced. In the latter case, however, the ordinary normal equations will simplify considerably if x is replaced by $x - \bar{x}$ in (12) because then $\sum (x - \bar{x})^m = 0$ for m odd. The normal equations (13) will then reduce to two sets of equations. Thus, if $k = 5$, the six normal equations will reduce to two sets of three equations each. The odd-numbered equations will involve only the unknowns c_0, c_2, and c_4, whereas the even-numbered equations will involve only the unknowns c_1, c_3, and c_5. The technique of how to use orthogonal polynomials may be found in one of the exercises at the end of this chapter.

7.2 Other Regression Functions

In the preceding section it was pointed out that when there are no theoretical reasons for preferring a certain type of regression function polynomials are selected because of their simplicity and convenience. There are numerous situations, however, in which the nature of the relationship between two variables is known from theoretical considerations. In such situations the fundamental regression problem is to obtain estimates of the parameters that are needed to determine the equation of the curve that represents the relationship. For example, the equation

$$pv^\gamma = \text{constant}$$

represents the relation between the pressure and volume of an ideal gas undergoing adiabatic change. Here γ is a parameter whose value depends on the particular gas and for which an estimate may be obtained from experimental data.

Another example of a nonpolynomial regression function is the function often used in studying simple growth phenomena. If it is assumed that the rate of growth of a biological population is proportional to its size, then the regression function is a simple exponential function. This fact comes from solving the differential equation

$$\frac{dy}{dt} = cy$$

which has as its solution $y = be^{ct}$. This function was also studied in Chapter 3 in connection with radioactive decay.

Suppose, now, that one is given a set of n points (t_1, y_1), (t_2, y_2), $\cdots$, (t_n, y_n) representing the size of a growing population at the times $t_1, t_2, \cdots, t_n$. If the parameters b and c are to be estimated by least squares, it is necessary to minimize the function

$$G(b, c) = \sum_{i=1}^{n} [y_i - be^{ct_i}]^2 .$$

Calculating the partial derivatives with respect to b and c and equating them to 0 will yield the normal equations

$$\sum [y_i - be^{ct_i}][-e^{ct}] = 0$$
$$\sum [y_i - be^{ct_i}][-be^{ct_i}t_i] = 0 .$$

These equations simplify to

(14)
$$b \sum e^{2ct_i} = \sum y_i e^{ct_i}$$
$$b \sum t_i e^{2ct_i} = \sum y_i t_i e^{ct_i} .$$

The solution of these equations is very difficult and requires tedious numerical methods. This example illustrates what frequently occurs, namely, that the method of least squares for nonpolynomial regression often gives rise to normal equations that are difficult to solve.

There are numerous other methods of fitting a curve to a set of points that can be employed when least squares gives rise to computational difficulties. One such method is to introduce new variables that are functions of the old variables in an effort to obtain a more tractable relationship. Thus, in the preceding illustration, it is convenient to work with the variable $z = \log y$ rather than with y itself. If logarithms, to the base e, of both sides of $y = be^{ct}$ are taken, then one obtains

$$\log y = \log b + ct .$$

Next, letting $z = \log y$ and $a = \log b$, this relationship reduces to the linear relationship

$$z = a + ct .$$

The problem has now been reduced to the problem of fitting a straight line to a set of points in the t, z plane and thus to a simple problem in least squares. These least squares estimates for c and a may then be used to yield estimates for c and b. The estimates for c and b obtained in this manner differ, of course, from those obtained by solving the original least squares equations (14), however, the differences are usually quite small.

In studying the problem of determining the accuracy of estimates of regression parameters, it is essential to know how the errors of estimation are distributed. The type of assumption made about their distribution will often determine whether to use direct least squares or to use least squares on a modification of the relationship. The problem of the accuracy of least squares estimates is considered in later chapters; however, it is mentioned here to point out that least squares applied to a modification of a regression relationship may sometimes be preferred to least squares applied to the original relationship and therefore that such a modification does not necessarily yield inferior estimates.

For some types of regression functions it is not possible to introduce changes of variables that will reduce the problem to one for which the least squares equations become tractable. For example, in studying growth phenomena of a somewhat more complex nature than the preceding one, the modified exponential function $y = a + be^{ct}$ is often used as a regression model. Taking logarithms will not help here because of the parameter a. For functions such as this, other fitting procedures are often used. The simplest procedure is to select three points which appear to represent the trend of the data-points and pass the curve through those points. The three equations resulting from having the coordinates of the points satisfy the equation would suffice to determine the three parameters. There are other more refined methods that could also be used here.

8 LINEAR DISCRIMINANT FUNCTION

A problem that arises quite often in certain branches of science is that of discriminating between two groups of individuals or objects on the basis of several properties of those individuals or objects. For example, a botanist might wish to classify a set of plants, some of which belong to one species and the rest to a second species, into their proper species by means of three or four measurements taken on each plant. If the two species were fairly similar with respect to all those measurements, it might not be possible to classify the plants correctly by means of any single measurement because of a fairly large amount of overlap in the distributions of this measurement for the two species; however, it might be possible to find a linear combination of those various measurements whose distributions for the two species would possess very little overlap. This linear combination could then be used to yield a type of index number by means of which plants of two species could be differentiated with a high percentage of success. The procedure for discriminating would consist in finding a critical value of the index such that any plant whose index value fell below the critical value would be classified as belonging to one species, otherwise to the other species.

The principal difference between a *linear discrimination function* and an ordinary linear regression function arises from the nature of the dependent variable. A linear regression function uses values of the dependent variable to determine a linear function that will predict the values of the dependent variable, whereas the discriminant function possesses no such values or variable but uses instead a two-way classification of the data to determine the linear function.

Consider a set of k variables X_1, X_2, $\cdots$, X_k, by means of which it is desired to discriminate between two groups of individuals. Let

$$(15) \qquad Z = \lambda_1 X_1 + \lambda_2 X_2 + \cdots + \lambda_k X_k$$

represent a linear combination of these variables. The problem then is to determine the λ's by means of some criterion that will enable Z to serve as an index for differentiating between members of the two groups. For the purpose of simplifying the geometrical discussion of the problem, consider two

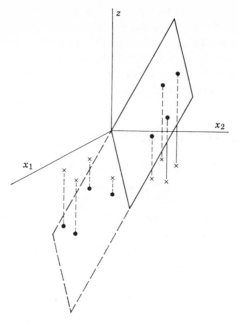

Fig. 5.   Example of a discriminating plane.

variables with n_1 and n_2 individuals, respectively, in the two groups. The equation

$$z = \lambda_1 x_1 + \lambda_2 x_2$$

then represents a plane in three dimensions passing through the origin and having direction numbers λ_1, λ_2, and -1. Figure 5 shows two sets of four points each, designated by x's, that are separated by a plane of this type. The corresponding z values are indicated by heavy dots. If two sets of points corresponding to two groups of individuals are such that they can be separated by means of a plane through the origin, such as shown in Fig. 5, it is clear that the values of z corresponding to the two groups will assume increasingly large negative and positive values as the separating plane approaches perpendicularity to the x_1, x_2 plane. At the same time, however, the variations of the values of z within a group becomes increasingly large for both groups; consequently the increase in the separation of the values of z for the two groups occurs at the expense of an increase in the separation of the values of z within each group. This situation corresponds to that in which the means of two distributions are separating but for which the standard deviations are increasing to such an extent that greater discrimination between the two distributions does not necessarily result. It would be desirable, therefore, to choose a plane that separates the values of z for the two groups as widely as possible relative to the variation of the values of z within the two groups. As a measure of the separation of the two groups it is convenient to use $(\bar{z}_1 - \bar{z}_2)^2$, where $\bar{z}_1$ and $\bar{z}_2$ are the means of the two groups. As a measure of the variation of the values of z within the two groups it is convenient to use

$$\sum_{i=1}^{2} \sum_{j=1}^{n_i} (z_{ij} - \bar{z}_i)^2.$$

Here z_{ij} denotes the z value of the jth individual in the ith group, where $i = 1$ or 2. Then the desired plane will be that plane for which the λ's are determined to maximize the function

(16)
$$G = \frac{(\bar{z}_1 - \bar{z}_2)^2}{\sum_{i=1}^{2} \sum_{j=1}^{n_i} (z_{ij} - \bar{z}_i)^2}.$$

Although the arguments leading to (16) were elucidated by means of two variables and three-dimensional geometry, they hold equally well for k variables; consequently the solution of the problem will be carried out for the general case.

Let x_{pij} represent the value of X_p for the jth individual in the ith group and let $\bar{x}_{pi}$ represent the mean value of X_p for the n_i individuals in that group. Then from (15) it follows that

(17)
$$\bar{z}_1 - \bar{z}_2 = \lambda_1(\bar{x}_{11} - \bar{x}_{12}) + \cdots + \lambda_k(\bar{x}_{k1} - \bar{x}_{k2}),$$

and

(18)
$$z_{ij} - \bar{z}_i = \lambda_1(x_{1ij} - \bar{x}_{1i}) + \cdots + \lambda_k(x_{kij} - \bar{x}_{ki}).$$

If $d_p = \bar{x}_{p1} - \bar{x}_{p2}$, it follows from (17) that

$$(\bar{z}_1 - \bar{z}_2)^2 = (\lambda_1 d_1 + \cdots + \lambda_k d_k)^2$$
$$= \sum_{p=1}^{k} \sum_{q=1}^{k} \lambda_p \lambda_q d_p d_q.$$

If $S_{pq} = \sum_{i=1}^{2} \sum_{j=1}^{n_i} (x_{pij} - \bar{x}_{pi})(x_{qij} - \bar{x}_{qi})$, it follows from (18) that

$$\sum_{i=1}^{2} \sum_{j=1}^{n_i} (z_{ij} - \bar{z}_i)^2 = \sum_{i=1}^{2} \sum_{j=1}^{n_i} [\lambda_1(x_{1ij} - \bar{x}_{1i}) + \cdots + \lambda_k(x_{kij} - \bar{x}_{ki})]^2$$
$$= \sum_{i=1}^{2} \sum_{j=1}^{n_i} \sum_{p=1}^{k} \sum_{q=1}^{k} \lambda_p \lambda_q (x_{pij} - \bar{x}_{pi})(x_{qij} - \bar{x}_{qi})$$
$$= \sum_{p=1}^{k} \sum_{q=1}^{k} \lambda_p \lambda_q \sum_{i=1}^{2} \sum_{j=1}^{n_i} (x_{pij} - \bar{x}_{pi})(x_{qij} - \bar{x}_{qi})$$
$$= \sum_{p=1}^{k} \sum_{q=1}^{k} \lambda_p \lambda_q S_{pq}.$$

When these values are inserted in (16), it reduces to

$$(19) \qquad G = \frac{\displaystyle\sum_{p=1}^{k}\sum_{q=1}^{k}\lambda_p\lambda_q d_p d_q}{\displaystyle\sum_{p=1}^{k}\sum_{q=1}^{k}\lambda_p\lambda_q S_{pq}} = \frac{A}{B}.$$

Since the λ's are to be determined to make G a maximum, it is necessary that $\partial G/\partial\lambda_r = 0$ for $r = 1, \cdots, k$ at the maximizing point. This requirement may be expressed in the form

$$\frac{\partial G}{\partial\lambda_r} = \frac{B\dfrac{\partial A}{\partial\lambda_r} - A\dfrac{\partial B}{\partial\lambda_r}}{B^2} = 0, \qquad r = 1, \cdots, k$$

which is equivalent to

$$(20) \qquad \frac{\partial B}{\partial\lambda_r} = \frac{1}{G}\frac{\partial A}{\partial\lambda_r}, \qquad r = 1, \cdots, k.$$

For ease of differentiating, it is convenient to write out B in the form

$$B = \lambda_1\lambda_1 S_{11} + \cdots + \lambda_1\lambda_r S_{1r} + \cdots + \lambda_1\lambda_k S_{1k}$$
$$\cdots\cdots\cdots\cdots\cdots\cdots\cdots\cdots\cdots\cdots$$
$$\lambda_r\lambda_1 S_{r1} + \cdots + \lambda_r\lambda_r S_{rr} + \cdots + \lambda_r\lambda_k S_{rk}$$
$$\cdots\cdots\cdots\cdots\cdots\cdots\cdots\cdots\cdots\cdots$$
$$\lambda_k\lambda_1 S_{k1} + \cdots + \lambda_k\lambda_r S_{kr} + \cdots + \lambda_k\lambda_k S_{kk}.$$

It will be observed that λ_r occurs as a common factor of both the rth row and the rth column. Since $S_{ij} = S_{ji}$, it therefore follows that

$$\frac{\partial B}{\partial\lambda_r} = 2(\lambda_1 S_{r1} + \cdots + \lambda_k S_{rk}).$$

Similarly,

$$\frac{\partial A}{\partial\lambda_r} = 2(\lambda_1 d_r d_1 + \cdots + \lambda_k d_r d_k)$$
$$= 2(\lambda_1 d_1 + \cdots + \lambda_k d_k)d_r.$$

If these expressions are inserted in (20), it will reduce to

$$(21) \qquad \lambda_1 S_{r1} + \lambda_2 S_{r2} + \cdots + \lambda_k S_{rk} = cd_r, \qquad r = 1, \cdots, k$$

where $c = [\lambda_1 d_1 + \cdots + \lambda_k d_k]/G$ is independent of r.

Since

$$S_{pq} = \sum_{i=1}^{2}\sum_{j=1}^{n_i}(x_{pij} - \bar{x}_{pi})(x_{qij} - \bar{x}_{qi})$$

and

$$d_p = \bar{x}_{p1} - \bar{x}_{p2}$$

are numerical quantities in any given problem, the necessary conditions (21) constitute a set of k linear equations in the λ's. The solution of these equations determines the λ's except for the unknown factor c. Although c is actually a function of the λ's here, so that the solution of these equations expresses each λ_i as a constant times this function, the factor c cancels out from numerator and denominator of G when these values are substituted in (19). Thus there is no unique set of λ's maximizing G, and any multiple of a set of λ's satisfying equations (21) will do just as well. From (15) it is clear that such a multiple can be ignored because the two sets of z's would merely be multiplied by this constant factor and thus would be equivalent as far as discriminating between the two groups is concerned. As a matter of fact, it is usually convenient to choose $c = 1$, solve the equations, and then reduce (15) to the form in which one of the λ's, say λ_1, is unity.

As an illustration of the use of this function, consider the data of Table 4

TABLE 4

Race A	x_1	6.36	5.92	5.92	6.44	6.40	6.56	6.64	6.68	6.72	6.76	6.72
	x_2	5.24	5.12	5.36	5.64	5.16	5.56	5.36	4.96	5.48	5.60	5.08
Race B	x_1	6.00	5.60	5.64	5.76	5.96	5.72	5.64	5.44	5.04	4.56	5.48	5.76
	x_2	4.88	4.64	4.96	4.80	5.08	5.04	4.96	4.88	4.44	4.04	4.20	4.80

on the mean numbers of teeth found on the proximal (x_1) and distal (x_2) combs of two races of insects. The problem here is to discriminate between members of the two races by means of the two indicated variables.

Computations give $S_{11} = 2.68$, $S_{12} = 1.29$, $S_{22} = 1.75$, $d_1 = .915$, $d_2 = .597$; consequently if c is chosen equal to 1, (21) becomes

$$2.68\lambda_1 + 1.29\lambda_2 = .915$$

$$1.29\lambda_1 + 1.75\lambda_2 = .597 .$$

The solution of these equations is $\lambda_1 = .274$ and $\lambda_2 = .139$. If these values are used, the linear discriminant function (15) becomes

$$z = .274x_1 + .139x_2 .$$

For the purpose of computing values of z, it is more convenient to choose c so that either λ_1 or λ_2 equals 1. If c is chosen to make λ_1 equal 1, this discriminant function reduces to

$$z = x_1 + .507x_2 .$$

The values of z corresponding to the various members of the two races given in Table 4 are as follows:

Race A	9.02 8.52 8.64 9.30 9.02 9.38 9.36 9.19 9.50 9.60 9.30
Race B	8.47 7.95 8.15 8.19 8.54 8.28 8.15 7.91 7.29 6.61 7.61 8.19

It will be noted that the two races are separated by means of z except for the slight overlap found in the second entry for Race A and the fifth entry for Race B.

As presented in this section, a linear discriminant function is constructed for the purpose of classifying future observations into their proper group. Thus the problem is essentially one of estimating the λ's. This function could be used as a device for testing the hypothesis that the two groups differ in the manner described earlier; however, there are other more natural methods for treating the latter problem.

EXERCISES

Sec. 1

1. Calculate the value of r for the following artificial data

x	1	2	3	4	5	6
y	2	2	3	5	4	6

2. Calculate the value of r for the following data on the heights (X) and weights (Y) of 12 college students

x	63	72	70	68	66	69	74	70	63	72	65	71
y	124	184	161	164	140	154	210	164	126	172	133	150

3. How large a correlation coefficient is needed for a sample of size 30 before one is justified in claiming that $\rho \neq 0$?

4. Test the hypothesis that $\rho = .8$ if a sample of size 50 gave $r = .7$.

Sec. 3

5. What interpretation would you make if told that the correlation between the number of automobile accidents per year and the age of the driver is $r = -.60$ if only drivers with at least one accident are considered?

6. What would be the effect on the value of r for the correlation between height

and weight of males of all ages if only males in the 20–25 age group were sampled? Observe what effect this restriction would have on the scatter diagram.

7. Explain why it would not be surprising to find a high correlation between the density of traffic on Wall Street and the height of the tide in Maine if observations were taken every hour from 6:00 A.M. to 10:00 P.M. and high tide occurred at 8:00 A.M. Plot a scatter diagram to assist in the explanation.

8. Prove that $s_{x-y}^2 = s_x^2 + s_y^2 - 2rs_xs_y$.

9. Find an expression for s^2 for the first n positive integers by using a familiar formula for the sum of the squares of those integers.

10. If X and Y denote the ranks of an individual with respect to two properties for a group of n individuals, derive the formula $r = 1 - 6\Sigma(X - Y)^2/n(n^2 - 1)$ for the correlation of the two ranked variables by using the results of problems 8 and 9. Calculate r by means of this formula for the data of Table 1 and compare with the regular r value of .66. Replace tied ranks by the mean of the ranks involved.

Sec. 5

11. For the data of problem 1, find the equation of the regression line of Y on X.

12. For the data of problem 2, find the equation of the regression line of Y on X.

13. Derive the least squares equations for fitting a curve of the type $y = \alpha x + \beta/x$ to a set of n points.

14. Suppose it is known that the regression curve of y on x is a straight line that must pass through the origin. Find the equation of the least squares line that will be used to estimate this regression line.

15. Derive the least squares equations for fitting a curve of the type $y = axe^{-b(x-c)^2}$ to a set of n points.

16. Show that the equation of the simple regression line can be written in the form $y' - \bar{y} = r(s_y/s_x)(x - \bar{x})$.

17. Use the formula of problem 16 to show that the correlation between the regression line values y'_i and the observed values y_i is equal to r.

18. Derive the equations that would need to be solved if one were to estimate a and b in $y = a + bx$ by requiring that the sum of the squares of the perpendicular distances to the line be a minimum.

Sec. 6

19. The following data are for the three variables, honor points (Y), general intelligence scores (X_1), and hours of study (X_2). Find the equation of the regression plane of y on x_1 and x_2, where these letters denote variables measured from their sample means, given

$$\Sigma x_1^2 = 250, \quad \Sigma x_1 x_2 = 33, \quad \Sigma x_2^2 = 36, \quad \Sigma x_1 y = 106, \quad \Sigma x_2 y = 22 .$$

Sec. 7

20. Derive the least squares equations for fitting a modified exponential $y = c + ae^{bx}$ to a set of n points and indicate why these equations would be difficult to solve.

21. Suppose a regression curve of the form $y = \alpha\beta^x$ is to be fitted to a set of

points. By taking logarithms this becomes linear. Use the least squares technique on the linear version to estimate α and β for the following data

x	1	2	3	4	5	6	7	8
y	1.0	1.2	1.7	2.6	3.5	4.8	6.8	10.0

22. Two polynomials $P_i(x)$ and $P_j(x)$ of degrees i and j are said to be orthogonal on a set of points $x_1, \cdots, x_n$ provided that $\sum_{k=1}^{n} P_i(x_k)P_j(x_k) = 0$, $i \neq j$. $P_i(x)$ is said to be normalized on this set if $\sum_{k=1}^{n} P_i^2(x_k) = 1$. For the points $x = 0, 1, 2, 3, 4$, find an orthonormal set of polynomials $P_0(x)$, $P_1(x)$, $P_2(x)$.

23. Assuming the properties defined in problem 22, obtain the least squares equations for the coefficients of the polynomial

$$y = a_0 P_0(x) + a_1 P_1(x) + \cdots + a_k P_k(x)$$

and show that their solution for a particular coefficient a_i is the same regardless of the degree of the polynomial, provided that $i < k$.

24. The following data are for intelligence-test scores, grade-point averages, and reading-rate scores of students. Find the equation of the regression plane of G.P.A. on I.T. scores and R.R. scores.

G.P.A.	2.4	.6	.2	0	1	.6	1	.4	0	2.6
I.T.	295	152	214	171	131	178	225	141	116	173
R.R.	41	18	45	29	28	38	25	26	22	37

G.P.A.	2.6	0	1.8	0	.4	1.8	.8	1	.2	2.8
I.T.	230	195	174	177	210	236	198	217	143	186
R.R.	39	38	24	32	26	29	34	38	40	27

G.P.A.	1.4	.2	.4	1.4	0	.8	.8	.8	.4	2.2
I.T.	233	136	183	223	106	184	134	211	151	231
R.R.	44	32	26	50	24	32	48	18	20	26

G.P.A.	1.4	1.2	1.4	1.4	1.4	.8	.8	2.6	2.6	.2
I.T.	135	146	227	204	223	142	176	238	268	163
R.R.	26	19	35	26	18	22	23	27	40	33

25. For linear regression involving more than two variables the multiple correlation coefficient is defined to be the correlation between the y_i and the $y_i{}'$. It is designed to measure the extent to which the linear regression function can predict the variable y. Calculate the value of the multiple correlation for the data of problem 24 if the answer to problem 24 is $y' = .012x_1 - .007x_2 - .97$.

Sec. 8

26. Classify the individuals of problem 24 into one of two groups on the basis of having a G.P.A. less than or greater than .9. (a) Using the remaining variables, find the equation of the discriminant function for classifying individuals into the proper G.P.A. group. (b) Calculate the values of Z for the individuals and note whether this function does much better than either variable alone.

CHAPTER 8

General Principles for Statistical Inference

1 INTRODUCTION

Chapter 4 introduced some of the fundamental concepts of statistical inference but in a rather informal and incomplete manner. This chapter is concerned with pursuing those ideas in a more systematic and precise fashion. As was stated in section 6, Chapter 4, although statistical inference in its broadest sense is a type of decision making based on probability, a large share of the inferences made by statisticians are concerned with either the estimation or hypothesis testing of a parameter of a probability density function. Methods for solving these two classes of problems in an efficient manner will therefore constitute the principal motivation for this chapter.

2 ESTIMATION

An introduction to the problem of estimating parameters of density functions was given in Chapter 4. In that treatment the method of moments was introduced as a simple method for obtaining point estimates of parameters. In this chapter properties to be desired in point estimates will be studied and estimation by means of intervals will be introduced.

2.1 Unbiased Estimates

Perhaps the first property of an estimate that one would think of as being desirable is the property of the estimate converging, in some sense, to the value of the parameter as the sample size becomes increasingly large. Since almost any reasonable estimate will possess such a property, a closely related property that is somewhat more restrictive is often imposed instead. This is the property of being unbiased. For the purpose of defining this term, consider a random variable X whose density function depends on a parameter θ. Let $X_1, \cdots, X_n$ represent a random sample of size n of X and let

$t(X_1, \cdots, X_n)$ be any function of the sample that is being contemplated as an estimate of θ. A function such as this is called a *statistic*. It is a random variable whose value is completely determined by the sample values. Now for a given set of observational values $x_1, \cdots, x_n$ the estimate $t(x_1, \cdots, x_n)$ is merely a number obtained from calculations made on the observational values; however, from the point of view of procedure, an estimate is a function of the observational values. For example, the function $t(x_1, \cdots, x_n) = (x_1 + \cdots + x_n)/n$ is a typical estimate. It is customary for some writers to use the word *estimator* for the function of the corresponding random variables and the word estimate for the value of the function after the observational values have been substituted. Thus, $(X_1 + \cdots + X_n)/n$ would be called an estimator of θ, whereas its numerical valu in any given problem would be called an estimate of θ. Other writers, however, use the word estimate both for the function and its numerical value. In view of these remarks, the property of being unbiased may be defined as follows.

(1) DEFINITION: *The statistic* $t = t(X_1, \cdots, X_n)$ *is called an unbiased estimate* (*or estimator*) *of the parameter* θ *if* $E[t] = \theta$ *for all* θ.

The property of being unbiased merely states that the random variable t possesses a distribution whose mean is the parameter θ being estimated. This property was shown in section 5 of Chapter 5 to hold, for example, for $t = \bar{X}$ when estimating the mean μ of a distribution.

As an illustration of how the bias in a statistic may sometimes be determined by means of expected value formulas, consider the expected value of a sample variance $S^2 = 1/n \sum_{i=1}^{n} (X_i - \bar{X})^2$ based on a random sample of size n. From properties of E, and the definition of σ^2, it follows that

(2)
$$E[S^2] = E\left[\frac{1}{n}\sum_{i=1}^{n}(X_i - \bar{X})^2\right]$$

$$= E\left\{\frac{1}{n}\sum_{i=1}^{n}[(X_i - \mu) - (\bar{X} - \mu)]^2\right\}$$

$$= E\left[\frac{1}{n}\sum_{i=1}^{n}(X_i - \mu)^2 - (\bar{X} - \mu)^2\right]$$

$$= \frac{1}{n}\sum_{i=1}^{n}E(X_i - \mu)^2 - E(\bar{X} - \mu)^2$$

$$= \frac{1}{n}\sum_{i=1}^{n}\sigma^2 - \sigma_{\bar{X}}^2$$

$$= \sigma^2 - \frac{\sigma^2}{n}$$

$$= \frac{n-1}{n}\sigma^2 .$$

This shows that S^2 is not an unbiased estimate of σ^2, which means that if repeated samples of size n are taken and the resulting sample variances are averaged the average will not approach the true variance in value but will be consistently too small by the factor of $(n-1)/n$. For small samples this factor becomes important; consequently, one must be careful how he combines samples in making an estimate of the true variance when an unbiased estimate is desired. In order to overcome the bias in S^2, it is merely necessary to multiply S^2 by $n/(n-1)$ and use the resulting quantity as the estimate of σ^2. Then, because of (2),

$$E\left(\frac{n}{n-1}\,S^2\right) = \frac{n}{n-1}\,E[S^2] = \sigma^2\,.$$

Since

$$\frac{n}{n-1}\,S^2 = \frac{\sum\limits_{i=1}^{n}(X_i - \bar{X})^2}{n-1}$$

it is clear that one can avoid the bias in estimating variances by dividing the sum of squares of deviations by $n-1$ rather than by n, as was the practice in the preceding chapters. It is because of this property that some authors define the sample variance as $\sum (X_i - \bar{X})^2/(n-1)$.

As a second illustration, consider the problem of how to combine several sample variances to obtain a single unbiased estimate of the population variance. Such a problem would arise, for example, in quality-control work if one wished to obtain an unbiased estimate of the variability of a manufacturing process as measured by σ^2 and had available a number of daily estimates of the variability. Let $S_1^2, \cdots, S_k^2$ denote k sample variances based on random samples of sizes $n_1, \cdots, n_k$, respectively. Then, if each sample variance is weighted with the size of the sample on which it is based, the proper weighted average to use for estimating σ^2 is given by

$$t = \frac{n_1 S_1^2 + \cdots + n_k S_k^2}{a}$$

where a is chosen to make this estimate unbiased. From properties of E and the result in (2), it follows that

$$E[t] = \frac{1}{a}\,[(n_1 - 1)\sigma^2 + \cdots + (n_k - 1)\sigma^2]$$

$$= \frac{\sigma^2}{a}\,(n_1 + \cdots + n_k - k)\,.$$

In order that t be unbiased, it is therefore necessary to choose $a = n_1 + \cdots + n_k - k$. Thus the desired estimate of σ^2 is given by

(3)
$$\frac{n_1 S_1^2 + \cdots + n_k S_k^2}{n_1 + \cdots + n_k - k}.$$

As an illustration of how formula (3) is applied, suppose an unbiased estimate of the variability of a manufactured product is desired and on four different occasions samples of sizes 4, 6, 10, and 8 were obtained with resulting sample values of $s_1^2 = 30$, $s_2^2 = 40$, $s_3^2 = 36$, and $s_4^2 = 42$. A direct application of formula (3) then gives the unbiased estimate

$$\frac{4(30) + 6(40) + 10(36) + 8(42)}{4 + 6 + 10 + 8 - 4} = 44.$$

If the simple mean of the individual sample variances had been used, the estimate would have been 37.

As an exercise to illustrate the convenience of using the operator E for calculating mean values and at the same time to derive a useful formula, consider the problem of expressing the variance of a linear combination of a set of variables in terms of the variances and correlations of the variables. This was done for independent variables in Chapter 5; however, here the variables are not assumed to be independent. Let

$$Z = a_1 X_1 + \cdots + a_k X_k$$

be the linear function whose variance is desired. Then

$$E[Z] = a_1\mu_1 + \cdots + a_k\mu_k$$

and

$$Z - E[Z] = a_1(X_1 - \mu_1) + \cdots + a_k(X_k - \mu_k).$$

Hence, from the definition of the variance of a variable and this result,

$$\begin{aligned}
\sigma_Z^2 &= E[Z - E(Z)]^2 \\
&= E[a_1(X_1 - \mu_1) + \cdots + a_k(X_k - \mu_k)]^2 \\
&= E\left[\sum_{i=1}^{k} a_i^2(X_i - \mu_i)^2 + 2\sum_{i<j} a_i a_j(X_i - \mu_i)(X_j - \mu_j)\right] \\
&= \sum_{i=1}^{k} a_i^2 E(X_i - \mu_i)^2 + 2\sum_{i<j} a_i a_j E(X_i - \mu_i)(X_j - \mu_j).
\end{aligned}$$

Denoting the variance of the variable X_i by σ_i^2 and the correlation coefficient between X_i and X_j by ρ_{ij}, it follows from (12) and (14), Chapter 6, that

$$E(X_i - \mu_i)(X_j - \mu_j) = \rho_{ij}\sigma_i\sigma_j$$

and hence that

(4)
$$\sigma_Z{}^2 = \sum_{i=1}^{k} a_i{}^2 \sigma_i{}^2 + 2 \sum_{i<j} a_i a_j \rho_{ij} \sigma_i \sigma_j .$$

When the variables $X_1, \cdots, X_k$ are uncorrelated, this formula reduces to the formula

(5)
$$\sigma_Z{}^2 = \sum_{i=1}^{k} a_i{}^2 \sigma_i{}^2$$

which was discussed in section 7, Chapter 5. Formulas (4) and (5) are very useful for determining the accuracy of estimate of means of populations when these estimates are constructed as linear combinations of other estimates.

2.2   Best Unbiased Estimates

Although the property of being unbiased is a desirable one to seek in an estimate, it is not nearly so important as the property of an estimate being close in some sense to the parameter being estimated. Thus, if an estimate t is consistently closer to θ than another estimate t' in repeated samples of the same size, then t would certainly be preferred to t', even if t were biased and t' were unbiased. Because of the difficulty or impossibility of determining whether one of two estimates is closer than the other to θ for any reasonable definition of closeness, it is customary to substitute a measure of the variability of t about θ in place of closeness. Since the variance, or the standard deviation, has been used to measure variability throughout the preceding chapters, one would naturally think of selecting one or the other of these measures; however, unless θ happens to be the mean of the distribution of t, the variance will not measure the variability about θ. This difficulty can be overcome by using the second moment about θ as the desired measure. When θ is the mean of t, that is, when t is an unbiased estimate of θ, this measure reduces to the variance of t.

If now t_1 and t_2 are two estimates of θ that are to be compared, this can be done by comparing their second moments about θ. In this connection, a statistic t_1 will be said to be better than the statistic t_2 for estimating θ provided that $E(t_1 - \theta)^2 \leq E(t_2 - \theta)^2$ for all possible values of θ and provided that strict inequality holds for at least one value of θ.

The problem of deciding whether an estimate is a good one in comparison with all other possible estimates is not quite so simple. The difficulty is that a trivial estimate such as $t = c$, where c is some constant, will be better as an estimate of the mean θ of a normal population than $\bar{X}$ when θ happens to be equal to c. Thus one cannot expect to find a reasonable estimate such as $\bar{X}$ to possess a second moment about θ that is a minimum for all possible values of θ. In order to avoid such paradoxical results, it is customary to limit the discussion of the goodness of an estimate to unbiased estimates. Since the property of being unbiased is required to hold for all possible values of θ, trivial estimates such as $t = c$ are automatically eliminated from consideration. In view of the preceding discussion, the following definition is introduced as a basis for choosing a good estimate.

(6) DEFINITION: *A statistic $t = t(X_1, X_2, \cdots, X_n)$ will be called a best unbiased estimate (estimator) of the parameter θ if it is unbiased and if it possesses minimum variance among all unbiased estimates (estimators).*

This property must hold for all possible values of θ, that is, regardless of what the true value of the parameter may be. Although there are other definitions of a best estimate in use, the preceding definition is one that is frequently used. It should be realized that the variance was selected in (6) because it was considered to measure the concentration of the distribution of t about θ. Since it is easy to construct an example of a distribution in which most of the distribution is heavily concentrated about θ, yet for which the second moment is extremely large, one must appreciate that the second moment is not foolproof for giving the comparison of estimates that one originally had in mind. Nevertheless, similar criticisms can be leveled at any other substitute; furthermore, experience and theory have shown that (6) is a very useful definition.

As an application of the preceding ideas, consider the problem of determining whether some weighted average of a random sample from a population can yield a better unbiased estimate of the population mean than the sample mean. Let the two competing estimates be written

$$t_1 = a_1 X_1 + \cdots + a_n X_n$$

and

$$t_2 = \overline{X}.$$

The unknown a's in t_1 are selected to make t_1 unbiased and to minimize $E(t_1 - \theta)^2$. In order to determine the bias in t_1, calculate

$$E(t_1) = a_1 E(X_1) + \cdots + a_n E(X_n)$$
$$= a_1 \mu + \cdots + a_n \mu$$
$$= (a_1 + \cdots + a_n)\mu.$$

The statistic t_1 will be unbiased if the a's are restricted to satisfy

$$a_1 + \cdots + a_n = 1.$$

This merely states that the sum of the coefficients in t_1 must be 1. With this restriction on the a's, t_1 will be unbiased; therefore the problem now is to choose a set of weights, summing to 1, that will minimize the variance of t_1. But it was demonstrated in section 7, Chapter 5, that this variance will be minimized if the weights are chosen inversely proportional to the variances of the corresponding variables. Since the X_i all have the same variance, this implies that all the weights should be equal, and therefore that the best unbiased linear combination is obtained by choosing $a_i = 1/n, i = 1, \cdots, n$. Thus, no linear combination of the sample can yield a better unbiased estimate than the sample mean $\bar{X}$. If the variable X is normally distributed, it can be shown that $\bar{X}$ is not only the best linear combination of the sample values to use but is the best unbiased function of any kind to use, that is, $\bar{X}$ minimizes $E(t' - \mu)^2$, where t' is any unbiased estimate of μ. A proof of this fact is given in the appendix as an application of a formula that is derived there to enable one to determine whether a particular estimate satisfies definition (6) for being a best unbiased estimator.

2.3 Maximum Likelihood Estimates

In Chapter 4 the method of moments was introduced as a method for estimating parameters. That method was used because of its intuitive appeal and simplicity. There is no assurance, however, that estimates obtained by the method of moments possess desirable properties. For example, if t is an estimate of θ obtained by the method of moments, there is no assurance that the value of $E(t - \theta)^2$ will be as small as the corresponding value for some other estimate.

A method of estimation that is known to possess several desirable properties but which is somewhat more complicated than that of moments is the *method of maximum likelihood*. It is defined as follows. Let $f(x; \theta)$ be the density function of the random variable X, where θ is the parameter to be estimated. Suppose that n observations are to be made of the variable X. Let $X_1, \cdots, X_n$ denote the random variables corresponding to these n observations. Then the function given by

$$(7) \qquad L(x_1, \cdots, x_n; \theta) = \prod_{i=1}^{n} f(x_i; \theta)$$

defines a function of the random sample values $x_1, \cdots, x_n$ and the parameter θ which is known as the *likelihood function*.

For the purpose of interpreting this function, suppose that the observational values are obtained from n independent trials of an experiment for which $f(x; \theta)$ is the density function of a discrete random variable X. Then, for any particular set of observational values, the likelihood function gives the probability of obtaining that set of values, including their order of occurrence. If, however, X is a continuous variable, the likelihood function gives the probability density at the sample point $(x_1, x_2, \cdots, x_n)$, where the sample space is thought of as being n dimensional.

Using the notation and terminology of the preceding paragraphs, the method of maximum likelihood estimation may be defined in the following manner:

(8) DEFINITION: *A maximum likelihood estimate $\hat{\theta}$ of the parameter θ in the density function $f(x; \theta)$ is an estimate that maximizes the likelihood function $L(x_1, \cdots, x_n; \theta)$ as a function of θ.*

If the x_i are treated as fixed, the likelihood function becomes a function of θ only, say $L(\theta)$; consequently, the problem of finding a maximum likelihood estimate is the problem of finding the value of θ that maximizes $L(\theta)$. This maximizing value of θ is, of course, a function of the x_i that have been treated as fixed. In discussing the maximum likelihood estimator it is necessary, however, to write $\hat{\theta} = \hat{\theta}(X_1, X_2, \cdots, X_n)$ to show that the estimator is a function of the random variables $X_1, X_2, \cdots, X_n$ rather than of their numerical values.

Maximum likelihood estimates can usually be obtained by calculus methods because the relative maximum of the likelihood function obtained by differentiating $L(x_1, \cdots, x_n; \theta)$ with respect to θ and setting the derivative equal to zero is usually an absolute maximum.

As an illustration of the calculus technique for finding maximum likelihood estimates, consider the problem of estimating the parameter θ in the exponential density function $f(x; \theta) = \theta e^{-\theta x}$, $x > 0$, $\theta > 0$, if five observations on X yielded the values $x_1 = .9$, $x_2 = 1.7$, $x_3 = .4$, $x_4 = .3$, and $x_5 = 2.4$.

From (7), the likelihood function is

$$L = \theta e^{-\theta x_1} \cdot \theta e^{-\theta x_2} \cdots \theta e^{-\theta x_n}$$
$$= \theta^n e^{-\theta \sum\limits_{i=1}^{n} x_i}$$

Since the value of θ that maximizes L will be the same as the value that maximizes $\log L$, and the latter is easier to treat, $\log L$ is first calculated giving

$$\log L = n \log \theta - \theta \sum x_i .$$

Then differentiating with respect to θ, and setting the derivative equal to 0, one obtains the equation

$$\frac{n}{\theta} - \sum x_i = 0 .$$

The solution of this equation gives the desired maximum likelihood estimate of θ, namely,

(9) $$\hat{\theta} = \frac{n}{\sum x_i} .$$

By calculating the second derivative it is easily shown that θ does maximize L because $L \to 0$ if θ approaches a boundary. It will be observed that this estimate is merely the reciprocal of the arithmetic mean of the x_i.

In order to find the maximum likelihood estimate for the given observations, it is merely necessary to choose $n = 5$ and insert the five given observational values in (9). Computations yield the estimate $\theta = .88$.

As a second illustration, let p be the probability that an event A will occur when an experiment is performed and let the experiment be repeated until A does occur. Further, let X denote the number of experiments that are required before A occurs. Here the density function of X is the geometric density

$$(10) \qquad f(x; p) = (1 - p)^{x-1}p$$

because $x - 1$ successive failures followed by a success for the event A must occur if the event A is to occur the first time on experiment number x. The problem is to find the maximum likelihood estimate of p. Now the function given by (10) is also the likelihood function because only one experiment is being performed; therefore its maximum with respect to the parameter p must be found. It is convenient here to take logarithms and then maximize the $\log f(x; p)$ by calculus methods. The value of p that maximizes $\log f(x; p)$ will be the same as the value that maximizes $f(x; p)$. Thus

$$\log f(x; p) = (x - 1) \log (1 - p) + \log p .$$

Hence,

$$\frac{\partial \log f(x; p)}{\partial p} = -\frac{x - 1}{1 - p} + \frac{1}{p} .$$

If this derivative is set equal to 0, it will be found that the value of p which satisfies the resulting equation is given by

$$\hat{p} = \frac{1}{x} .$$

Here also the second derivative test will show that $\hat{p}$ maximizes L. Thus, if A were the event of getting a 1 to turn up in rolling a die, the estimate of p, whose value is $\frac{1}{6}$ here, would be the reciprocal of the number of rolls needed before a 1 appeared. An average of 6 rolls agrees with one's intuition.

As a slight generalization of this problem, suppose a set of n such experiments is carried out. Let $X_1, X_2, \cdots, X_n$ denote the number of trials of the experiment required before A occurs in each group of experiments. Each of the X_i possesses the density function given in (10); therefore the likelihood function now is

$$L = \prod_{i=1}^{n} f(x_i; p)$$
$$= (1 - p)^{x_1-1}p \cdot (1 - p)^{x_2-1}p \cdots (1 - p)^{x_n-1}p$$
$$= (1 - p)^{\Sigma x_i - n}p^n .$$

As before, the maximum is easier to find if one first takes logarithms. Thus

$$\log L = (\sum x_i - n) \log (1 - p) + n \log p .$$

Hence,

$$\frac{\partial \log L}{\partial p} = -\frac{\sum x_i - n}{1 - p} + \frac{n}{p} .$$

The solution of the equation obtained by setting this derivative equal to zero is given by

$$\hat{p} = \frac{n}{\sum x_i} .$$

The similarity of this result with that given in (9) should not tempt one to generalize about the nature of maximum likelihood estimates.

As an illustration of a problem for which calculus methods are inappropriate, consider the problem of finding the maximum likelihood estimate of the parameter θ for the uniform density $f(x; \theta) = 1/\theta$, $0 \le x \le \theta$, based on the random sample values $x_1, \cdots, x_n$.

Here

$$L(\theta) = \prod_{i=1}^{n} f(x_i; \theta) = \left(\frac{1}{\theta}\right)^n .$$

This function will be maximized by choosing θ as small as possible, subject to the restriction $0 \le x_i \le \theta$, $i = 1, \cdots, n$. The smallest value of θ that satisfies these inequalities is clearly the largest value of the x_i. Thus, the maximum likelihood estimate of θ is given by

$$\theta = \max \{x_1, x_2, \cdots, x_n\} .$$

Although the discussion of estimation has been limited to that of estimating a parameter of a density function, there are methods available for estimating various properties of a density function, such as its maximum value. In addition, there are methods for estimating the density function itself.

The method of maximum likelihood is easily the most popular of all methods for finding point estimates of parameters. This popularity is based partly on the ease with which such estimates are often obtained but more importantly on the desirable properties possessed by such estimates.

Among the desirable features of maximum likelihood estimates is their property of often yielding best estimates in the sense of being unbiased and possessing minimum variance. Examples can be found for which the maximum likelihood estimate is a poor one; however, for most familiar applications it is either a best estimate or very nearly so.

A second desirable feature of maximum likelihood estimates is their excellent large sample properties. If θ denotes a maximum likelihood estimator, and if some mild restrictions are placed on the density function $f(x; \theta)$, it can be shown that the random variable

$$(11) \qquad \frac{\hat{\theta} - \theta}{\dfrac{a}{\sqrt{n}}}$$

possesses a distribution approaching that of a standard normal variable as $n \to \infty$. The constant a in the denominator depends on $f(x; \theta)$. The situation here is very similar to that in section 6, Chapter 5, where it was shown that $(\bar{X} - \mu)\sqrt{n}/\sigma$ possesses a distribution approaching that of a standard normal variable. It is customary to call such limiting distributions *asymptotic distributions*. Thus the maximum likelihood estimator θ is said to be asymptotically normally distributed. The quantity $a/\sqrt{n}$ in the denominator of (11) is called the asymptotic standard deviation of θ. Now it can be shown that among all estimators that are asymptotically normally distributed, the maximum likelihood estimator possesses minimum asymptotic variance. Thus, in the sense of possessing minimum asymptotic variance, one can say that among all asymptotically normally distributed estimators the maximum likelihood estimator is a best estimator.

It will be found that maximum likelihood estimators are often biased; hence, if an unbiased estimate is desired, it may be necessary to multiply the maximum likelihood estimator by a constant that depends on n, such as was done with S^2, in order to obtain an unbiased estimate. In some problems it is not possible to adjust a maximum likelihood estimator in this manner.

The preceding properties are the principal ones that justify the popularity of maximum likelihood estimation.

2.4 Confidence Intervals

Thus far, only point estimates of parameters have been considered. In many problems of estimation, however, one prefers an interval estimate that will express the accuracy of the estimate as well. If the sample is sufficiently large and the estimate is a maximum likelihood estimate, one can use normal curve methods as indicated in the preceding section to find such an interval;

however, in order to be able to treat more general problems, a more general method is needed for constructing interval estimates. Such a method, known as the method of *confidence intervals*, is now described by means of a particular example.

Suppose that a random sample of size 100 has been taken from a population that is known to be normal and whose variance is known to be equal to 16. Suppose, further, that the mean of this sample is 30. Then the problem is to estimate the population mean μ by the use of an interval of values of X. Since $\sigma^2 = 16$, $\sigma_{\bar{X}} = \sigma/\sqrt{n} = .4$. From the theory of Chapter 5, it follows that $\bar{X}$ will be normally distributed with mean μ and standard deviation .4. Hence, one can write

(12) $$P\{\mu - .8 < \bar{X} < \mu + .8\} = .95 .$$

By performing some algebra on the inequality inside the braces, it can be rearranged to produce the equivalent inequality

$$\bar{X} - .8 < \mu < \bar{X} + .8 .$$

Hence, one can just as well write

(13) $$P\{\bar{X} - .8 < \mu < \bar{X} + .8\} = .95 .$$

From a frequency interpretation point of view, (12) states that in the long run of experiments of this type 95 percent of them will yield a value of $\bar{X}$ that lies inside the interval determined by $\mu \pm .8$. The frequency interpretation of (13) is that in the long run of experiments of this type 95 percent of them will yield an interval that contains the unknown mean μ. These interpretations are represented geometrically in Fig. 1.

Each point represents an $\bar{X}$ based on a sample of 100. The upper diagram corresponds to the case in which μ is assumed known and a probability statement is made concerning $\bar{X}$'s. The lower diagram corresponds to the case in which μ is assumed unknown and the variable intervals $\bar{X} \pm .8$ are plotted. If a point lies inside the 95 percent band of the upper diagram, its interval in the lower diagram must necessarily cover μ, and not otherwise.

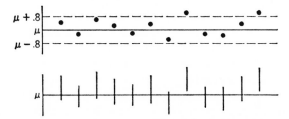

Fig. 1.   Illustration of confidence interval methods.

In practice, only one such $\bar{X}$ is available, so that only the first point and its corresponding interval of $30 \pm .8$ is available. On the basis of this one experiment, the claim will be made that the interval $30 \pm .8$ contains the population mean μ. If for each such experiment the same claim were made for the interval corresponding to that experiment, then for these experiments 95 percent of such claims would be true in the long run. It is in this sense that correct probability statements can be made concerning population parameters. The interval $30 \pm .8$ is called a 95 percent *confidence interval* for μ. The end points of a confidence interval for a parameter are called *confidence limits* for the parameter.

It should be clearly understood that one is merely betting on the correctness of the rule of procedure when applying the confidence interval technique to a given experiment. It is obviously incorrect to make the claim that the probability is .95 that the interval $30 \pm .8$ contains μ. The latter probability is either 1 or 0, depending on whether μ does or does not lie in this fixed interval. It is only when the random interval $\bar{X} \pm .8$ is considered that one can make correct probability statements of the type desired.

The preceding analytical method for finding confidence intervals is used extensively in the following chapters for finding confidence intervals for the more common statistical parameters. An examination of this illustration and those in the following chapters will reveal that the method for finding confidence intervals consists in first finding a random variable, call it Z, that involves the desired parameter θ but whose distribution does not depend on any unknown parameters. Thus $Z = \bar{X} - \mu$ is such a variable. Next, two numbers, z_1 and z_2, are chosen such that

$$P\{z_1 < Z < z_2\} = 1 - \alpha$$

where $1 - \alpha$ is the desired *confidence coefficient*, such as .95. Then this inequality is solved so that this probability statement assumes the form

$$P\{\underline{\theta} < \theta < \bar{\theta}\} = 1 - \alpha$$

where $\underline{\theta}$ and $\bar{\theta}$ are random variables depending on Z but not involving θ. Finally, one substitutes the sample values in $\underline{\theta}$ and $\bar{\theta}$ to obtain a numerical interval which is then the desired confidence interval. The preceding technique does not always lead to a confidence interval because the rearrangement of the probability inequality may not yield an interval. It is also clear that any number of confidence intervals can be constructed for a parameter by choosing z_1 and z_2 differently each time or by choosing different random variables of the Z type.

3   TESTING HYPOTHESES

The fundamental concepts for testing hypotheses were introduced in Chapter 4. There it was stated that most of the statistical hypotheses met in practice are assertions made concerning a parameter of a density function and that the test of such a hypothesis is usually based on a sample estimate of that parameter. Thus, the statistic $\bar{X}$ was used to test a hypothesis about the parameter μ of a normal distribution, and the statistic X/n, where X is the total number of successes in n trials of an experiment, was used to test a hypothesis concerning the parameter p of a binomial distribution. Now, not only was the statistic for a given type of problem selected on intuitive grounds, but the critical region for the test, such as the two tails of the distribution of the statistic, was also selected on an intuitive basis rather than on the principle of good test construction introduced in section 6.3, Chapter 4. In this chapter the ideas introduced earlier on testing hypotheses will be studied more carefully and extended somewhat to include a broader class of problems.

In all the problems of testing hypotheses that have been considered thus far the procedure for deciding whether to accept or reject the hypothesis has consisted in selecting a statistic based on a sample of fixed size n, calculating the value of the statistic for the sample, and then rejecting the hypothesis if and only if the value of the statistic corresponded to a point in the chosen critical region.

A more general procedure that possesses striking advantages in many situations is one in which the random sample is obtained by selecting one individual at a time until a sufficiently large sample has been accumulated to arrive at a reliable decision. This method of sampling, called sequential sampling, often arrives at a decision some time before the fixed-size sample, with the same size type I and type II errors, is exhausted, and thus it often decreases the cost of sampling. In the sequential procedure one must decide at every stage of the sampling whether to accept the hypothesis, to reject the hypothesis, or to continue sampling. The fixed-size sample procedure does not permit any conclusions to be drawn until the entire sample has been taken and does not permit additional sampling. A sequential method for testing hypotheses is discussed in Chapter 13; hence only fixed-size sample procedures are considered in this chapter.

3.1   Kinds of Tests

The statistical hypotheses that will be considered in this chapter are assertions about the parameters of a density function. For the purpose of describing them, let $f(x; \theta_1, \theta_2, \cdots, \theta_k)$ denote a known density function that depends on k parameters. A statistical hypothesis then becomes an assertion about the k parameters. In studying hypotheses of this kind it is convenient to classify them into one of two types by means of the following definition.

(14) DEFINITION: *If a hypothesis specifies the values of all the parameters of a density function, it is called a simple hypothesis; otherwise, it is called a composite hypothesis.*

As an illustration, suppose the density function is

$$f(x; \theta_1, \theta_2) = \frac{e^{-\frac{1}{2}\left(\frac{x-\theta_1}{\theta_2}\right)^2}}{\sqrt{2\pi}\,\theta_2}.$$

If the hypothesis is $H_0: \theta_1 = 10$, $\theta_2 = 2$, then H_0 is a simple hypothesis. If, however, the hypothesis is $H_0: \theta_1 = 10$, $\theta_2 < 2$, then H_0 is composite because the value of θ_2 has not been specified.

The theory of how to design good tests for simple hypotheses is much simpler than that for composite hypotheses. In the next two sections two methods for constructing good tests are discussed. The first method is directly applicable to simple hypotheses only, although it sometimes solves composite problems also, whereas the second method is applicable to both simple and composite hypotheses.

3.2 Best Tests for Simple Hypotheses

In this section a method is given for constructing best tests, in the sense of the principle introduced in Chapter 4, for simple hypotheses. In discussing the relative merits of different tests, this principle requires that only tests with an agreed upon type I error size, denoted by α, be considered. Then a best test is defined as a test in this set that minimizes the size of the type II error, denoted by β. The method of constructing a best test depends on the use of a theorem that was first proved and used by the two statisticians after whom it is named. The theorem, called the Neyman-Pearson lemma, will be proved for a density function, $f(x; \theta)$, of a single continuous variable and a single parameter; however, by merely thinking of x and θ as vectors, the proof will be seen to hold for any number of random variables and parameters. The variables $x_1, x_2, \cdots, x_n$ occurring in the theorem are understood to represent the values of a random sample of size n from the population whose density function is $f(x; \theta)$. The theorem is concerned with a simple hypothesis $H_0: \theta = \theta_0$ and a simple alternative $H_1: \theta = \theta_1$. This is the type of problem discussed and illustrated in Chapter 4 beginning with the illustration in section 6.2, Chapter 4. One should review that material before studying the following. In particular one should recall that the phrase "critical region of size α" means that the critical region is one for which the probability of making a type I error is α. In terms of this language, the theorem may be expressed as follows:

(15) NEYMAN-PEARSON LEMMA: *If there exists a critical region A of size α and a constant k such that*

$$\frac{\displaystyle\prod_{i=1}^{n} f(x_i; \theta_1)}{\displaystyle\prod_{i=1}^{n} (x_i; \theta_0)} \geq k \text{ inside } A$$

and

$$\frac{\displaystyle\prod_{i=1}^{n} f(x_i; \theta_1)}{\displaystyle\prod_{i=1}^{n} f(x_i; \theta_0)} \leq k \text{ outside } A$$

then A is a best critical region of size α.

To prove this lemma, let A^* be any other critical region of size α. The regions A and A^* may be represented geometrically as the regions interior to the indicated closed surfaces in Fig. 2. For simplicity of notation, let

$$L_0 = \prod_{i=1}^{n} f(x_i; \theta_0)$$

denote the density function for the variables $X_1, X_2, \cdots, X_n$ when H_0 is true, and let L_1 denote this function when H_1 is true. Further, to simplify the

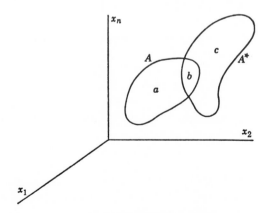

Fig. 2. Two critical regions.

notation, write

$$\int_A \cdots \int \prod_{i=1}^{n} f(x_i; \theta_0) \, dx_1 \cdots dx_n = \int_A L_0 \, dx$$

with a similar expression for L_1.

Since A and A^* are both critical regions of size α,

(16)
$$\int_A L_0 \, dx = \int_{A^*} L_0 \, dx \,.$$

But from Fig. 2 it is clear that the integral over b, which is the common part of A and A^*, will cancel from both sides of (16) and reduce it to the form

(17)
$$\int_a L_0 \, dx = \int_c L_0 \, dx \,.$$

Now calculate the size of the type II error for both A and A^*. Since the size of the type II error is the probability that the sample point will fall outside the critical region when H_1 is true, which in turn is equal to 1 minus the probability that it will fall inside the critical region when H_1 is true, these errors may be written in the form

$$\beta^* = 1 - \int_{A^*} L_1 \, dx$$

and

$$\beta = 1 - \int_A L_1 \, dx \,.$$

Consequently

$$\beta^* - \beta = \int_A L_1 \, dx - \int_{A^*} L_1 \, dx \,.$$

If the integral over the common part b is canceled, this difference will reduce to

(18)
$$\beta^* - \beta = \int_a L_1 \, dx - \int_c L_1 \, dx \,.$$

Since region a lies in A, it follows from the definition of A given in (15) that every point of a satisfies the inequality

$$L_1 \geq k L_0 \,;$$

hence,

$$\int_a L_1 \, dx \geq k \int_a L_0 \, dx \,.$$

Similarly, since c lies outside A, every point of c satisfies the second inequality in (15), namely,

$$L_1 \leq kL_0 \; ;$$

hence,

$$\int_c L_1 \, dx \leq k \int_c L_0 \, dx .$$

When these two results are used in (18), it follows that

$$\beta^* - \beta \geq k \int_a L_0 \, dx - k \int_c L_0 \, dx .$$

But from (17) the right side must be equal to zero; hence,

$$\beta^* \geq \beta .$$

Since β^* is the size of the type II error for any critical region of size α, other than A, the preceding analysis proves that A is a best critical region of size α, where best is understood to mean a critical region with a minimum size type II error.

The constant k of this lemma is chosen to make A a critical region of size α. In most problems, as k goes from 0 to infinity, the size of A decreases from 1 to 0, thus making it possible to determine a value of k that will yield a critical region of size α for any α satisfying $0 < \alpha < 1$.

The usefulness and meaning of this lemma is best explained by means of illustrations. Consider first the problem that was discussed in section 6.3, Chapter 4. For that problem

$$f(x; \theta) = \theta e^{-\theta x}, \qquad x \geq 0 .$$

In order to discuss a somewhat more general problem let the hypothesis be $H_0 : \theta = \theta_0$ and the alternative be $H_1 : \theta = \theta_1 < \theta_0$ and assume that a sample of size n is to be taken. The corresponding likelihood functions are

$$L_0 = \prod_{i=1}^{n} f(x_i; \theta_0) = \theta_0^{\,n} e^{-\theta_0 \sum\limits_{i=1}^{n} x_i}$$

and

$$L_1 = \prod_{i=1}^{n} f(x_i; \theta_1) = \theta_1^{\,n} e^{-\theta_1 \sum\limits_{i=1}^{n} x_i}.$$

According to (15), the region A is the region in which

$$\frac{\theta_1^{\,n} e^{-\theta_1 \Sigma x_i}}{\theta_0^{\,n} e^{-\theta_0 \Sigma x_i}} \geq k .$$

This inequality may be written in the form

$$e^{(\theta_1 - \theta_0)\Sigma x_i} \le \frac{1}{k}\left(\frac{\theta_1}{\theta_0}\right)^n .$$

Taking logarithms, the inequality becomes

$$(\theta_1 - \theta_0) \sum x_i \le \log \frac{1}{k}\left(\frac{\theta_1}{\theta_0}\right)^n .$$

Since H_1 specifies that $\theta_1 < \theta_0$, dividing both sides by $\theta_1 - \theta_0$ will reverse the inequality and yield

$$(19) \qquad \sum x_i \ge \frac{\log \frac{1}{k}\left(\frac{\theta_1}{\theta_0}\right)^n}{\theta_1 - \theta_0} .$$

Now the problem of Chapter 4 has $n = 1$, $\theta_0 = 2$, and $\theta_1 = 1$; hence for that problem the best critical region, as given by (19), would be that part of the x axis to the right of the point

$$x_0 = \frac{\log \dfrac{\theta_1}{k\theta_0}}{\theta_1 - \theta_0} = \log 2k$$

where k is chosen to make the probability .135 that X will exceed x_0. Thus the right tail, which was shown in Chapter 4 to be better than the left tail for that problem, is now shown to be the best possible critical region for that problem.

The derivation that led to (19) does not depend on the particular value of θ_1, provided that $\theta_1 < \theta_0$. Thus the same critical region is used whatever the value of θ_1, as long as $\theta_1 < \theta_0$. The value of k necessary to produce the same x_0 for (19) will, of course, depend on the value of θ_1. This discussion shows that (19) gives the best critical region for testing the hypothesis $H_0 : \theta = \theta_0$ against the composite alternative $H_1 : \theta < \theta_0$. Thus the Neyman-Pearson lemma, although designed to test a simple hypothesis against a simple alternative, can sometimes be used to solve a problem in which the alternative hypothesis is composite. From this result it follows that the critical region selected for the problem discussed in Chapter 4 is the best critical region for testing $H_0 : \theta = 2$ against $H_1 : \theta < 2$. This form of the alternative hypothesis would undoubtedly be much more realistic and satisfying to the experimenter than the original alternative $H_1 : \theta = 1$.

As a second illustration, consider the problem of testing whether a normal population with unit variance has a mean $\theta = \theta_0$ or a mean $\theta = \theta_1 < \theta_0$.

Here

$$f(x; \theta) = \frac{e^{-\frac{1}{2}(x-\theta)^2}}{\sqrt{2\pi}}.$$

Then

$$L_0 = \prod_{n=1}^{n} f(x_i; \theta_0) = (2\pi)^{-\frac{n}{2}} e^{-\frac{1}{2} \sum_{i=1}^{n} (x_i-\theta_0)^2}$$

and

$$L_1 = \prod_{i=1}^{n} f(x_i; \theta_1) = (2\pi)^{-\frac{n}{2}} e^{-\frac{1}{2} \sum_{i=1}^{n} (x_i-\theta_1)^2}.$$

The region A in (15) is therefore the region in which

$$\frac{e^{-\frac{1}{2}\Sigma(x_i-\theta_1)^2}}{e^{-\frac{1}{2}\Sigma(x_i-\theta_0)^2}} = e^{\frac{1}{2}[\Sigma(x_i-\theta_0)^2-\Sigma(x_i-\theta_1)^2]} \geq k.$$

If logarithms are taken, this inequality will reduce to

$$\sum (x_i - \theta_0)^2 - \sum (x_i - \theta_1)^2 \geq 2 \log k.$$

Simplification of the left side will produce the form

$$2(\theta_1 - \theta_0) \sum x_i \geq 2 \log k + (\theta_1^2 - \theta_0^2)n.$$

If both sides are divided by $2n(\theta_1 - \theta_0)$, which is a negative number because it was assumed that $\theta_1 < \theta_0$, this inequality will finally reduce to

$$(20) \qquad \bar{x} \leq \frac{2 \log k + (\theta_1^2 - \theta_0^2)n}{2n(\theta_1 - \theta_0)}.$$

By choosing k properly, the quantity on the right can be made to have a value $\bar{x}_0$ such that the probability that $\bar{X}$ will be less than $\bar{x}_0$ when H_0 is true will be equal to, say, $\alpha = .05$. Thus the best critical region here is the left tail of the $\bar{X}$ distribution.

As in the first illustration of this section, it will be observed that the critical region obtained by applying (15) is the same for all alternative values θ_1, provided that $\theta_1 < \theta_0$, and is the best critical region for the more general composite alternative $H_1: \theta < \theta_0$.

If $\theta_1 > \theta_0$, inequality (20) will be reversed; consequently the best critical region will consist of the right tail of the $\bar{X}$ distribution. This critical region will also be best for the composite alternative $H_1: \theta > \theta_0$.

If one wished to test $H_0: \theta = \theta_0$ against $H_1: \theta \neq \theta_0$, there would be no best critical region for all possible alternative values θ_1 because when $\theta_1 < \theta_0$ the left tail will be best, whereas when $\theta_1 > \theta_0$ the right tail will be best. The preceding result is typical; best critical regions usually exist only if the alternative values of the parameter are suitably restricted.

As a final illustration, consider a discrete variable problem. Although lemma (15) was proved for continuous variables, the same proof will apply to discrete variables if one replaces integrals by sums. A certain difficulty arises with discrete variable problems in that there may be very few, or no other, critical regions having the same value of α as that for a selected critical region. If this were true, it would be academic to say that a certain critical region is a best critical region of size α. These possibilities are considered in the following illustration.

Let X possess a Poisson distribution with mean μ and let the hypothesis $H_0: \mu = \mu_0$ be tested against the alternative hypothesis $H_1: \mu = \mu_1 < \mu_0$. By proceeding as for continuous variables,

$$\frac{L_1}{L_0} = \frac{\prod_{i=1}^{n} e^{-\mu_1} \dfrac{\mu_1^{x_i}}{x_i!}}{\prod_{i=1}^{n} e^{-\mu_0} \dfrac{\mu_0^{x_i}}{x_i!}} = e^{n(\mu_0-\mu_1)} \left(\frac{\mu_1}{\mu_0}\right)^{\Sigma x_i}.$$

The inequality

$$e^{n(\mu_0-\mu_1)} \left(\frac{\mu_1}{\mu_0}\right)^{\Sigma x_i} \geq k$$

is equivalent to the inequality

$$\sum x_i \log \frac{\mu_1}{\mu_0} \geq \log k + n(\mu_1 - \mu_0).$$

Since $\log \mu_1/\mu_0 < 0$ because it was assumed that $\mu_1 < \mu_0$, the preceding inequality can be written

$$\sum x_i \leq \frac{\log k + n(\mu_1 - \mu_0)}{\log \mu_1 - \log \mu_0}.$$

It was shown in Chapter 5 that the sum of independent Poisson variables is a Poisson variable with its mean equal to the sum of the means; it therefore follows that the variable $Z = \sum X_i$ is a Poisson variable with mean $n\mu$. The critical region determined by the preceding inequality is therefore equivalent to a critical region of the type $z \leq z_0$ for the Poisson variable Z where z_0 is chosen to make the region one of size α.

This is where the difficulty with discrete variable problems arises. Since the sample space for a Poisson variable consists of the points $z = 0, 1, 2, \cdots$, the critical region $z \leq z_0$ is constructed by starting with the point $z = 0$ and adding successive points $z = 1$, $z = 2$, etc., until the sum of the probabilities for those points under H_0 is equal to α. But it is unlikely that this sum will exactly equal a previously specified α value. This unsatisfactory state of affairs can be overcome by employing what is known as a *randomization* device. Suppose, for example, that $\alpha = .05$ and that the Poisson probabilities under H_0 corresponding to $z = 0$, 1, 2, $\cdots$ are .018, .072, .144, $\cdots$. Choosing $z = 0$ as the critical region makes $\alpha = .018$, whereas choosing it to consist of the two points $z = 0$ and $z = 1$ makes $\alpha = .018 + .072 = .091$. The randomization device that will yield a value of $\alpha = .05$ consists in agreeing to reject H_0 when $Z = 0$ but to reject H_0 only a certain proportion of the time when $Z = 1$. The proper proportion here is p, where p satisfies the equation $.018 + .072p = .05$. The solution of this equation is $p = \frac{4}{9}$; consequently, in carrying out the test, one would consult a table of random numbers, or use some game of chance that would yield successes $\frac{4}{9}$ of the time, to determine whether to place $Z = 1$ in the critical region when the value $Z = 1$ is obtained. By using such randomization devices, it is possible to discuss best tests and to apply lemma (15) to discrete variable problems in much the same manner as for continuous variables. In practical applications with discrete variables one usually dispenses with these devices and chooses a critical region whose size is feasible and close to the desired α value.

3.3 Likelihood Ratio Tests

When the Neyman-Pearson lemma fails to yield a best test, or when the hypothesis is composite rather than simple, it is necessary to place further restrictions on the class of tests and then attempt to find a best test in this restricted class or to introduce some other principle for obtaining good tests. In this section a second principle for constructing good tests is introduced and discussed. Since any method for testing composite hypotheses will include the testing of simple hypotheses as a special case, this principle is introduced from the point of view of composite hypotheses.

Suppose that the variable X has a density function $f(x; \theta_1, \cdots, \theta_k)$ that depends on k parameters. Let the composite hypothesis to be tested be denoted by $H_0: \theta_i = \theta_i'(i = 1, 2, \cdots, k)$, where θ_i' may or may not denote a numerical value. Thus, if there are two parameters, H_0 might be the hypothesis that $\theta_1 = 10$ with θ_2 unspecified; then $\theta_1' = 10$ and $\theta_2' = \theta_2$. As a second illustration, H_0 might be the hypothesis that $\theta_1 = \theta_2$; then $\theta_1' = \theta_1$ and $\theta_2' = \theta_1$. With the aid of this notation, $f(x; \theta_1', \cdots, \theta_k')$ will denote the density function of X when H_0 is true.

Let $\hat{\theta}_i$ denote the maximum likelihood estimator of θ_i for the likelihood function $L(X, \theta) = \prod_{i=1}^{n} f(X_i; \theta_1, \cdots, \theta_k)$, where here the likelihood function is treated as a function of the parameters and the random variables $X_1, \cdots, X_n$. Similarly, let $\hat{\theta}_i'$ denote the maximum likelihood estimator of

θ_i when H_0 is true; that is, for the likelihood function $L(X, \theta') = \prod_{i=1}^{n} f(X_i; \theta'_1, \cdots, \theta'_k)$. Now, form the ratio

$$(21) \qquad \lambda = \frac{L(X, \theta')}{L(X, \theta)} .$$

This is the ratio of the two likelihood functions when their parameters have been replaced by their maximum likelihood estimators. Since the maximum likelihood estimators are functions of the random variables $X_1, X_2, \cdots, X_n$, the ratio λ is a function of $X_1, X_2, \cdots, X_n$ only and is therefore an observable random variable, that is, a statistic.

The denominator of λ is the maximum of the likelihood function with respect to all the parameters, whereas the numerator is the maximum only after some or all of the parameters have been restricted by H_0; consequently it is clear that the numerator cannot exceed the denominator in value and therefore that λ can assume values between 0 and 1 only. Now the value of the likelihood function gives the probability density (or probability in the case of a discrete variable) at the sample point $x_1, x_2, \cdots, x_n$. Therefore, if λ has a value close to 1, it follows that the probability density (or probability) of the sample point could not be increased much by allowing the parameters to assume values other than those possible under H_0; consequently, a value of λ near 1 corresponds intuitively to considerable belief in the reasonableness of the hypothesis H_0. If, however, the value of λ is close to 0, it implies that the probability density (or probability) of the sample point is very low under H_0 as contrasted to its value under certain other possible values of the parameters not permitted under H_0, and therefore a value of λ near 0 corresponds to considerable belief in the unreasonableness of the hypothesis. If increasing values of λ are treated as corresponding to increasing degrees of belief in the truth of the hypothesis, then λ may serve as a statistic for testing H_0, with small values of λ leading to the rejection of H_0.

Now suppose that H_0 is true and the density function of the random variable λ, say $g(\lambda)$, has been found. This is theoretically possible if the explicit form of $f(x; \theta'_1, \cdots, \theta'_k)$ is known. Suppose, further, that $g(\lambda)$ does not depend on any unknown parameters. Then one can find a value of λ, call it λ_0, such that

$$(22) \qquad \int_0^{\lambda_0} g(\lambda)\, d\lambda = \alpha .$$

The critical region of size α for testing H_0 by means of the statistic λ then is chosen to be the interval $0 \leq \lambda \leq \lambda_0$.

The preceding explanation of how likelihood ratio tests are constructed may be summarized in the following form.

(23) LIKELIHOOD RATIO TESTS: *To test a hypothesis H_0, simple or composite, use the statistic λ given by (21) and reject H_0 if, and only if, the sample value of λ satisfies the inequality $\lambda \leq \lambda_0$, where λ_0 is given by (22).*

There is a great deal of similarity between the techniques used to obtain a best test and a likelihood ratio test. They both use the ratio of the two likelihood functions as a basis for making decisions. This similarity may be observed by comparing (15) and (21).

Although the use of λ as a statistic for testing hypotheses has been justified largely on intuitive grounds, it can be shown that such tests possess several very desirable properties. These properties will be discussed briefly after a few illustrations have been given on how to construct likelihood ratio tests.

Consider the second illustration of the preceding section, namely, the problem of testing the hypothesis $H_0: \theta = \theta_0$, where

$$f(x; \theta) = \frac{e^{-\frac{1}{2}(x-\theta)^2}}{\sqrt{2\pi}}.$$

For this problem

$$L(x, \theta) = (2\pi)^{-\frac{n}{2}} e^{-\frac{1}{2} \sum_{i=1}^{n} (x_i - \theta)^2}.$$

The hypothesis H_0 is a simple hypothesis; nevertheless it will be tested by using the likelihood ratio method to see how that test compares with a test obtained by using the Neyman-Pearson lemma. For the present, H_1 will not be specified. A second illustration will consider a problem in which H_0 is composite. Since $L(x, \theta)$ will be maximized if $\log L(x, \theta)$ is maximized, it will suffice to maximize $\log L(x, \theta)$. But

$$\frac{\partial \log L(x, \theta)}{\partial \theta} = \sum_{i=1}^{n} (x_i - \theta) \, ;$$

hence $\hat{\theta} = \bar{x}$, and therefore

$$L(x, \hat{\theta}) = (2\pi)^{-\frac{n}{2}} e^{-\frac{1}{2} \sum_{i=1}^{n} (x_i - \bar{x})^2}.$$

Since there are no parameters to be estimated under H_0,

$$L(x, \hat{\theta}') = L(x, \theta') = (2\pi)^{-\frac{n}{2}} e^{-\frac{1}{2} \sum_{i=1}^{n} (x-\theta_0)^2}.$$

Then λ, as given by (21) for a set of sample values, becomes

$$\lambda = e^{-\frac{1}{2} \left[\sum_{i=1}^{n} (x_i - \theta_0)^2 - \sum_{i=1}^{n} (x_i - \bar{x})^2 \right]}.$$

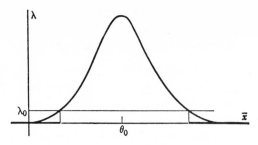

Fig. 3.   Relationship between λ and $\bar{x}$.

Upon simplifying the exponent, λ reduces to

(24)
$$\lambda = e^{-\frac{n}{2}(\bar{x}-\theta_0)^2}$$

Now n and θ_0 are known constants; hence (24) expresses a relationship between λ and $\bar{x}$. By means of this relationship the critical value λ_0 can be determined without finding $g(\lambda)$. The nature of the relationship expressed by (24) is most easily seen graphically, as in Fig. 3.   To each value of λ correspond two values of $\bar{x}$, which are symmetrical with respect to $\bar{x} = \theta_0$. There are therefore two critical values of $\bar{x}$ corresponding to the critical value of $\lambda = \lambda_0$. Figure 3 also shows that increasingly small values of λ correspond to increasingly large values of $|\bar{x} - \theta_0|$.   Therefore the 5 percent critical region for λ, consisting of the interval $0 \leq \lambda \leq \lambda_0$, will correspond to the two $2\frac{1}{2}$ percent tails of the normal $\bar{X}$ distribution.   Thus the 5 percent critical region for the likelihood ratio test is equivalent to the two equal tails of the $\bar{X}$ distribution given by the familiar inequality $|\bar{X} - \theta_0|\sqrt{n} > 1.96$.

The preceding illustration was concerned with testing a simple hypothesis and was selected for the purpose of comparing the result of applying the Neyman-Pearson lemma for best tests with that obtained by the likelihood ratio approach.   It will be recalled that a best test exists for this problem only for one-sided alternatives, $\theta_1 > \theta_0$ or $\theta_1 < \theta_0$; hence the likelihood ratio test cannot be a best test.   It serves as a compromise test when there is no restriction placed on the alternative values of θ.   It should be noted that the likelihood ratio test is concerned only with the hypothesis H_0 and that no mention is made of H_1. This implies that the alternative hypothesis should be $H_1 : \theta \neq \theta_0$.

In the preceding illustration it was not necessary to find the distribution of λ because it turned out that λ was a simple function of $\bar{X}$ whose distribution is known.   In general, however, there is no assurance that some such nice relationship to a familiar variable will exist.   Then one must use whatever tools he has available in an effort to find the distribution of λ. Fortunately, for large samples there is a good approximation to the distribution of λ

which eliminates the necessity for finding the exact distribution. This result from the advanced theory of statistics may be expressed in the form of a theorem.

(25) THEOREM: *Under certain regularity conditions, the random variable* $-2 \log_e \lambda$, *where* λ *is given by* (21), *has a distribution that approaches that of a* χ^2 *variable as n becomes infinite, with its degrees of freedom equal to the number of parameters that are determined by the hypothesis* H_0.

Since small values of λ correspond to large values of $-2 \log_e \lambda$, it follows that the critical region for a test based on $-2 \log_e \lambda$ will consist of large values of this variable. If the borderline of a critical region for the χ^2 variable $-2 \log_e \lambda$ is denoted by χ_0^2, then χ_0^2 must be a number such that $P\{\chi^2 > \chi_0^2\} = \alpha$. Thus, in order to determine the critical region for this approximate likelihood ratio test, it suffices to find a critical value of χ^2 in Table III in the appendix.

According to theorem (25), the number of degrees of freedom in the approximating χ^2 distribution for the illustration considered earlier is $v = 1$ because only a single parameter was determined by H_0. Furthermore, from (24) the value of $-2 \log_e \lambda$ is $n(\bar{X} - \theta_0)^2$. From (25) it therefore follows that the critical region for this approximate test is the region in which $n(\bar{X} - \theta_0)^2 > \chi_0^2$. This is the same critical region as that obtained from the exact likelihood ratio test. It should be noted that $\sqrt{n}(\bar{X} - \theta_0)$ is a standard normal variable because the standard deviation of $\bar{X}$ is $1/\sqrt{n}$ here; therefore from section 8, Chapter 5, its square is a χ^2 variable with 1 degree of freedom. Thus this theorem is seen to check with the known exact distribution here.

The preceding illustration is not typical of the kind of problems for which the likelihood ratio test was designed because the hypothesis was a simple hypothesis. To obtain an illustration that is a typical composite hypothesis testing problem, the preceding setup will be changed by assuming that the variance of the normal variable X is unknown. Thus, X is assumed to be normally distributed with mean μ and unknown variance σ^2 and the hypothesis to be tested is $H_0 : \mu = \mu_0$ against $H_1 : \mu \neq \mu_0$.

The likelihood function for a sample of size n is

$$L(x, \theta) = \frac{e^{-\frac{1}{2} \frac{\Sigma(x_i - \mu)^2}{\sigma^2}}}{(2\pi)^{\frac{n}{2}} \sigma^n} \cdot$$

When H_0 is true, this function is

$$L(x, \theta') = \frac{e^{-\frac{1}{2} \frac{\Sigma(x_i - \mu_0)^2}{\sigma^2}}}{(2\pi)^{\frac{n}{2}} \sigma^n} \cdot$$

The maximum likelihood estimates of the parameters in these two functions are $\hat{\mu} = \bar{x}$, $\hat{\sigma}^2 = (1/n) \sum (x_i - \bar{x})^2$, and $\hat{\hat{\sigma}}^2 = (1/n) \sum (x_i - \mu_0)^2$, respectively. Substitution of those values will give

$$L(x, \theta) = \frac{e^{-\frac{n}{2}}}{(2\pi)^{\frac{n}{2}}\hat{\sigma}^n} \quad \text{and} \quad L(x, \theta') = \frac{e^{-\frac{n}{2}}}{(2\pi)^{\frac{n}{2}}\hat{\hat{\sigma}}^n} .$$

As a result

$$\lambda = \frac{\hat{\sigma}^n}{\hat{\hat{\sigma}}^n} = \left[\frac{\sum (x_i - \bar{x})^2}{\sum (x_i - \mu_0)^2} \right]^{\frac{n}{2}} .$$

The inequality $\lambda \leq \lambda_0$ is equivalent to the inequality $\lambda^{-2/n} \geq \lambda_0^{-2/n}$; hence, using this last inequality, the critical region for the test is given by the inequality

$$\frac{\sum (x_i - \mu_0)^2}{\sum (x_i - \bar{x})^2} \geq c$$

where $c = \lambda_0^{-2/n}$. Writing

$$\sum (x_i - \mu_0)^2 = \sum [(x_i - \bar{x}) + (\bar{x} - \mu_0)]^2 = \sum (x_i - \bar{x})^2 + n(\bar{x} - \mu_0)^2$$

this inequality will assume the form

$$\frac{n(\bar{x} - \mu_0)^2}{\sum (x_i - \bar{x})^2} \geq c - 1 .$$

It is illuminating to write this as

$$\frac{(\bar{x} - \mu_0)^2}{s^2/n} \geq n(c - 1) .$$

Denoting $n(c - 1)$ by c', this is equivalent to

$$\left(\frac{\bar{x} - \mu_0}{s/\sqrt{n}} \right)^2 \geq c' .$$

If s were replaced by σ, the expression in parentheses would be a standard normal variable and the critical region would consist of the two equal tails of its distribution, or equivalently, the two equal tails of the $\bar{x}$ distribution. Since s will be a good estimate of σ for large n, the preceding critical region is in good agreement with what was to be expected for large samples.

The distribution of the random variable $(\bar{X} - \mu_0)\sqrt{n}/S$, which is needed here if the test is to be carried out exactly, will be derived in Chapter 10. For the present it will be necessary to resort to large sample methods if the test is to be applied. From the preceding theory it follows that one could calculate $-2 \log \lambda$ and treat it as a chi-square variable; however it can be shown that this is equivalent to applying the earlier large sample methods for testing a mean by replacing σ by s and using the two tails of the $\bar{X}$ distribution. In view of this equivalence the earlier technique will be used to solve problems of this type until Chapter 10 has been studied.

3.4 Power Function

The preceding methods have been concerned with how to choose the critical region for testing a hypothesis but, except for the case of a simple hypothesis versus a simple hypothesis, they did not consider the problem of determining how good a given test is. This requires studying the size of the type II error for composite hypotheses.

For these more general classes of alternatives, the size of the type II error β will depend on the particular alternative value of θ being considered. In order to determine how good the chosen test may be, compared to a competing test, it is therefore necessary to compare the type II errors for all possible alternative values of θ rather than for just one alternative value as before. For this purpose, it is necessary to consider the calculation of the size of the type II error as a function of θ. The size of this error is denoted by $\beta(\theta)$. Now $\beta(\theta)$ is the probability that the sample point will fall in the noncritical region when θ is the true value of the parameter. It is usually more convenient to work exclusively with the critical region; therefore it is customary to calculate $1 - \beta(\theta)$, which is the probability that the sample point will fall in the critical region when θ is the true value of the parameter. The function $1 - \beta(\theta)$ is called the power function and may be defined formally as follows.

(26) DEFINITION: *The power function $P(\theta)$ of a test is the function of the parameter that gives the probability that the sample point will fall in the critical region of the test when θ is the true value of the parameter.*

Since $P(\theta) = 1 - \beta(\theta)$, seeking for a test that minimizes the type II error $\beta(\theta)$ is equivalent to seeking for one that maximizes the power $P(\theta)$.

The first problem that was considered in the preceding section will be used to illustrate how the power function can assist one in comparing tests when there is more than a single alternative value of the parameter. For that illustration, let the hypothesis to be tested be $H_0: \theta = 2$, as it was before, but let the alternative hypothesis now be $H_1: \theta < 2$ rather than $H_1: \theta = 1$. As before, let $x > 1$ and $x < .07$ be the respective critical regions of size $\alpha = .135$ that were obtained for the two competing tests in section 3.2, and let $P_1(\theta)$ and $P_2(\theta)$ denote the power functions for the two tests. From (26), the power functions of these tests are given by integrating the density function

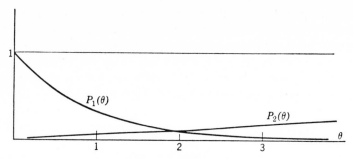

Fig. 4.　Two competing power curves.

over the respective critical regions; hence

(27)
$$P_1(\theta) = \int_1^\infty \theta e^{-\theta x}\, dx = e^{-\theta}$$

and

$$P_2(\theta) = \int_0^{.07} \theta e^{-\theta x}\, dx = 1 - e^{-.07\theta}.$$

The graphs of $P_1(\theta)$ and $P_2(\theta)$, which are called the power curves for the two tests, are shown in Fig. 4. These curves must intersect at the point $(2, .135)$ because the power function gives the probability that the sample point will fall in the critical region and this probability has been chosen equal to $\alpha = .135$ when $H_0: \theta = 2$ is true. Since the power curve for the first test lies above the power curve for the second test for all values of $\theta < 2$ and the only alternative values permitted in the problem are those given by $H_1: \theta < 2$, it follows that the first test is superior to the second for the problem under discussion. This, of course, was to be expected because the first test is the test obtained by applying the Neyman-Pearson lemma.

Not only is the power function useful for assisting one in comparing tests and finding best tests when more than one alternative value of the parameter exists, but it is also useful for determining the effectiveness of a given test for making the correct decision as a function of the parameter value. For example, the power function $P_1(\theta)$ given by (27) shows that the probability is .37 of making the correct decision, namely, rejecting H_0 when $\theta = 1$, and that this probability rises to .61 when $\theta = \frac{1}{2}$. By studying the power function, or power curve, of a test the experimenter can determine his chances of detecting various possible alternative values of the parameter that may occur and thus determine whether his experiment is large enough to give him the confidence that he would like in whatever decision will be made by the test.

As a second illustration of how to calculate power functions, consider the second problem of section 3.2. There the best critical region was determined to be a region of the form $\bar{x} \leq c$. Suppose a sample of size $n = 16$ is to be taken and the hypotheses under consideration are $H_0: \mu = 20$

and $H_1: \mu < 20$. Since X is a normal variable with $\sigma = 1$, $\bar{X}$ is normally distributed with mean μ and variance $1/n = \frac{1}{16}$; consequently the density of $\bar{X}$ is given by

$$f(\bar{x}; \mu) = \frac{4e^{-\frac{16}{2}(\bar{x}-\mu)^2}}{\sqrt{2\pi}}.$$

If $\alpha = .05$ it follows from the illustration in section 3.2 that the best critical region is the region where $\bar{x} < \mu_0 - 1.64\sigma_{\bar{x}}$, which is the region where $\bar{x} < 20 - 1.64/4$, or $\bar{x} < 19.59$. The power function is therefore given by the integral

$$P(\mu) = \int_{-\infty}^{19.59} \frac{4e^{-\frac{16}{2}(\bar{x}-\mu)^2}}{\sqrt{2\pi}}\, d\bar{x}.$$

Letting $t = 4(\bar{x} - \mu)$ will reduce this integral to the form

$$P(\mu) = \int_{-\infty}^{4(19.59-\mu)} \frac{e^{-\frac{t^2}{2}}}{\sqrt{2\pi}}\, dt.$$

Values of $P(\mu)$ can be obtained by evaluating the upper limit of this integral and using Table II. A few such values of $P(\mu)$ are sufficient to enable one to sketch the power curve. For example, such calculations yielded the following values:

μ	20	19.8	19.59	19.4	19.2
$4(19.59 - \mu)$	-1.64	$-.84$	0	.76	1.56
$P(\mu)$	.05	.20	.50	.78	.94

The graph of the power function for $\mu \leq 20$, which includes the only values of μ that need to be considered here since the alternative hypothesis is $H_1: \mu < 20$, is shown in Fig. 5.

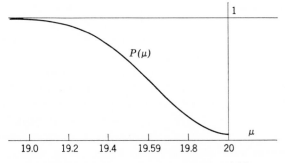

Fig. 5. The power curve for a normal variable.

EXERCISES

Sec. 2.1

1. Given that X is normally distributed and given the following three sample values of the variance, (a) combine them to yield an unbiased estimate of σ^2, (b) show that $(2s_1^2 + 2s_2^2 + s_3^2)/5$ is not an unbiased estimate of σ^2. The sample values are $s_1^2 = 8$, $s_2^2 = 10$, $s_3^2 = 14$ with $n_1 = 10$, $n_2 = 7$, $n_3 = 5$.

2. Show that $2\bar{X}$ is an unbiased estimate of θ for $f(x; \theta) = 1/\theta$, $0 < x < \theta$.

3. Show that the distribution function of $Z = \max\{X_1, X_2, \cdots, X_n\}$ is given by $F(z) = (z/\theta)^n$ when X has the density of problem 2. In this connection see problem 21, Chapter 5. Use the preceding result to show that $(1 + 1/n)Z$ is also an unbiased estimate of θ in problem 2.

4. Using the expected value operator, derive an expression for the correlation between $u = a_1X_1 + \cdots + a_kX_k$ and $v = b_1X_1 + \cdots + b_kX_k$, where the a's and b's are constants and the variables $X_1, \cdots, X_k$ are independent.

5. A fisheries investigator catches fish from a lake until he has obtained k fish of a certain species. His total catch then is N. Assuming that the lake has a very large number of fish, show that the density function of the variable N is given by $\binom{N-1}{k-1}p^k(1-p)^{N-k}$, $N = k$, $k+1$, $\cdots$, where p is the proportion of this species in the lake. Use this result to show that $(k-1)/(N-1)$ is an unbiased estimate of p.

Sec. 2.2

6. Show that $\sum (X - c)^2$ is a minimum when $c = E[X]$.

7. Consider the variable $Z = (a_1X_1 + \cdots + a_kX_k)/(a_1 + \cdots + a_k)$, where the variables $X_1, \cdots, X_k$ are independently distributed with 0 means and variances $\sigma_1^2, \cdots, \sigma_k^2$. Prove that the variance of Z will be minimized if the weight a_i is chosen inversely proportional to σ_i^2.

8. Given $\sigma_1^2 = 1$, $\sigma_2^2 = 2$, $\sigma_3^2 = 3$, $\sigma_4^2 = 4$, $\sigma_5^2 = 5$, calculate the variance of Z in problem 7 when (a) a_i is chosen inversely proportional to σ_i^2, (b) a_i is chosen equal to $1/k$. (c) Comment on the advantage of the weighting used in (a).

9. Why is it reasonable to take the ratio of the variances rather than the ratio of the standard deviations of two estimates to compare the estimates?

10. Compare the variances of the two estimates of θ obtained in problems 2 and 3.

11. If X is the number of successes in n trials with probability p of success, and if $Z_1 = X/n$ and $Z_2 = (X + 1)/(n + 2)$, are two estimates of p, for what values of p will the second estimate posess a smaller mean squared error than the first estimate?

Sec. 2.3

12. Given the density $f(x; \theta) = e^{-\frac{1}{2}(x-\theta)^2}/\sqrt{2\pi}$, find the maximum likelihood estimator of θ based on a sample of size n.

13. Given the density $f(x; \theta) = e^{-\theta}\theta^x/x!$, $x = 0, 1, \cdots$, and given the six observed values 4, 10, 4, 8, 7, and 7, find the maximum likelihood estimate of θ.

14. Find the maximum likelihood estimator of θ based on n observations for the density $f(x; \theta) = (1 + \theta)x^\theta$, $\theta > -1$, $0 < x < 1$.

15. Given the density $f(x; \theta) = e^{-x^2/2\theta^2}/\theta\sqrt{2\pi}$, find the maximum likelihood estimator of θ based on a sample of size n.

16. In problem 15 treat θ^2 as the parameter to be estimated and write the density as $f(x; \phi) = e^{-x^2/2\phi}/\sqrt{2\pi\phi}$, where $\phi = \theta^2$. Now find the maximum likelihood estimator of ϕ and compare with the result for problem 15.

17. A box contains 8 balls of which the proportion p are white. Let X equal the number of white balls obtained by drawing 2 balls from the box. Find the density $f(x; p)$ and then find the values of p that will maximize $f(x; p)$. Here the only values that p can assume are $p = i/8$, $i = 0, 1, \cdots, 8$.

18. Show that the likelihood function $L(\theta)$ will be maximized when $\log L(\theta)$ is maximized if standard calculus methods may be used to obtain the maximum.

19. Find the maximum likelihood estimator of p for a binomial distribution based on a total of n trials.

20. Find the maximum likelihood estimator of p for a binomial distribution based on N experiments of n trials each and with successes $x_1, x_2, \cdots, x_N$.

21. Find the maximum likelihood estimator of q for $f(x; q) = pq^x$, $p = 1 - q$, $x = 0, 1, 2, \cdots$, if n experiments yielded the observations $x_1, x_2, \cdots, x_n$.

22. Given $f(x; \theta) = \theta e^{-\theta x}$, $x > 0$, (a) find the maximum likelihood estimator for θ, (b) find the maximum likelihood estimator for $E[X]$.

23. Find the maximum likelihood estimators of μ and σ for a normal distribution.

24. A lake contains N fish. A netting experiment yielded x fish, which were tagged and released. A second experiment yielded y fish of which z were found to be tagged. If y is small compared with N, show that the maximum likelihood estimate of N is given, approximately, by $\hat{N} = xy/z$.

Sec. 2.4

25. Find an 80 percent confidence interval for the mean of a normal variable if $\sigma = 2$ and if a sample of size 8 gave the values 9, 14, 10, 12, 7, 13, 11, 12.

26. Assume that X possesses a Poisson distribution with unknown mean μ. If 10 observations yielded the values 20, 23, 17, 16, 21, 22, 19, 19, 25, 18, find an approximate 80 percent confidence interval for μ. Use a normal approximation and base your interval on $\bar{x}$ only.

27. Assuming that n is large enough to justify a normal approximation to the binomial distribution, show that a 95 percent confidence interval for binomial p is given by $p_1 < p < p_2$ where p_1 and p_2 are solutions of the quadratic equation

$$(p - p^2)(1.96)^2 = n(\hat{p} - p)^2.$$

Sec. 3.1

28. Let p denote the probability of success in an experiment. The experiment is

to be performed twice. Let $H_0:p = \frac{1}{2}$ and $H_1:p = 1$. Construct all possible tests for this problem and calculate the values of α and β for each test.

29. In testing $H_0:\mu = 20$ for a normal variable, what is the probability that you will accept H_0 when the mean is actually $\sigma/3$ units above 20, if a sample of size 16 is taken and the critical region is chosen as the 5 percent right tail of the $\bar{X}$ distribution?

30. Let X denote the number of successes occurring in a Poisson process in a 5-minute period. Calculate α and β if you wish to test $H_0:\mu = 2$ against $H_1:\mu = 3$ where μ is the mean for a 5-minute period and $X \geq 4$ is used as the critical region.

Sec. 3.2

31. Suppose that you are testing $H_0:\mu = 3$ against $H_1:\mu = 2$ for the Poisson distribution by means of a sample of size 2. Indicate by means of a sketch in the sample space x_1, x_2 the part of that space you would choose for the critical region. Justify your choice.

32. Estimate the size of the type II error if the type I error is chosen to be $\alpha = .16$, if you are testing $H_0:\mu = 6$ against $H_1:\mu = 5$ for a normal variable with $\sigma = 2$ by means of a sample of size 25, and if the proper tail of the $\bar{X}$ distribution is used as critical region.

33. Under the three possible hypotheses H_1, H_2, and H_3, a discrete variable X has the following distributions:

x	1	2	3	4	5	6	7	8	9	10
$f(x \mid H_1)$	0	.58	.02	.05	.03	.11	.01	.07	.04	.09
$f(x \mid H_2)$	.60	0	.06	.08	.03	.01	.04	.12	.02	.04
$f(x \mid H_3)$	.54	0	.10	.03	.12	.06	.04	.01	.08	.02

(a) Choose $\alpha = .10$ and find a best critical region for testing H_1 against H_2.
(b) Determine whether there is a best critical region for testing H_1 against both H_2 and H_3 with $\alpha = .10$.

34. Forty pairs of runners have been matched with respect to ability. Each member of a pair is given a pill, with one member receiving a stimulant in his pill and the other not. Races are run between each pair. Let X denote the races won by the stimulated runners. Construct a test to test $H_0:p = \frac{1}{2}$ against $H_1:p > \frac{1}{2}$, where p is the probability the stimulated runner will win. Choose α to be approximately .10. Calculate the value of β for this test if $p = .6$.

35. If X is normally distributed with $\sigma = 5$ and it is desired to test $H_0:\mu = 105$, how large a sample should be taken if the probability of accepting H_0 when H_1 is true is to be .02 and if a critical region of size .05 is used?

36. By means of the Neyman-Pearson lemma, prove that the best test for testing $H_0:\sigma = \sigma_0$ against $H_1:\sigma = \sigma_1 > \sigma_0$ for X normally distributed with 0 mean is given by choosing as critical region the region where $\sum_1^n X_i^2 > c$, where c is the proper constant.

37. Use the Neyman-Pearson lemma to determine the nature of a best critical region based on a sample of size n for testing $H_0: \theta = \theta_0$ against $H_1: \theta = \theta_1 < \theta_0$ if $f(x; \theta) = (1 + \theta)x^\theta$, $0 < x < 1$.

38. Can the Neyman-Pearson lemma be applied to testing $H_0: \theta = 1$ against $H_1: \theta = 2$ if $f(x; \theta) = 1/\theta$, $0 < x < \theta$ and a sample of one is to be taken?

39. Given $f(x; p) = pq^x$, $p = 1 - q$, $x = 0, 1, 2, \cdots$, find a best test based on a sample of size n for testing $H_0: q = q_0$ against $H_1: q = q_1 > q_0$. Is this test also best for $H_1: q > q_0$?

40. Derive a formula for determining the size sample needed to produce given values of α and β in testing the hypothesis $H_0: \mu = \mu_0$ against $H_1: \mu = \mu_1 > \mu_0$ if the random variable X is normal and σ is known.

41. Given $X \sim N(\mu_X, 100)$ and $Y \sim N(\mu_Y, 225)$, how large an equal size sample should be taken from X and Y to make $\alpha = .05$ and $\beta = .20$ if the critical region for testing $H_0: \mu_Y = \mu_X$ against $H_1: \mu_Y = \mu_X + 5$ is of the form $\bar{Y} - \bar{X} > c$. In this notation the variances are 100 and 225, respectively.

42. You are given the density $f(x) = \theta^x(1 - \theta)^{1-x}$, $x = 0, 1$, and 0 elsewhere. Let $H_0: \theta = \frac{1}{10}$, $H_1: \theta < \frac{1}{10}$. If a sufficiently large sample is taken to justify using the central limit theorem, what critical region of size .05 would you select for this test?

Sec. 3.3

43. Given that X is normal with mean 0 and variance σ^2, find the expression for λ for the likelihood ratio test for testing $H_0: \sigma = 1$.

44. Work problem 43 if the mean is μ rather than 0, with μ unknown.

45. Construct a likelihood ratio test for testing $H_0: \theta = 1$, given that $f(x; \theta) = \theta e^{-\theta x}$, $x > 0$. Carry your solution to the stage of obtaining λ as a function of $\bar{x}$.

46. Construct a likelihood ratio test for testing $H_0: p = p_0$ by means of n observations of a binomial variable with probability p. Is this a best test for some alternative value?

47. Construct a likelihood ratio test for problem 37 with $H_1: \theta \neq \theta_1$.

48. Using the fact that $-2 \log_e \lambda$ possesses an approximate χ^2 distribution with 1 degree of freedom for the likelihood ratio test of problem 45, use the result of problem 45 and the following sample values to test the hypothesis $H_0: \theta = 1$. The sample values are 1.5, 2, .8, 1.3, 2.8, .9, 1.6, .6, 4.2, 3.1, 1.4, 2.2, .7, 1.6, .8.

49. Show that the likelihood ratio test for testing the simple hypothesis $H_0: \theta = \theta_0$ against $H_1: \theta = \theta_1$ for the normal variable X with σ given is equivalent to the test based on the Neyman-Pearson lemma.

50. Show that the likelihood ratio test for testing $H_0: \sigma_X = \sigma_Y$ for independent $X \sim N(\mu_X, \sigma_X^2)$ and $Y \sim N(\mu_Y, \sigma_Y^2)$ based on equal size random samples of size n for X and Y can be based on the random variable

$$F = \frac{\sum\limits_{i=1}^{n} (X_i - \bar{X})^2}{\sum\limits_{i=1}^{n} (Y_i - \bar{Y})^2}$$

Sec. 3.4

51. Graph the power function by plotting a few points and drawing a smooth curve through them for testing the hypothesis $H_0: \mu = 0$ when using the two $2\frac{1}{2}$ percent tails of the $\bar{X}$ normal distribution if $\sigma = 1$ and $n = 9$.

52. If p denotes the probability that an event A will occur in a single trial of an experiment, then $f(x; p) = p(1 - p)^{x-1}$ is the density function for X, the number of trials needed for A to occur. Find the power function for testing $H_0: p = \frac{1}{4}$ if the critical region consists of the points $x = 1, 2, 3$. Criticize this choice.

53. Find an expression for the power function of the test when testing $H_0: \mu = 20$ against $H_1: \mu > 20$ for a normal variable with unit variance. Use the right tail of the $\bar{X}$ distribution as critical region with $n = 16$ and $\alpha = .10$.

54. An experiment is to be conducted 100 times to determine whether a possible outcome has probability $p = .4$ or $p > .4$. If X denotes the number of outcomes and if $X \geq 48$ is chosen as the critical region for testing $H_0: p = .4$, find an expression for the power function of the test.

55. In problem 54 use the normal approximation to find an expression for the power function.

56. Given $f(x) = \dfrac{e^{-\mu}\mu^x}{x!}$, $x = 0, 1, \cdots$, $H_0: \mu = 1$, $H_1: \mu < 1$. Let $Z = X_1 + \cdots + X_{10} \leq 4$ be the C.R. for this test based on a sample of size 10. Find an expression for the power function of this test.

CHAPTER 9

Testing Goodness of Fit

A problem that arises frequently in statistical work is the testing of the compatibility of a set of observed and theoretical frequencies. For example, if Mendelian inheritance suggests that four kinds of plants should occur in the proportions $9:3:3:1$ and if a sample of 240 plants yielded 120, 40, 55, 25 in the four categories, one would like to know whether these frequencies are compatible with those expected under Mendelian inheritance.

This type of problem has already been discussed and solved for the special case in which there are only two pairs of frequencies to be compared. Then the binomial distribution may be applied as shown in the first illustration of section 9.3, Chapter 5. When more than two pairs of frequencies are to be compared a distribution that is a generalization of the binomial distribution, is needed.

1 MULTINOMIAL DISTRIBUTION

The binomial distribution is capable of solving only those successive trials problems in which each outcome can be classified as either a success or a failure. One can always reduce more than two categories to only two by combining them, but this procedure is likely to throw away much valuable information; therefore it would be desirable to have a distribution that takes account of all such categories. Such a distribution exists in what is known as the *multinomial distribution*. It is obtained in the following manner.

Consider an experiment in which there are k mutually exclusive possible outcomes $A_1, A_2, \cdots, A_k$. Let p_i be the probability that event A_i will occur at a trial of the experiment and let n trials be made. Then the probability that event A_1 will occur x_1 times, event A_2 will occur x_2 times, etc., where $\sum_1^k x_i = n$, may be calculated by using the same reasoning as that used in deriving the binomial distribution. In this connection, consider the particular sequence of events given by

$$\overbrace{A_1, \cdots, A_1}^{x_1}, \overbrace{A_2, \cdots, A_2}^{x_2} \cdots, \overbrace{A_k, \cdots, A_k}^{x_k}.$$

Since the trials are independent, the probability of obtaining this particular sequence of events is

$$(1) \qquad p_1^{x_1} p_2^{x_2} \cdots p_k^{x_k}.$$

Now every arrangement of the preceding set of A's has this same probability of occurring and satisfies the conditions of the problem; consequently, it is necessary to count the number of arrangements. But this is merely the number of permutations of n things of which x_1 are alike, x_2 are alike, etc., which by (18), Chapter 2, is equal to

$$(2) \qquad \frac{n!}{x_1! \, x_2! \cdots x_k!}.$$

Since all these arrangements are the mutually exclusive ways in which the desired event can occur and since each of them has the probability given by (1), the desired probability is obtained by multiplying the quantities given in (1) and (2). This result may be summarized as follows:

(3) MULTINOMIAL DISTRIBUTION:

$$f(x_1, x_2, \cdots, x_k) = \frac{n!}{x_1! \, x_2! \cdots x_k!} \, p_1^{x_1} p_2^{x_2} \cdots p_k^{x_k}.$$

This name is given to the distribution because (3) represents the general term in the expansion of the multinomial function

$$(p_1 + p_2 + \cdots + p_k)^n$$

just as the binomial density function represents the general term in the expansion of the binomial function $(q + p)^n$.

As it stands, the multinomial distribution is not very convenient for calculating probabilities unless n is small. The problem of finding an approximation here is considerably more difficult than in the case of the binomial distribution.

2 THE χ^2 TEST

The problem that is to be considered here can be formulated quite generally in terms of the preceding notation. Because of tradition, let n_i be used in place of X_i to represent the random variable of which x_i is the observed value.

Then n_i is a binomial variable which gives the number of successes that will be obtained in n trials of an experiment for which p_i is the probability of success in a single trial. Hence, the expected value of n_i is given by

$$e_i = E[n_i] = np_i, \qquad i = 1, \cdots, k\,.$$

In terms of this notation, the problem is to determine whether a set of observed values of $n_1, \cdots, n_k$ is compatible with the expected values $e_1, \cdots, e_k$.

An analysis of the preceding discussion will show that the problem is really one of testing a hypothesis because it is assumed that the multinomial distribution is the proper model here and interest centers on whether the postulated p's are correct. Thus the problem can be treated as a problem of testing the hypothesis

(4) $$H_0 : p_i = p_{i0}, \qquad i = 1, 2, \cdots, k$$

where the p_{i0}'s are the postulated values of the probabilities of a multinomial distribution.

The hypothesis expressed in (4) is a simple hypothesis, but unless alternative values of the p's are specified the alternative hypothesis is composite. As a result, lemma (15) of Chapter 8 is not applicable; consequently the likelihood ratio test is the natural test to employ here. Now it will be found that the expression for λ in this test is so complicated that it is not feasible to find its distribution; therefore only the large sample approximation given by theorem (25), Chapter 8, for a general likelihood ratio test is ordinarily used.

If the various steps involved in evaluating λ in (21), Chapter 8, are carried out, it will be found that

(5) $$-2 \log_e \lambda = 2 \sum_{i=1}^{k} n_i \log_e \frac{n_i}{e_i}\,.$$

Now according to theorem (25), Chapter 8, this quantity possesses an approximate χ^2 distribution when n is large. The number of degrees of freedom here is given by $\nu = k - 1$ because the multinomial distribution is determined by only $k - 1$ parameters in view of the restriction that $\sum_1^k p_i = 1$ and all of those parameters are determined under H_0. The test of hypothesis H_0 therefore consists in choosing as critical region the right tail of the χ^2 distribution with $k - 1$ degrees of freedom.

Although (5) does yield a valid large sample test for the hypothesis (4), this test is not the one that is customarily employed here. A modification of it, which is based on approximating the right side of (5), is more commonly used. This approximation is obtained by expanding the logarithms and retaining only the dominating terms. Since the error term converges to zero as $n \to \infty$, the results of such manipulations produce the following theorem.

THEOREM: *If* $n_1, n_2, \cdots, n_k$ *and* $e_1, e_2, \cdots, e_k$ *represent the observed and expected frequencies, respectively, for the k possible outcomes of an experiment that is to be performed n times, then, as n becomes infinite, the distribution of the random variable*

(6)
$$\sum_{i=1}^{k} \frac{(n_i - e_i)^2}{e_i}$$

will approach that of a χ^2 *variable with* $k - 1$ *degrees of freedom.*

The test procedure here is the same as for the test based on (5). Thus, after calculating the value of the quantity given by (6), one determines whether this value exceeds the critical value χ_0^2 that is obtained from the table of critical values of the χ^2 distribution given in Table III in the appendix. Although this test was derived here as an approximate likelihood ratio test, it was obtained by other methods many years before likelihood ratio tests were introduced. Since statisticians were already familiar with the preceding theorem when the test based on (5) was introduced, they continued using the test based on (6), known as the χ^2 *test for goodness of fit.*

The derivation of the theorem given in connection with (6) as an application of theorem (25), Chapter 8, is sketched in the appendix.

As a simple illustration of how to apply this theorem, consider a typical problem. Suppose that a gambler's die is rolled 60 times and a record is kept of the number of times each face comes up. If the die is an "honest" die, each face will have the probability $\frac{1}{6}$ of appearing in a single roll. Therefore, each face would be expected to appear 10 times in an experiment of this kind. Suppose that the experiment produced the following results, where the row labeled n_i represents the observed frequencies and the row labeled e_i represents the expected frequencies.

Face	1	2	3	4	5	6
n_i	15	7	4	11	6	17
e_i	10	10	10	10	10	10

As explained in an earlier paragraph, the problem is one of testing a hypothesis about a multinomial distribution, namely, the hypothesis

$$H_0 : p_1 = \cdots = p_6 = \tfrac{1}{6} .$$

Since $v = k - 1$ and $k = 6$ here, $v = 5$. If a critical region of size .05 is chosen, it will consist of those values of the approximate χ^2 variable given in (6) that exceed the value χ_0^2 which cuts off 5 percent of the right tail of the

χ^2 distribution with five degrees of freedom. From Table III it will be found that $\chi_0{}^2 = 11.1$. Now calculations show that

$$\sum_{i=1}^{6} \frac{(n_i - e_i)^2}{e_i} = \frac{(15 - 10)^2}{10} + \frac{(7 - 10)^2}{10} + \frac{(4 - 10)^2}{10} + \frac{(11 - 10)^2}{10}$$

$$+ \frac{(6 - 10)^2}{10} + \frac{(17 - 10)^2}{10} = 13.6 .$$

Since this value exceeds the critical value $\chi_0{}^2 = 11.1$, it lies in the critical region and therefore the hypothesis H_0 is rejected. Thus one would conclude that the gambler's die is "dishonest." The error introduced in using the approximate χ^2 distribution here would be very small because n is fairly large; consequently the χ^2 test based on the theorem in (6) may be applied with confidence to this problem.

3 LIMITATIONS ON THE χ^2 TEST

Since the χ^2 distribution is only an approximation to the exact distribution of the quantity $\sum (n_i - e_i)^2/e_i$, care must be exercised that the χ^2 test is used only when the approximation is good. Experience and theoretical investigations indicate that the approximation is usually satisfactory, provided that the $e_i \geq 5$ and $k \geq 5$. If $k < 5$, it is best to have the e_i slightly larger than 5. This limitation is similar to that placed on the use of the normal curve approximation to the binomial density function in which np and nq were required to exceed 5.

If the expected frequency of a cell does not exceed 5, this cell should be combined with one or more other cells until the above condition is satisfied. For example, suppose that the gambler's die of the preceding section had been rolled only 24 times and the following results had been obtained:

Face	1	2	3	4	5	6
n_i	6	5	2	3	0	8
e_i	4	4	4	4	4	4

Here none of the expected frequencies exceeds 5; therefore, to satisfy the preceding restriction it is necessary to combine each cell with some other cell. If successive pairs of cells are combined, the preceding empirical rule will be satisfied and the following table of values will be obtained:

Face	1 or 2	3 or 4	5 or 6
n_i	11	5	8
e_i	8	8	8

The application of the χ^2 test will now yield a value of $\chi^2 = 2.25$ with $\nu = 2$. From a theoretical point of view it is legitimate to combine cells in any desired manner, provided that one is not influenced by the observed frequencies. In many applications, however, there are practical reasons for combining neighboring cells as in the preceding impractical illustration.

4 APPLICATIONS

In experiments on the breeding of flowers of a certain species, an experimenter obtained 120 magenta flowers with a green stigma, 48 magenta flowers with a red stigma, 36 red flowers with a green stigma, and 13 red flowers with a red stigma. Theory predicts that flowers of these types should be obtained in the ratios $9:3:3:1$. Are these experimental results compatible with the theory?

This is a problem of testing the hypothesis

$$H_0 : p_1 = \tfrac{9}{16}, \quad p_2 = \tfrac{3}{16}, \quad p_3 = \tfrac{3}{16}, \quad p_4 = \tfrac{1}{16}$$

for a multinomial distribution involving four cells and for which $n = 217$. Under H_0, the expected frequencies, correct to the nearest integer, are those in the second row of the following table.

n_i	120	48	36	13
e_i	122	41	41	14

Calculations give

$$\chi^2 = \frac{(120 - 122)^2}{122} + \frac{(48 - 41)^2}{41} + \frac{(36 - 41)^2}{41} + \frac{(13 - 14)^2}{14} = 1.9 .$$

From Table III the 5 percent critical value of χ^2 for three degrees of freedom is $\chi_0^2 = 7.8$; consequently the result is not significant. The hypothesis H_0 is acceptable here and thus there is no reason on the basis of this test for doubting that the theory is applicable to these data.

As a second application, consider the following problem. On the basis of extensive experience with trainees, a training station determined four scores in marksmanship so that equal numbers of trainees would be located in the resulting five categories of skill. A new group of 200 trainees is given the marksmanship test with the following results:

Category	I	II	III	IV	V
n_i	54	44	40	35	27
e_i	40	40	40	40	40

If the five categories are listed according to increasing ability, would you be justified in claiming that the 200 trainees represent an inferior group of trainees with respect to marksmanship? This problem may be treated as a problem of testing the hypothesis

$$H_0:p_1 = \cdots = p_5 = \tfrac{1}{5}$$

for a multinomial distribution with $n = 200$. Calculations give $\chi^2 = 10.1$. Since $\chi_0^2 = 9.5$ for $\nu = 4$, this result is significant; hence one is justified in claiming that the new group of trainees is not typical of past trainees. Because of the excess frequencies at the lower end of the scale, the new trainees are very likely inferior marksmen.

5 GENERALITY OF THE χ^2 TEST

In the preceding applications the expected frequencies for the various cells were known because the cell probabilities were assumed known; however, many applications involve situations in which the cell probabilities are functions of some unknown parameters. For example, suppose that one is interested in studying the sex distribution of children in families having eight children. If it is assumed that the probability is p that a child selected at random from a family with eight children will be a son, and if N such families are selected then the expected frequencies for the nine cells corresponding to $0, 1, 2, \cdots, 8$ sons will be given by the successive terms in the expansion of the binomial $N(q + p)^8$. Here the probabilities for the various cells depend on the unknown parameter p. Except for crude work, the difficulty cannot be overcome by assuming that the two sexes are equally divided because experience has shown that p is slightly larger than $\tfrac{1}{2}$.

Fortunately, the χ^2 test possesses a remarkable property that permits it to be applied even when the cell probabilities depend on unknown parameters as in this problem. This property, although very difficult to prove, is very simple to state. It may be expressed as follows:

(7) PROPERTY: *The χ^2 test is applicable when the cell probabilities depend on unknown parameters, provided that the unknown parameters are replaced by their maximum likelihood estimates and provided that one degree of freedom is deducted for each parameter estimated.*

It is assumed, of course, that the cell frequencies are large enough to justify the use of the regular χ^2 test. Since $v = k - 1$ when there are k cells and the cell probabilities are known, it follows that $v = k - 1 - l$ when the cell probabilities depend on l parameters. The preceding property enables the χ^2 test to be applied to a wide variety of problems involving the comparison of observed and expected frequencies. Some of these applications are considered in the next few sections.

6 DENSITY CURVE FITTING

If a density function has been fitted to an empirical distribution, the question of whether the fit is satisfactory naturally arises. For example, in Chapter 4 it was assumed on the basis of inspecting Fig. 4 that the length of a telephone conversation could be treated as a normal variable. Although the justification was based on mentally fitting a normal curve to the histogram of Fig. 4, such a fitting could have been carried out analytically by choosing $\mu = \bar{x}$ and $\sigma = s$ in a normal density and then graphing the resulting normal curve, adjusted to give the correct area, on Fig. 4. When a normal curve is fitted to a histogram, it is usually assumed that the data represent a sample selected at random from a normal population and that the fitted normal curve is an approximation to the population curve. Thus, the question whether a fit is satisfactory can be answered only if one knows what sort of histograms will be obtained in random samples from a normal population.

Now, the χ^2 test can be employed to give a partial answer to this question. Since the χ^2 test is concerned only with comparing sets of observed and expected frequencies, it is capable of testing only those features of the fitted distribution that affect a lack of agreement in the compared sets of frequencies. For example, the χ^2 test is not capable of distinguishing between the two curves shown in Fig. 1, in which the x axis has been divided into six intervals to give six cells for the χ^2 test and in which the areas under the two curves for each of the six intervals are equal.

With this understanding of the capabilities of the χ^2 test, consider the problem of testing the adequacy of a normal curve fit to the histogram of Fig. 4, Chapter 4. Since calculations there gave $\bar{x} = 475$ and $s = 151$, the

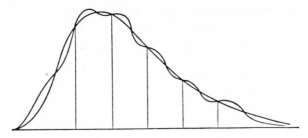

Fig. 1. Two χ^2 equivalent density functions.

equation of the normal curve that would be chosen to fit that histogram is

$$y = \frac{ce^{-\frac{1}{2}\left(\frac{x-475}{151}\right)^2}}{151\sqrt{2\pi}}$$

where c is a constant that will make the area under the curve the same as the area of the histogram. This function was integrated between successive pairs of the boundaries of the intervals of Fig. 4 to obtain expected frequencies corresponding to the observed frequencies for those intervals. This was accomplished by calculating the value of $z = (x - 475)/151$ for each such boundary point and then using Table II of the standard normal variable. Since one can work with proportions here and multiply the total number of observations, namely 1000, by each area proportion to obtain an expected frequency, it is not necessary to know the value of c in carrying out these computations.

Calculations of this type produced the following table of values corresponding to the ten intervals of Fig. 4. It will be found that the value of χ^2

n_i	6	28	88	180	247	260	133	42	11	5
e	6.4	28.0	88.6	185.5	255.1	230.3	138.0	52.3	13.3	2.2

here is 10.4. Since there are two parameters being estimated from the data and there are 10 pairs of frequencies to be compared, the number of degrees of freedom for this problem is $\nu = 9 - 2 = 7$. From Table III it will be found that the .05 critical value of χ^2 for $\nu = 7$ is $\chi_0^2 = 14.1$. Since $10.4 < 14.1$, the hypothesis that the data were obtained from sampling a normal population is substantiated, as far as compatibility of corresponding pairs of frequencies is concerned. Thus, the normal curve fitting here would be considered satisfactory from this point of view.

If a binomial distribution is fitted to an empirical frequency distribution by estimating the two parameters p and n in $f(x) = n!/[x!(n - x)!] \times p^x q^{n-x}$

from the data, the number of degrees of freedom in the χ^2 test will be $\nu = k - 1 - 2 = k - 3$ just as in normal curve fitting; however, it often happens in binomial problems that one or more of the parameters will be specified from other considerations. For example, suppose that one were interested in studying the sex distribution in families of eight children. Here $n = 8$ is known; hence it is not obtained as a maximum likelihood estimate from the data. Consequently the number of degrees of freedom would be $k - 2$. If one were to assume that $p = \frac{1}{2}$ rather than estimate p from the observations, the number of degrees of freedom would be $k - 1$.

Since the fitting of a Poisson distribution involves only the parameter μ, the χ^2 test will possess $k - 2$ or $k - 1$ degrees of freedom, depending on whether μ is replaced by $\bar{x}$ or is known from other considerations.

Property (7) of the χ^2 test requires that the unknown parameters be estimated by the method of maximum likelihood using the likelihood function $p_1{}^{n_1} p_2{}^{n_2} \cdots p_k{}^{n_k}$, corresponding to the multinomial distribution. If one calculates the maximum likelihood estimates of μ and σ from this likelihood function when fitting a normal curve to a histogram, the resulting estimates will not be exactly equal to $\bar{x}$ and s, which are the maximum likelihood estimates of μ and σ for ungrouped data; however, they will ordinarily be very nearly the same. Thus, although theoretically speaking one should calculate the maximum likelihood estimates for the multinomial situation, it usually suffices to use the well-known maximum likelihood estimates for the continuous situation.

7 CONTINGENCY TABLES

Another very useful application of the χ^2 test occurs in connection with testing the compatibility of observed and expected frequencies in two-way tables. Such two-way tables are usually called *contingency tables*. Table 1, in

TABLE 1

Marriage-Adjustment Score

		Very low	Low	High	Very high	Totals
Education	College	18 (27)	29 (39)	70 (64)	115 (102)	232
	High school	17 (13)	28 (19)	30 (32)	41 (51)	116
	Grades only	11 (6)	10 (9)	11 (14)	20 (23)	52
	Totals	46	67	111	176	400

which are recorded the frequencies corresponding to the indicated classifications for a sample of 400, is an illustration of a contingency table.

A contingency table is usually constructed for the purpose of studying the relationship between the two variables of classification. In particular, one may wish to know whether the two variables are related. By means of the χ^2 test it is possible to test the hypothesis that the two variables are independent. Thus, in connection with Table 1, the χ^2 test can be used to test the hypothesis that there is no relationship between an individual's educational level and his adjustment to marriage.

Before considering how the χ^2 test may be applied to this particular problem, consider a general contingency table containing r rows and c columns. Let p_{ij} be the probability that an individual selected at random from the population under consideration will be a member of the cell in the ith row and jth column of the contingency table. Let $p_{i.}$ be the probability that the individual will be a member of the ith row and let $p_{.j}$ be the probability that the individual will be a member of the jth column. Then the hypothesis that the two variables are independent can be written in the form

$$H_0 : p_{ij} = p_{i.}p_{.j}, \quad \begin{pmatrix} i = 1, \cdots, r \\ j = 1, \cdots, c \end{pmatrix}.$$

If a sample of n individuals is selected and n_{ij} of them are found in the cell in the ith row and jth column, then χ^2 as defined by (6) will assume the form

$$\chi^2 = \sum_{i=1}^{r} \sum_{j=1}^{c} \frac{(n_{ij} - np_{ij})^2}{np_{ij}}.$$

But under the hypothesis H_0, this expression will become

(8)
$$\chi^2 = \sum_{i=1}^{r} \sum_{j=1}^{c} \frac{(n_{ij} - np_{i.}p_{.j})^2}{np_{i.}p_{.j}}.$$

Since the $p_{i.}$ and $p_{.j}$ are unknown, it is necessary to estimate them from the sample. If the estimates are maximum likelihood estimates, the theory discussed in section 5 will permit the χ^2 test to be applied here, provided that one degree of freedom is deducted for each parameter so estimated. Since $\sum_1^r p_{i.} = 1$ and $\sum_1^c p_{.j} = 1$, there are $r - 1 + c - 1 = r + c - 2$ parameters that need to be estimated; hence the proper number of degrees of freedom for testing independence in a contingency table of r rows and c columns is given by $v = k - 1 - l = rc - 1 - (r + c - 2) = (r - 1)(c - 1)$.

In order to complete the discussion, it is necessary to find the maximum likelihood estimates of the $p_{i.}$ and $p_{.j}$. For this purpose let $n_{i.}$ denote the sum of the frequencies in the ith row and let $n_{.j}$ denote the sum of the frequencies in the jth column. Since the variables n_{ij} are discrete, the likelihood of the sample is the probability of obtaining the sample in the order in which it occurred. Thus the likelihood of the sample is given by

$$L = \prod_{i=1}^{r} \prod_{j=1}^{c} p_{ij}^{n_{ij}}.$$

But, because of H_0 and the definitions of $n_{i \cdot}$ and $n_{\cdot j}$, this will reduce to

$$L = \prod_{i=1}^{r} \prod_{j=1}^{c} (p_{i \cdot} p_{\cdot j})^{n_{ij}}$$

$$= \prod_{i=1}^{r} \prod_{j=1}^{c} p_{i \cdot}^{n_{ij}} \prod_{i=1}^{r} \prod_{j=1}^{c} p_{\cdot j}^{n_{ij}}$$

$$= \prod_{i=1}^{r} p_{i \cdot}^{\sum_{j=1}^{c} n_{ij}} \prod_{j=1}^{c} p_{\cdot j}^{\sum_{i=1}^{r} n_{ij}}$$

$$= \prod_{i=1}^{r} p_{i \cdot}^{n_{i \cdot}} \prod_{j=1}^{c} p_{\cdot j}^{n_{\cdot j}}.$$

Before differentiating L with respect to $p_{i \cdot}$ for maximizing purposes, it is necessary to express one of the $p_{i \cdot}$'s, say $p_{r \cdot}$, in terms of the remaining ones through the relation $\sum_{i=1}^{r} p_{i \cdot} = 1$. If this is done, L will assume the form

$$L = \left(1 - \sum_{1}^{r-1} p_{i \cdot}\right)^{n_{r \cdot}} \prod_{i=1}^{r-1} p_{i \cdot}^{n_{i \cdot}} \prod_{j=1}^{c} p_{\cdot j}^{n_{\cdot j}}.$$

Taking logarithms,

$$\log L = n_{r \cdot} \log \left(1 - \sum_{1}^{r-1} p_{i \cdot}\right) + \sum_{i=1}^{r-1} n_{i \cdot} \log p_{i \cdot} + K$$

where K does not involve the variable $p_{i \cdot}$. Now, differentiating with respect to $p_{i \cdot}$ and setting the derivative equal to 0 for a maximum,

$$\frac{\partial \log L}{\partial p_{i \cdot}} = -\frac{n_{r \cdot}}{1 - \sum_{1}^{r-1} p_{i \cdot}} + \frac{n_{i \cdot}}{p_{i \cdot}} = 0.$$

Since $1 - \sum_{1}^{r-1} p_{i \cdot} = p_{r \cdot}$, this equation is equivalent to

$$p_{i \cdot} = \frac{p_{r \cdot}}{n_{r \cdot}} n_{i \cdot} = c n_{i \cdot}.$$

where c does not depend on the index i. Since this must hold for $i = 1$, $2, \cdots, r$ and since

$$1 = \sum_{1}^{r} p_{i \cdot} = c \sum_{1}^{r} n_{i \cdot} = cn$$

it follows that $c = 1/n$ and that the maximum likelihood estimate of $p_i.$ is

$$\hat{p}_{i\cdot} = \frac{n_{i\cdot}}{n}.$$

As in the case of earlier problems, the value of a parameter obtained by setting the derivative of L with respect to that parameter equal to 0 and solving the equation does maximize L. In solving problems of this type, no attempt will be made to prove by second derivative tests or by other methods that the necessary conditions for a maximum also turn out to be sufficient conditions for a maximum. By symmetry the maximum likelihood estimate of $p_{\cdot j}$ is

$$\hat{p}_{\cdot j} = \frac{n_{\cdot j}}{n}.$$

If $p_{i\cdot}$ and $p_{\cdot j}$ in (8) are replaced by their maximum likelihood estimates, χ^2 will become

(9)
$$\chi^2 = \sum_{i=1}^{r} \sum_{j=1}^{c} \frac{\left(n_{ij} - \dfrac{n_{i\cdot}n_{\cdot j}}{n} \right)^2}{\dfrac{n_{i\cdot}n_{\cdot j}}{n}}.$$

According to the theory of section 5, this quantity may be treated as possessing a χ^2 distribution with $(r - 1)(c - 1)$ degrees of freedom, provided that n is sufficiently large and H_0 is true.

Now, consider the application of (9) to testing independence in Table 1. To calculate the values of the $n_{i\cdot}n_{\cdot j}/n$ in the ith row, it is merely necessary to multiply the column totals $n_{\cdot j}$ by the fraction $n_{i\cdot}/n$. Thus the values of $n_{i\cdot}n_{\cdot j}/n$ for the first row of Table 1 are obtained by multiplying the column totals by 232/400 and similarly for the remaining rows. These values, correct to the nearest integer, are inserted in parentheses in Table 1. The calculation of χ^2 is now like that for (6), with the values in parentheses treated as the e_i. It will be found that $\chi^2 = 20.7$. Since $\chi_0^2 = 12.6$ for $(3 - 1)(4 - 1) = 6$ degrees of freedom, this result is significant and the hypothesis H_0 of independence is therefore rejected. An inspection of Table 1 shows that individuals with some college education appear to adjust themselves to marriage more readily than those with less education.

8 HOMOGENEITY TESTS

A problem that differs very little from that of testing the independence of the two variables of classification in a contingency table is the problem of testing homogeneity in such a table. It arises when a set of k similar experiments is carried out and the experimenter wishes to know whether they represent independent versions of the same experiment. For example, suppose daily samples of the same size are taken of a manufactured product and the number of defectives is recorded. It would then be of interest to know whether a set of such numbers is compatible with the assumption that the probability of getting a defective is the same from day to day. Or, suppose an anthropologist wishes to compare several tribes of aborigines to determine whether they are the same with respect to blood types. Since the study of blood types uses four groupings, he would require a contingency table having four rows and as many columns as there are tribes to be tested to record his experimental results.

The essential difference between a contingency table of this type and the kind that arises in testing independence is that here the experimental outcomes are obtained a column at a time and the column totals, which represent the number of observations taken in those individual experiments, are usually fixed in advance of experimentation.

For the purpose of describing this type of problem, consider the simple case in which the same number of observations, n, is taken in each experiment, in which each observation is recorded as a success or failure, and in which k such experiments are performed. Let $x_1, \cdots, x_k$ represent the observed number of successes that occurred in those k experiments. The results of the experiments can be conveniently recorded in a $2 \times k$ contingency table as follows:

Successes	x_1	$n - x_2 -$	$\cdots$	x_k
Failures	$n - x_1$	$n - x_1$	$\cdots$	$n - x_k$

If $p_1, \cdots, p_k$ denote the probabilities of success for these k experiments, then the homogeneity hypothesis becomes the hypothesis

$$H_0 : p_1 = p_2 = \cdots = p_k .$$

This is a composite hypothesis because the common value of these probabilities is unknown; therefore a likelihood ratio test should be constructed. As in the case of the test for independence in a contingency table, λ is so complicated that it is necessary to resort to the asymptotic χ^2 distribution of $-2 \log \lambda$ in order to arrive at a feasible solution. Even then, the expression for $-2 \log \lambda$ is so messy that one becomes discouraged with the computations needed to carry out the test. Fortunately, it can be shown that the asymptotic χ^2 distribution of $-2 \log \lambda$ for this test is precisely the same as for the test of independence. Thus, one can forget about the differences in these two

TABLE 2

													Totals
Successes	19 (15)	6 (15)	9 (15)	18 (15)	15 (15)	13 (15)	14 (15)	15 (15)	16 (15)	20 (15)	22 (15)	14 (15)	181
Failures	71 (75)	84 (75)	81 (75)	72 (75)	75 (75)	77 (75)	76 (75)	75 (75)	74 (75)	70 (75)	68 (75)	76 (75)	899
Totals	90	90	90	90	90	90	90	90	90	90	90	90	1080

contingency tables and treat the homogeneity testing problem as though it were an independence testing problem. The number of degrees of freedom in the preceding problem is $k - 1$ because that is the number of degrees of freedom when testing for independence.

As an illustration of a simple homogeneity problem, consider the following data giving the number of infected plants per plot of 90 plants for 12 plots: 19, 6, 9, 18, 15, 13, 14, 15, 16, 20, 22, 14. The problem here is to determine whether it is reasonable to assume that the rate of infection is the same over the 12 plots. Formally, this is a problem of testing the hypothesis

$$H_0 : p_1 = p_2 = \cdots = p_{12}$$

where p_i denotes the probability that a plant selected at random from the 90 plants in the ith plot will be infected.

The expected frequencies listed in parenthesis in Table 2 were obtained by multiplying each column total, which is 90, by the ratio of the corresponding row sum to the total of the row sums and rounding off to the nearest integer. This is the same technique as that employed in the illustration of section 7. Calculations give $\chi^2 \doteq 18$. For 11 degrees of freedom the .05 critical value is $\chi_0^2 = 19.7$; consequently H_0 is accepted here.

Problems of testing homogeneity that involve more than two rows are solved in the same manner as those for testing independence in a contingency table.

EXERCISES

Sec. 1

1. Calculate the probability of getting a total of 5 if three dice are thrown simultaneously.

2. A game of chance consists in tossing a ball into boxes numbered 1, 2, 3, and 4. If the probabilities for landing in those boxes are $\frac{12}{25}$, $\frac{6}{25}$, $\frac{4}{25}$, and $\frac{3}{25}$, respectively, and one receives in dollars the number on the box, what is the probability of winning at least 6 dollars when taking 2 tosses?

Sec. 2

3. In a breeding experiment it was expected that ducks would be hatched in the ratio of 1 duck with a white bib to every 3 ducks without bibs. Of 80 ducks hatched, 14 had white bibs. Are these data compatible with expectation?

4. According to Mendelian inheritance offspring of a certain crossing should be colored red, black, or white in the ratios $9:3:4$. If an experiment gave 70, 36, and 38 offspring in those categories, would you claim the theory is not applicable here?

5. The number of individuals possessing the four blood types should be in the proportions $q^2 : p^2 + 2pq : r^2 + 2qr : 2pr$, where $p + q + r = 1$. Given the observed frequencies 180, 360, 132, 98, test for compatibility with $p = .4, q = .5$, and $r = .1$.

6. Toss a coin 100 times and apply the χ^2 test to see whether the coin is unbiased.

7. By integration verify the .05 critical value of χ^2 given in Table III for $\nu = 2$.

8. Use the table of random numbers to sample from the population given by

x	0	1	2
$f(x)$	.4	.4	.2

Take samples of 25 each and perform 20 (or more) such sampling experiments. For each sample of 25 calculate the value of χ^2 for observed and expected frequencies in the above three cells. Classify the 20 values of χ^2 into a frequency table and compare its histogram with the χ^2 curve with $\nu = 2$. As a class exercise this should make the χ^2 theory seem plausible.

Sec. 5

9. According to the Hardy-Weinberg formula, the number of flies of certain types resulting from certain crossings should be in the proportions $q^2:2pq:p^2$, where $p + q = 1$. If an experiment gave the frequencies 40, 50, 20 would the results be compatible with this formula (a) if $q = .5$, (b) if q is estimated from the data by using the maximum likelihood estimate

$$\hat{q} = \frac{n_1 + n_2/2}{n_1 + n_2 + n_3} ?$$

10. Prove that the estimate used in problem 9 for q is the maximum likelihood estimate based on the multinomial distribution.

11. On the basis of a given hypothesis, indicate by inspection of the χ^2 table why, if an experiment yields a value of $\chi^2 = \chi_1^2$ slightly less than the critical value for ν degrees of freedom and if the experiment is repeated with approximately the same results, the two combined experiments will yield a degree of confidence in the hypothesis different from that given by the first experiment alone.

Sec. 6

12. Apply the χ^2 test to the normal curve fit for the following 500 determinations of the width of a spectral band of light. Here e denotes the fitted normal curve frequencies obtained by estimating all the parameters.

o	5	12	43	61	105	103	89	54	19	7	2
e	5	14	36	71	102	109	85	50	21	7	2

13. Given the following data, state how many degrees of freedom you would probably use in the χ^2 test if you attempted to fit the histogram with a (a) normal function, (b) Poisson function, (c) binomial function with theory suggesting that $n = 10$

x	0	1	2	3	4	5	6	7
f	2	4	10	15	19	12	8	3

14. Apply the χ^2 test for goodness of fit to the results of problem 39, Chapter 3.

15. Apply the χ^2 test for goodness of fit to the results of problem 27, Chapter 3.

Sec. 7

16. A certain drug is claimed to be effective in curing colds. In an experiment on 164 people with colds half were given the drug and half were given sugar pills. The patients' reactions to the treatment are recorded in the following table. Test the hypothesis that the drug is no better than the sugar pills.

	Helped	Harmed	No Effect
Drug	50	10	22
Sugar	42	12	28

17. In an epidemic of a certain disease 920 children contracted the disease. Of these 400 received no treatment and of those 98 suffered after-effects. Of the remainder who did receive treatment, 162 suffered after-effects. Test the hypothesis that the treatment was not effective and comment about the paradoxical conclusion.

18. Is there any relation between the mentality and weight of criminals as judged by the following data?

Weight

Mentality	90–120	120–130	130–140	140–150	150–
Normal	21	51	94	106	124
Weak	15	18	34	15	15

19. The following data are for school children in a city in Scotland. Test to see whether hair color and eye color are independently distributed. Comment on the usefulness of a test for such large samples.

Eye \ Hair	Fair	Red	Medium	Dark	Black
Blue	1368	170	1041	398	1
Light	2577	474	2703	932	11
Medium	1390	420	3826	1842	33
Dark	454	255	1848	2506	112

20. Show that for a 2 × 2 contingency table with cell frequencies a, b, c, and d, respectively,

$$\chi^2 = \frac{(a + b + c + d)(ad - bc)^2}{(a + b)(a + c)(b + d)(c + d)}$$

21. Show that the method of Chapter 5 for testing the difference of proportions is equivalent to the χ^2 test when applied to the 2 × 2 contingency table of successes and failures. It is assumed that p is estimated from the combined sample in the difference of proportions method.

Sec. 8

22. The number of automobile accidents per week in a certain city were 12, 8, 20, 2, 14, 10, 15, 6, 9, 4. Test the homogeneity of these frequencies over this 10-week period.

23. Five boxes of different brands of canned salmon containing 24 cans each were examined for high quality specifications. The number of cans below specification were, respectively, 3, 10, 5, 3, 9. Can one conclude that the five brands are of comparable quality?

24. The following data give the number of colonies of bacteria that developed on 15 different plates from the same dilution. Is one justified in claiming that the dilution technique is satisfactory in the sense that the bacteria behave as though they were randomly distributed in the dilution? The number of colonies were, respectivley, 193, 168, 161, 153, 183, 152, 171, 156, 159, 140, 151, 152, 133, 164, 157.

Small Sample Distributions

Many of the statistical techniques considered in the preceding chapters are applicable only when large samples are available. For example, the method used in section 9.2, Chapter 5, for testing the hypothesis that two population means are equal assumes that the samples are so large that population variances may be replaced by their sample estimates without appreciably affecting the validity of the test. In this chapter methods are developed that do not require the assumption of large samples. Although such methods are called small sample methods, they obviously apply to large samples as well and might better have been called exact methods. Some small sample methods require more information or assumptions than the corresponding large sample methods; consequently, small sample techniques cannot completely displace the techniques designed for large samples.

1 DISTRIBUTION OF A FUNCTION OF A RANDOM VARIABLE

The procedure for obtaining a distribution that eliminates the necessity for estimating unknown parameters is based on a change of variable technique. The simplest of these arises when one wishes to apply normal distribution theory to a problem but discovers that the random variable X involved does not possess a normal distribution. The question then arises, is it possible to find a change of variable, say $Y = h(X)$, such that the density function of Y will be approximately normal? If one thinks of what this means geometrically, one would surmise that the answer is yes. In this connection, compare the graph of a particular density function, $f(x)$, given in Fig. 1, with that of a standard normal variable, $g(y)$, shown in the same sketch.

To any value of x, say x_0, one can find a corresponding value of y, denoted by y_0, such that the areas to the left of these values under the corresponding

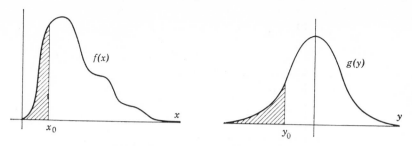

Fig. 1.   Graphs of two density functions.

curves will be equal. If one chooses a large number of values of x and obtains the corresponding values of y by means of Table II, then this set of x and y values will yield a functional relationship which may serve as a change of variable that transforms the non-normal $f(x)$ into the normal $g(y)$. This relationship is a numerical one and therefore is only an approximation of the complete relationship. If the complete relationship $y = h(x)$ were known, one could transform any x value over to its corresponding y value and treat it as an observation taken from a standard normal variable population.

One can reverse the preceding process by starting with a given density function $f(x)$ and a given change of variable $Y = h(X)$ and ask for the density function $g(y)$ of the new variable. If $g(y)$ should turn out to be a normal density, or approximately so, then the transformation $Y = h(X)$ would have accomplished the desired objective. Now it is relatively easy to find $g(y)$ if the function $h(x)$ in the transformation is an increasing function of x, or a decreasing function, throughout the range of x values. The technique for doing this, which is now demonstrated, is based upon finding the distribution function of Y.

It follows from formula (31), Chapter 2, that the distribution function of Y, which is denoted by $G(y)$, satisfies the relations

(1) $$G(t) = P\{Y \le t\} = P\{h(X) \le t\}$$

where t is any desired value. Now the inequality $h(X) \le t$ can be expressed as an inequality on X. The relationship between y and x where $h(x)$ is an increasing function is like that shown in Fig. 2. For such a relationship there is a unique value of x to each value of y. Here the value of x corresponding to the value t for y has been denoted by τ; consequently, since $h(x) \le t$ if, and only if, $x \le \tau$,

$$P\{h(X) \le t\} = P\{X \le \tau\} = \int_{-\infty}^{\tau} f(x)\, dx .$$

Thus from (1)

$$G(t) = \int_{-\infty}^{\tau} f(x)\, dx .$$

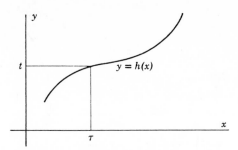

Fig. 2.   The graph of an increasing function.

Now, as shown in Chapter 2, a continuous density function can be obtained by differentiating the corresponding distribution function. In view of the fact that τ is a function of t, it follows from the calculus formula for differentiating an integral with respect to its upper limit that

(2) $$\frac{dG(t)}{dt} = \frac{dG(t)}{d\tau}\frac{d\tau}{dt} = f(\tau)\frac{d\tau}{dt}.$$

This formula is valid at any point τ where $f(\tau)$ is continuous. It will be assumed here that $f(x)$ is a continuous function of x. Since t and τ were any pair of corresponding values of y and x, respectively, and were introduced to keep from confusing upper-limit variables with dummy variables of integration, this relationship may be rewritten

$$\frac{dG(y)}{dy} = f(x)\frac{dx}{dy}.$$

But in view of the relationship between a distribution function and its density function, the left side is the density function of Y; hence the desired formula is

$$g(y) = f(x)\frac{dx}{dy}.$$

If one follows through this derivation for a change of variable $Y = h(X)$ in which $h(x)$ is a decreasing function, he will obtain the negative of this result. Since dx/dy will be negative in this case, a formula that is valid for both cases will be given by

(3) $$g(y) = f(x)\left|\frac{dx}{dy}\right|.$$

Before this formula can be applied, it is necessary to replace x in $f(x)$ by its value in terms of y, which means that it is necessary to solve the relation $y = h(x)$ for x in terms of y. One can calculate dx/dy from this inverse relationship, or else calculate dy/dx from the original relationship $y = h(x)$ and take its reciprocal.

Because of the importance and usefulness of this formula for later work, the result of this derivation is expressed formally.

(4) CHANGE OF VARIABLE TECHNIQUE: *If $y = h(x)$ is an increasing or decreasing function and $f(x)$ is the density function of X then $g(y)$, the density function of Y, is given by the formula*

$$g(y) = f(x) \left| \frac{dx}{dy} \right|$$

in which x is to be replaced by its value in terms of y by means of the relation $y = h(x)$.

Although formula (4) will not be valid unless $h(x)$ is either an increasing or a decreasing function, the procedure used to derive the formula can be applied to more complicated problems.

As an illustration of the use of formula (4), consider the problem of finding the density function of Y if $Y = \sqrt{X}$ and $f(x) = e^{-x}$, $x \geq 0$. Since $y = \sqrt{x}$ is an increasing function of x, formula (4) may be applied without the absolute value signs. The inverse relationship is $x = y^2$; hence

$$g(y) = e^{-y^2} \frac{dx}{dy} = 2ye^{-y^2}, \qquad y > 0.$$

The relationship between these two density functions is shown geometrically in Fig. 3. Incidentally, it will be observed that $g(y)$ is considerably more like a normal curve in appearance than is $f(x)$.

As a second illustration, consider the problem of finding the density function of the kinetic energy $Z = \frac{1}{2}mV^2$, given the distribution of the velocity V. The letter Z is used in place of E to avoid confusing an observed value of E with exponential e. The density function of V, the velocity for a gas molecule with mass m, is given by

$$f(v) = av^2 e^{-bv^2}$$

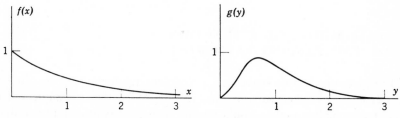

Fig. 3. Distribution of X and $Y = X^{\frac{1}{2}}$ for $f(x) = e^{-x}$, $x \geq 0$.

where $v > 0$, b is a constant depending on the gas, and a is determined to yield unit area. Since $z = \frac{1}{2}mv^2$ is an increasing function of v for positive values of v, formula (4) may be applied by choosing $X = V$ and $Y = Z$. Here the inverse relationship is $v = \sqrt{2z/m}$; therefore

$$g(z) = a\frac{2z}{m}e^{-\frac{b2z}{m}}\frac{1}{\sqrt{2mz}}$$

or

$$g(z) = \alpha z e^{-\beta z}, \qquad z > 0$$

where α and β are constants depending upon a, b, and m.

As an illustration of a problem which cannot be solved by a direct application of formula (4), consider the problem of finding the density of $Y = X^2$ where X is a standard normal variable. As in the derivation that led to (4), one first calculates the distribution function of Y. Thus,

$$G(y) = P\{Y \le y\} = P\{X^2 \le y\} = P\{-\sqrt{y} \le X \le \sqrt{y}\}$$

$$= \int_{-\sqrt{y}}^{\sqrt{y}} \frac{e^{-\frac{x^2}{2}}}{\sqrt{2\pi}}\,dx = 2\int_0^{\sqrt{y}} \frac{e^{-\frac{x^2}{2}}}{\sqrt{2\pi}}\,dx\ .$$

To obtain the density function of Y it is necessary to calculate $G'(y)$. This is done as in (2) by letting $\tau = \sqrt{y}$. Thus,

$$g(y) = \frac{dG(y)}{dy} = \frac{dG(y)}{d\tau}\frac{d\tau}{dy} = 2\frac{e^{-\frac{\tau^2}{2}}}{\sqrt{2\pi}}\frac{1}{2\sqrt{y}} = \frac{y^{-\frac{1}{2}}e^{-\frac{y}{2}}}{\sqrt{2\pi}}, \qquad y > 0\ .$$

Since $\Gamma(\frac{1}{2}) = \sqrt{\pi}$, a comparison of this density with the chi-square densities defined by (39), Chapter 3, shows that Y possess a chi-square distribution with one degree of freedom. This result was obtained in Theorem 4, Chapter 5, by means of moment generating function techniques. It is to be noted that $y = x^2$ is not an increasing or decreasing function of x over its entire range of values and therefore that the problem cannot be solved by a direct application of formula (4).

2   DISTRIBUTION OF A FUNCTION OF TWO RANDOM VARIABLES

The preceding section developed a technique for finding the distribution of a function of a random variable. In this section that technique will be extended to finding the distribution of a function of two random variables. Such functions are needed for developing small sample methods.

Let X and Y be two continuous variables with the density function $f(x, y)$ and consider the problem of finding the density function of the variable $Z = t(X, Y)$, where t is some function of interest. The particular functions that are of interest in this chapter are $t(X, Y) = Y - X$ and $t(X, Y) = Y/X$; however, it is desirable to have available a general method of attack for such problems.

One method of approach is to adapt the change of variable technique of the preceding section to functions of two variables by holding one of the variables fixed. Toward this end, suppose the value of X is fixed so that the relation $Z = t(X, Y)$ becomes a relation between the random variables Z and Y only. Assume that $t(X, Y)$ is an increasing, or decreasing, function of Y. Then, for X fixed, the relation $Z = t(X, Y)$ represents a change of variable from Y to Z to which formula (4) applies. If $g(y \mid x)$ and $k(z \mid x)$ are used to denote the conditional density functions of Y and Z respectively, for X fixed, then by that formula

(5)
$$k(z \mid x) = \frac{g(y \mid x)}{\left| \dfrac{\partial z}{\partial y} \right|} .$$

Next, write $f(x, y)$ in the factored form

$$f(x, y) = f(x)g(y \mid x) .$$

Similarly, if $h(x, z)$ denotes the joint density function of X and Z one can write

$$h(x, z) = f(x)k(z \mid x) .$$

Taking the ratio of these two joint density functions and using (5) will then yield the formula

(6)
$$h(x, z) = \frac{f(x, y)}{\left| \dfrac{\partial z}{\partial y} \right|} .$$

In this formula it is necessary to replace y by its value in terms of x and z by means of the relation $z = t(x, y)$.

Formula (6) gives the joint density function of X and Z in terms of that of X and Y. In order to obtain the density function of Z, it is therefore merely necessary to integrate $h(x, z)$ with respect to x over the entire range of x values for z fixed. This follows from formula (3), Chapter 6, for marginal distributions.

As an application of this technique, consider the problem of finding the density function of the ratio $Z = Y/X$ when X and Y are independently distributed. Since $f(x, y) = f(x)g(y)$ and $\partial z/\partial y = 1/x$ here, it follows directly from formula (6) that

$$h(x, z) = |x| f(x) g(y)$$
$$= |x| f(x) g(zx) .$$

The density function of Z, say $q(z)$, is therefore given by

$$(7) \qquad q(z) = \int |x| f(x) g(zx) \, dx$$

where the integration is over the range of x values for z fixed.

As a special case of (7), let $f(x) = e^{-x}$, $x > 0$, and $g(y) = e^{-y}$, $y > 0$.
Then (7) yields

$$q(z) = \int_0^\infty x e^{-x} e^{-xz} \, dx .$$

The substitution $w = x(1 + z)$ will lead to the result

$$q(z) = (1 + z)^{-2}, \qquad z > 0 .$$

As a second application of this general technique, consider the problem of finding the density function of the difference $Z = Y - X$. Here $\partial z / \partial y = 1$; consequently (6) reduces to

$$h(x, z) = f(x, y) = f(x, x + z) .$$

The density function of Z is therefore given by

$$(8) \qquad q(z) = \int f(x, x + z) \, dx$$

where the integration is over the range of x values for z fixed.

The only real difficulty in finding the density function of a variable $Z = t(X, Y)$ by means of the preceding technique lies in selecting the proper limits of integration when integrating the function $h(x, z)$ with respect to x. The following problem illustrates the nature of such difficulties.

Let $f(x, y) = 8xy$, $0 < x < 1$, $0 < y < x$, and let $Z = X + Y$. Then $\partial z / \partial y = 1$ and (6) reduces to

$$h(x, z) = f(x, y) = f(x, z - x) = 8x(z - x) .$$

Now when z is fixed, x can range over only those values that correspond to points of the sample space lying on the line whose equation is $x + y = z$ and whose graph is shown in Fig. 4. The sample space here is the triangle bounded by the lines $y = x$, $x = 1$, and $y = 0$. If z is fixed at any value satisfying $z < 1$, as indicated by line l_1 in Fig. 4, then the range of possible x values is $x = z/2$ to $x = z$. However, if $z > 1$, as shown by line l_2, then the range is $x = z/2$ to $x = 1$. As a consequence, the density function of Z is

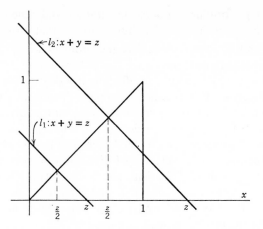

Fig. 4. Sample space corresponding to $Z = X + Y$.

given by the two formulas

$$k(z) = \int_{\frac{z}{2}}^{z} 8x(z - x)\,dx = \tfrac{2}{3}z^3, \qquad 0 < z \leq 1$$

$$= \int_{\frac{z}{2}}^{1} 8x(z - x)\,dx = -\tfrac{2}{3}z^3 + 4z - \tfrac{8}{3}, \qquad 1 < z < 2\,.$$

The graph of this density function is shown in Fig. 5.

A somewhat more general problem arises when the joint distribution of two functions of the basic variables, say $u = u(X, Y)$ and $v = v(X, Y)$, is desired. This corresponds to a change from the coordinate system x, y to the coordinate system u, v. This is a familiar procedure in calculus, for

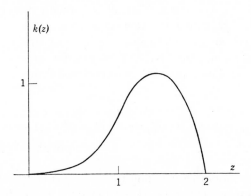

Fig. 5. Graph of $k(z)$ for $Z = X + Y$.

example, when performing a double integration and accomplishing it by shifting to polar coordinates. The functions in that case are given by $r = \sqrt{x^2 + y^2}$ and $\theta = \tan^{-1} y/x$. Here the problem would be to find out how r and θ are distributed when given the distribution of X and Y.

There exists a simple formula for finding the density function of such transformed variables. It is obtained by applying probability considerations to an advanced calculus formula for integration and involves the Jacobian function. This formula can be extended to any number of variables as well. The theory that is developed in this book does not require the use of these more general methods; however, a brief discussion of them is given in the appendix for the benefit of those who are familiar with advanced calculus methods and wish to become acquainted with the general methods.

The methods that have been explained in this section are now used to develop some of the theory of small samples.

3 THE χ^2 DISTRIBUTION

The chi-square distribution arose in Chapter 5 in connection with the sample variance and its use in making inferences about σ^2 for a normal variable, and it occurred in Chapter 8 in connection with the asymptotic distribution of a likelihood ratio test variable. In this section its role with respect to the first of these two types of problems will be extended further by showing that it is capable of handling inference problems with respect to σ^2 without the necessity of knowing the value of μ, which was required in the earlier treatment.

In solving some of the inference problems it is necessary to use an additive property of independent chi-square variables. Although this property is implicit in Theorem 4, Chapter 5, it will be formally demonstrated here by means of moment generating function techniques.

Let χ_1^2 and χ_2^2 possess independent chi-square distributions with ν_1 and ν_2 degrees of freedom, respectively. Consider the variable $W = \chi_1^2 + \chi_2^2$. From moment generating function properties and formula (40), Chapter 3, it follows that

$$M_W(\theta) = M_{\chi_1^2}(\theta)M_{\chi_2^2}(\theta) = (1 - 2\theta)^{-\frac{\nu_1}{2}}(1 - 2\theta)^{-\frac{\nu_2}{2}}$$
$$= (1 - 2\theta)^{-\frac{1}{2}(\nu_1 + \nu_2)} .$$

Since this is the moment generating function of a chi-square variable, the uniqueness theorem relating generating functions and distributions implies the following result.

THEOREM 1: *If χ_1^2 and χ_2^2 possess independent chi-square distributions with ν_1 and ν_2 degrees of freedom, respectively, then $\chi_1^2 + \chi_2^2$ will possess a chi-square distribution with $\nu_1 + \nu_2$ degrees of freedom.*

3.1 Distribution of S^2

Let X be normally distributed with mean μ and variance σ^2 and let $\bar{X}$ and S^2 be sample estimates of those parameters based on a random sample of size n.

Now if the mean μ were known, one could use the estimate $\sum (X_i - \mu)^2/n$ and Theorem 4, Chapter 5, to carry out inferences regarding σ^2. Since μ is not known here it is necessary to use $S^2 = \sum (X_i - \bar{X})^2/n$ for this purpose. This in turn can be done only if the distribution of the random variable S^2 can be found. The difficulty in finding the distribution of S^2 arises from the presence of $\bar{X}$ in place of μ, which prevents the terms in the sum of squares in S^2 from being independent variables. In order to make allowance for this substitution of $\bar{X}$ for μ, it is necessary to carry out certain algebraic manipulations. After these have been made, it can be shown by moment generating function methods that the chi-square distribution is still applicable.

Obvious algebraic operations will show that

$$nS^2 = \sum (X_i - \bar{X})^2 = \sum [(X_i - \mu) - (\bar{X} - \mu)]^2$$
$$= \sum (X_i - \mu)^2 - n(\bar{X} - \mu)^2 .$$

Because of the convenience of working with standard units, this relationship is divided by σ^2 and then written in the form

$$\frac{nS^2}{\sigma^2} + \left(\frac{\bar{X} - \mu}{\sigma/\sqrt{n}}\right)^2 = \sum \left(\frac{X_i - \mu}{\sigma}\right)^2$$

or symbolically as

$$J + K = L .$$

If the moment generating function of both sides is taken,

(9) $$M_{J+K}(\theta) = M_L(\theta) .$$

Now it can be shown that $\bar{X}$ and S^2 are independently distributed when the basic variable X is normally distributed. A proof of this property is given in the appendix. This fact is therefore assumed here. Since J is a function of S^2 and K is a function of $\bar{X}$, it follows from the independence of $\bar{X}$ and S^2 that J and K are independently distributed. The independence of J and K permits the left side of (9) to be factored; therefore (9) may be written in the form

$$M_J(\theta)M_K(\theta) = M_L(\theta) .$$

Since S^2 is the variable of interest here, this relationship is written in the form

$$(10) \qquad M_J(\theta) = \frac{M_L(\theta)}{M_K(\theta)}.$$

From Theorem 4, Chapter 5, it follows that L possesses a χ^2 distribution with n degrees of freedom. Now the variable $(\bar{X} - \mu)\sqrt{n}/\sigma$ is a normal variable with zero mean and unit variance; therefore it constitutes a random sample of size 1 from such a variable. The same theorem therefore shows that K possesses a χ^2 distribution with one degree of freedom.

The moment generating function of a χ^2 variable with ν degrees of freedom as given by formula (40), Chapter 3, is

$$(11) \qquad M_{\chi^2}(\theta) = (1 - 2\theta)^{-\frac{\nu}{2}}.$$

Application of this formula to (10) will yield

$$M_J(\theta) = \frac{(1 - 2\theta)^{-\frac{n}{2}}}{(1 - 2\theta)^{-\frac{1}{2}}} = (1 - 2\theta)^{-\frac{1}{2}(n-1)}.$$

Because a distribution is uniquely determined by its moment generating function, this result together with (11) proves the following theorem.

THEOREM 2: *If X is normally distributed with variance σ^2 and*

$$S^2 = \sum_{i=1}^{n}(X_i - \bar{X})^2/n$$

is the sample variance based on a random sample of size n, then nS^2/σ^2 has a χ^2 distribution with $n - 1$ degrees of freedom.

Although the name "degrees of freedom" is merely a name given to the parameter ν in the χ^2 distribution, it is well chosen because the parameter ν represents the number of independent variables whose sum of squares is a χ^2 variable. Thus L has $\nu = n$ because the n variables being squared and summed are independent, whereas the n variables being squared and summed in S^2 contain only $n - 1$ independent variables because the sum of the variables is 0.

3.2 Applications of the χ^2 Distribution

In this section Theorems 1 and 2 are used to test hypotheses about, and obtain confidence limits for, the variance of a normal variable.

As a first illustration, consider a problem of testing a hypothetical value of σ. If past experience with the quality of a manufactured product has shown that $\sigma = 7.5$ for the quality variable in question, and if the latest sample of size 25 gave a value of $s = 10$, would there be justification for believing that the variability of the quality had increased? This problem may be treated as a problem of testing the hypothesis

$$H_0 : \sigma = 7.5$$

against the alternative hypothesis

$$H_1 : \sigma > 7.5 \ .$$

From Theorem 2, nS^2/σ^2 possesses a χ^2 distribution with 24 degrees of freedom. If the right tail of the χ^2 distribution is chosen as the critical region for testing H_0 against H_1, it will be found from Table III that the critical value of χ^2 is given by $\chi_0^2 = 36.4$. Since here

$$\frac{nS^2}{\sigma^2} = \frac{25 \cdot 10^2}{(7.5)^2} = 44$$

the hypothesis H_0 is rejected in favor of H_1, which implies that there is justification for believing that the variability has increased.

The solution of one of the problems of Chapter 8 shows that the right tail of the χ^2 distribution is the best critical region for testing H_0 against H_1, provided that the mean of X is 0. If the mean were μ rather than 0, one would use $\sum (x_i - \mu)^2 > c$ in place of $\sum x_i^2 > c$ to define the best critical region. When the mean is not known, as in the problem just solved, it can be shown by methods somewhat more complicated than those used to solve that problem that $\sum (x_i - \bar{x})^2 > c$, where c is chosen properly, defines a restricted type of best critical region for the problem being discussed. Since $ns^2/\sigma^2 > \chi_0^2$ is equivalent to $\sum (x_i - \bar{x})^2 > c$, where $c = \sigma^2 \chi_0^2$, it follows that the test employed in solving this problem is a restricted type of best test.

If one were to test the hypothesis $H_0 : \sigma = \sigma_0$ against the alternative $H_1 : \sigma < \sigma_0$, one would use the left tail of the χ^2 distribution to obtain a best test; however, if the problem were one of testing $H_0 : \sigma = \sigma_0$ against $H_1 : \sigma \neq \sigma_0$, then methods like those of Chapter 8 will show that there does not exist a best critical region in this case. For this last type of alternative it is customary to use the two equal tails of the χ^2 distribution as the critical region.

As a second application of the χ^2 distribution, consider the problem of finding confidence limits for σ^2. Let X be normally distributed with variance σ^2, and let S^2 be the sample variance based on a random sample of size n. Then 95 percent confidence limits for σ^2 may be obtained by using the analytical methods explained in section 2.4, Chapter 8, in the following manner.

From Table III for $n - 1$ degrees of freedom find two values of χ^2, namely, χ_1^2 and χ_2^2, such that the probability is .975 that $\chi^2 > \chi_1^2$ and such that the probability is .025 that $\chi^2 > \chi_2^2$. Then it follows from Theorem 2 that the probability is .95 that

$$\chi_1^{\,2} < \frac{nS^2}{\sigma^2} < \chi_2^{\,2}$$

or that

(12) $$\frac{nS^2}{\chi_2^{\,2}} < \sigma^2 < \frac{nS^2}{\chi_1^{\,2}} \, .$$

These two numbers yield 95 percent confidence limits for σ^2. From the discussion in the section on confidence intervals it follows that in the long run 95 percent of the inequalities of this type that are computed will be true inequalities. This method, of course, is not restricted to 95 percent limits.

As a numerical illustration of the use of formula (12), consider once more the data for the first illustration of this section. Since the hypothetical value of $\sigma = 7.5$ was rejected, one would use the sample value $s^2 = 100$, or the unbiased version of it, as the point estimate of σ^2; however, if one were interested in an interval estimate, (12) would be used. Here $n = 25$ and $ns^2 = 2500$. A direct application of (12) and Table III will show that 96 percent confidence limits for σ^2 are given by

$$\frac{2500}{40.27} < \sigma^2 < \frac{2500}{11.99} \, .$$

This inequality is equivalent to

$$62 < \sigma^2 < 209 \, .$$

It is clear from this result that σ^2 cannot be estimated with much precision for such a small sample and such variable data.

As a third illustration, consider the problem of finding confidence limits for σ^2 when several sample variances are available. In particular, suppose the following sample values based on samples of size 5 each were obtained: $s_1^2 = 237$, $s_2^2 = 320$, $s_3^2 = 853$, $s_4^2 = 296$, and $s_5^2 = 141$. Since each of these variances is based on a random sample of size 5, Theorem 2 shows that the variables $n_i S_i^2/\sigma^2$ ($i = 1, \cdots, 5$), will possess independent χ^2 distributions with four degrees of freedom each. By Theorem 1 their sum, $\sum n_i S_i^2/\sigma^2$, will therefore possess a χ^2 distribution with 20 degrees of freedom. Since $\sum n_i s_i^2 = 9235$, formula (12) and Table III will then yield the following 96 percent confidence limits for σ^2:

$$\frac{9235}{35.02} < \sigma^2 < \frac{9235}{9.237}$$

or

$$264 < \sigma^2 < 1000 \, .$$

For data of the type just considered, the technique of combining several sample variances to obtain an estimate of σ^2 has certain advantages over the customary method of combining all the data to obtain a single direct estimate of σ^2. In the problem considered it may be that the variability of the product is unchanged from day to day but that the mean has changed. If all the data were combined, the change in the mean would tend to increase the value of S^2 over what it would be if the mean were stable from day to day. The sum of the daily values of S^2, however, would not be affected by such changes in the mean. Thus, by using the sums of daily variances, one may be able to obtain a valid estimate of σ^2, even though the product is not strictly under control. Here σ^2 is understood to be the population variance of the product when shifts in the mean do not occur.

The preceding methods for making inferences on σ^2 are valid only if the basic variable X possesses a normal distribution. If, for example, the fourth moment of X differs considerably from that of a normal variable, the distribution of nS^2/σ^2 may differ considerably from that of a chi-square variable and the foregoing inference techniques should not be applied. Under such circumstances it is better to apply large sample methods based on the central limit theorem, or to use a nonparametric technique, some of which will be discussed in Chapter 12.

4   STUDENT'S *t* DISTRIBUTION

Consider the data of Table 1 on the additional hours of sleep gained by 10 patients in an experiment with a certain drug. The problem is to determine whether these data justify the claim that the drug does produce additional sleep.

Assume that these patients may be treated as a random sample of size 10 from a population of such patients. Furthermore, assume that the number of additional hours of sleep that a patient obtains from the use of this drug

TABLE 1

Patient	1	2	3	4	5	6	7	8	9	10
Hours gained	0.7	−1.1	−0.2	1.2	0.1	3.4	3.7	0.8	1.8	2.0

is a normally distributed variable. The problem may then be treated as a problem of testing the hypothesis

$$H_0 : \mu = 0$$

against the alternative

$$H_1 : \mu > 0 .$$

If this problem were treated in the traditional large-sample manner of Chapter 5, the experimenter would use the data of Table 1 to obtain

$$\bar{x} = 1.24 \quad \text{and} \quad s = 1.45 .$$

Then he would calculate the value of the standard normal variable

$$Z = \frac{\bar{X} - \mu}{\sigma_{\bar{x}}} = \frac{\bar{X} - 0}{\sigma} \sqrt{n}$$

and approximate its value by replacing σ by s to obtain the numerical value

$$z \doteq \frac{1.24\sqrt{10}}{1.45} = 2.70 .$$

From Table II, the probability of obtaining a value of a standard normal variable exceeding 2.70 is .0035; consequently the hypothesis that $\mu = 0$ would be rejected here in favor of the alternative that $\mu > 0$. The drug undoubtedly has a beneficial effect with respect to sleep, even though it may be due to psychological factors affecting the patient.

This method of solving the problem is subject to one serious objection. For a sample as small as this, the sample standard deviation, S, will not be an accurate estimate of σ; consequently a serious error may be introduced in the value of Z in replacing σ by its sample estimate. In most applied problems the true standard deviation is unknown. In order to overcome this defect in the test, it is necessary to replace the random variable Z by a new random variable which involves the sample standard deviation rather than the population standard deviation. Such considerations will lead to what is known as *Student's t distribution*.

Although the t distribution is being introduced here to solve a particular problem, it has many other important applications. In its most general form a Student t variable is a variable of the type

(13)
$$T = \frac{U\sqrt{\nu}}{V}$$

where U is a standard normal variable and V^2 is a χ^2 variable with ν degrees of freedom distributed independently of U.

The density function of T can be obtained by finding the density functions of the numerator and denominator of T and then applying formula (7).

The numerator variable $U\sqrt{\nu}$, which is denoted by Y, is a normal variable with mean zero and variance ν because U is a standard normal variable; consequently the density function of Y, which is denoted by $k(y)$, is given by

$$(14) \qquad k(y) = \frac{e^{-\frac{y^2}{2\nu}}}{\sqrt{2\pi\nu}} .$$

The denominator variable V is the square root of a χ^2 variable; therefore its distribution can be found by using formula (4). Toward this end, let the variables x and y in that formula be set equal to $x = v^2$ and $y = v$. Then the required change of variable is given by the relationship $y = \sqrt{x}$. Application of (4) then yields

$$g(v) = f(v^2)2v .$$

But V^2 is a χ^2 variable with ν degrees of freedom whose density function is given by (39), Chapter 3; consequently

$$(15) \qquad g(v) = a(v^2)^{\frac{\nu}{2}-1} e^{-\frac{v^2}{2}} \cdot 2v$$

$$= 2av^{\nu-1} e^{-\frac{v^2}{2}} .$$

Here a is the χ^2 distribution constant $1/2^{\nu/2}\Gamma(\nu/2)$.

In order to apply formula (7) to (13), it is necessary to associate the variable v with x and the variable $u\sqrt{\nu}$ with y. The function $f(x)$ of (7) is therefore given by replacing v by x in (15). The function $g(y)$ of (7) is given by $k(y)$ in (14). Finally, it is necessary to associate the variable T with the variable Z. After these substitutions in notation have been made, formula (7) when applied to (13) will yield

$$q(t) = \int_0^\infty x \cdot 2ax^{\nu-1}e^{-\frac{x^2}{2}} \cdot \frac{e^{-\frac{(tx)^2}{2\nu}}}{\sqrt{2\pi\nu}} dx$$

$$= \frac{2a}{\sqrt{2\pi\nu}} \int_0^\infty x^\nu e^{-\frac{1}{2}x^2\left(1+\frac{t^2}{\nu}\right)} dx .$$

Now let $w = x^2(1 + t^2/\nu)/2$; then $dx = dw/\sqrt{2w}\sqrt{1 + t^2/\nu}$ and

$$q(t) = \frac{2^{\frac{\nu}{2}}a}{\sqrt{\pi\nu}}\left(1 + \frac{t^2}{\nu}\right)^{-\frac{1}{2}(\nu+1)} \int_0^\infty w^{\frac{1}{2}(\nu+1)} e^{-w} dw .$$

From (35), Chapter 3, it will be observed that this last integral is equal to $\Gamma[(\nu + 1)/2]$; consequently

$$q(t) = c\left(1 + \frac{t^2}{\nu}\right)^{-\frac{1}{2}(\nu+1)}$$

where c is the constant

(16)
$$c = \frac{2^{\frac{\nu}{2}} a}{\sqrt{\pi\nu}} \Gamma\left(\frac{\nu + 1}{2}\right) = \frac{\Gamma\left(\dfrac{\nu + 1}{2}\right)}{\sqrt{\pi\nu}\,\Gamma\left(\dfrac{\nu}{2}\right)}.$$

The preceding derivation proves the following theorem.

THEOREM 3: *If U is normally distributed with zero mean and unit variance and V^2 has a χ^2 distribution with ν degrees of freedom, and U and V are independently distributed, then the variable*

$$T = \frac{U\sqrt{\nu}}{V}$$

has a Student's t distribution with ν degrees of freedom given by

$$f(t) = c\left(1 + \frac{t^2}{\nu}\right)^{-\frac{1}{2}(\nu+1)}$$

where c is the constant given in (16).

Now consider once more the problem that was introduced at the beginning of this section in order to see how this theorem can remedy the defect in the large sample method of solution. Since X is normally distributed with 0 mean, the variable

$$U = \frac{\bar{X}}{\sigma_{\bar{x}}} = \frac{\bar{X}\sqrt{n}}{\sigma}$$

possesses the properties of U in Theorem 3. From Theorem 2 it follows that

$$V^2 = \frac{nS^2}{\sigma^2}$$

possesses the properties of V^2 in Theorem 3 with $\nu = n - 1$. Since it is known that $\bar{X}$ and S^2 are independently distributed, Theorem 3 may be applied to give the numerical value

$$t = \frac{\bar{x}\sqrt{n - 1}}{s} = \frac{1.24\sqrt{9}}{1.45} = 2.57, \qquad \nu = 9.$$

From Table IV it will be found that the probability is approximately .017

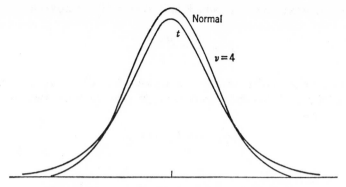

Fig. 6. Standard normal and Student's t distributions.

of obtaining a value of $T > 2.57$. This result is therefore significant at the 5 percent significance level.

A comparison of the probability of $P = .017$ with that of $P = .0035$ obtained by the use of large sample methods shows that the large sample method is not accurate for a sample as small as 10. It will be found that the large sample method gives probabilities that are consistently too small; consequently large sample methods will claim significant results more often than is justified. The explanation for this bias on the part of large sample methods is that the t distribution has a slightly larger dispersion than the standard normal distribution. The situation is shown graphically in Fig. 6, which gives the graphs of the standard normal distribution and Student's t distribution for four degrees of freedom.

The important feature of the t distribution is that it does not depend on any unknown population parameters, hence there is no necessity for replacing parameter values by questionable sample estimates as there is in the large sample normal curve method.

5 APPLICATIONS OF THE t DISTRIBUTION

5.1 Confidence Limits for a Mean

Let X be normally distributed with mean μ and variance σ^2. Let $\bar{X}$ and S^2 be their sample estimates based on a random sample of size n. Then, as before,

$$U = \frac{\bar{X} - \mu}{\sigma/\sqrt{n}}$$

and

$$V^2 = \frac{nS^2}{\sigma^2}$$

satisfy the requirements of U and V in Theorem 3; consequently,

$$(17) \qquad T = \frac{(\bar{X} - \mu)\sqrt{n - 1}}{S}$$

possesses a t distribution with $n - 1$ degrees of freedom. If $t_{.05}$ represents the value of T such that the probability is .05 that $|T| > t_{.05}$, then the probability is .95 that

$$\left| \frac{(\bar{X} - \mu)\sqrt{n - 1}}{S} \right| < t_{.05}$$

or that

$$(18) \qquad \bar{X} - t_{.05} \frac{S}{\sqrt{n - 1}} < \mu < \bar{X} + t_{.05} \frac{S}{\sqrt{n - 1}}.$$

This inequality determines a 95 percent confidence interval for μ. Since the probabilities heading the columns of Table IV are for one tail only, it is necessary to look in the column headed .025 in order to find the value of $t_{.05}$ needed in (18). If some probability other than .95 is desired, it is merely necessary to replace $t_{.05}$ by the corresponding value of t from Table IV, once more looking in the column headed with half the probability attached to t. The entries in the last row of Table IV, which are those for a standard normal variable, enable one to observe how rapidly Student's t distribution approaches that of a standard normal variable as the sample size increases. They also enable one to select the correct column in looking up critical values of t because of familiarity with large sample normal curve critical values such as 1.64 and 1.96.

5.2 Difference of Two Means

The t distribution may be used to eliminate the error in large sample methods when testing the difference of two means in the same manner as for testing one mean. Let X and Y be normally distributed with means μ_X and μ_Y and with the same variance σ^2. Let random samples of sizes n_X and n_Y be taken from these two populations. Denote the sample means and variances by $\bar{X}$, $\bar{Y}$, S_X^2, and S_Y^2. Then

$$U = \frac{(\bar{X} - \bar{Y}) - (\mu_X - \mu_Y)}{\sigma_{\bar{X} - \bar{Y}}}$$

$$= \frac{(\bar{X} - \bar{Y}) - (\mu_X - \mu_Y)}{\sigma \sqrt{\dfrac{1}{n_X} + \dfrac{1}{n_Y}}}$$

will possess the required properties of U in Theorem 3. Furthermore

$$V^2 = \frac{n_X S_X^2 + n_Y S_Y^2}{\sigma^2}$$

with $v = n_X + n_Y - 2$ degrees of freedom, is easily seen to possess the properties of V^2 in Theorem 3. This follows from Theorems 1 and 2 because $n_X S_X^2/\sigma^2$ and $n_Y S_Y^2/\sigma^2$ possess independent χ^2 distributions with $n_X - 1$ and $n_Y - 1$ degrees of freedom, respectively. Consequently

(19)
$$T = \frac{(\bar{X} - \bar{Y}) - (\mu_X - \mu_Y)}{\sqrt{n_X S_X^2 + n_Y S_Y^2}} \sqrt{\frac{n_X n_Y (n_X + n_Y - 2)}{n_X + n_Y}},$$
$$v = n_X + n_Y - 2$$

will have Student's t distribution with $n_X + n_Y - 2$ degrees of freedom. Then, to test the hypothesis that $\mu_X = \mu_Y$, it is merely necessary to calculate the value of T and use Table IV to see whether the sample value of T numerically exceeds the critical value.

It will be noted that the value of T does not depend on any population parameters as in the large sample method explained in section 9.2, Chapter 5. It will also be noted, however, that the t test is less general than the large sample method because here it is necessary to assume equality of the variances, which was not true for the large sample approach.

Formula (19) may also be used to determine confidence limits for $\mu_X - \mu_Y$. If it has been shown that the hypothesis $\mu_X = \mu_Y$ is not a reasonable one, it may be of interest to know how large or how small a difference is reasonable. For a given probability, confidence limits for $\mu_X - \mu_Y$ will give the desired answer.

As a numerical illustration, consider the data of Table 2 on the yield of corn in bushels per plot on 20 experimental plots of ground, half of which were treated with phosphorus as a fertilizer.

The problem is to decide whether the addition of phosphorus will improve the yield of corn. It may be treated as a problem of testing the hypothesis

$$H_0 : \mu_X = \mu_Y$$

against the alternative

$$H_1 : \mu_X > \mu_Y$$

TABLE 2

Treated	6.2	5.7	6.5	6	6.3	5.8	5.7	6	6	5.8
Untreated	5.6	5.9	5.6	5.7	5.8	5.7	6	5.5	5.7	5.5

where X and Y denote the yield on a treated and untreated plot, respectively. It will be assumed that all the plots were treated alike, except for the addition of phosphorus to half of them selected at random, and that the yield of corn on a plot may be treated as a normal variable. It will also be assumed that $\sigma_X = \sigma_Y$. These assumptions are sufficient to permit formula (19) to be applied to this problem. Calculations here give

$$\bar{x} = 6, \quad n_X s_Y^2 = .64$$
$$\bar{y} = 5.7, \, n_Y s_Y^2 = .24 .$$

When (19) is applied,

$$t = \frac{.3}{\sqrt{.64 + .24}} \sqrt{\frac{100(18)}{20}} = 3.03 , \quad \nu = 18 .$$

From Table IV the .005 critical value of t is $t = 2.878$, using only the right tail because of H_1; consequently, this result is certainly significant, and the hypothesis of no increase in mean yield is rejected.

If the assumptions of normality and equality of variances are reasonable so that the experimenter can justifiably claim that this significant difference is caused by a real difference in the population means, he will undoubtedly want confidence limits for $\mu_X - \mu_Y$. The same calculations as before give

$$t = \frac{.3 - (\mu_X - \mu_Y)}{.0989} .$$

Then, 95 percent confidence limits are given by

$$\left| \frac{.3 - (\mu_X - \mu_Y)}{.0989} \right| < 2.101$$

which reduces to

$$.092 < \mu_X - \mu_Y < .508 .$$

From this result it is clear that for a sample as small as 10 one cannot promise with any great degree of certainty more than about .092 unit increase in yield, which is only about a 2 percent increase in the mean yield of $\bar{y} = 5.7$, because of the addition of this amount of phosphorus.

The preceding methods are valid only under the assumption that $\sigma_X = \sigma_Y$. If $\sigma_X \neq \sigma_Y$ but the values of σ_X and σ_Y are known, one can test the hypothesis $\mu_X = \mu_Y$ by means of the standard normal variable

$$Z = \frac{(\bar{X} - \bar{Y}) - (\mu_X - \mu_Y)}{\sigma_{\bar{X} - \bar{Y}}}$$
$$= \frac{(\bar{X} - \bar{Y}) - (\mu_X - \mu_Y)}{\sqrt{\dfrac{\sigma_X^2}{n_X} + \dfrac{\sigma_Y^2}{n_Y}}} .$$

The values of the two variances are seldom known; therefore it is usually necessary to replace them by their sample estimates, just as was done for the large sample method in section 9.2, Chapter 5. The difficulty here is that only small samples are assumed to be available.

If $\sigma_X{}^2$ and $\sigma_Y{}^2$ are replaced by their unbiased sample estimates,

$$\tilde{\sigma}_X{}^2 = \frac{\sum\limits_{i=1}^{n_X}(X_i - \bar{X})^2}{n_X - 1} \qquad \text{and} \qquad \tilde{\sigma}_Y{}^2 = \frac{\sum\limits_{i=1}^{n_Y}(Y_i - \bar{Y})^2}{n_Y - 1}$$

the resulting variable

(20)
$$\tau = \frac{(\bar{X} - \bar{Y}) - (\mu_X - \mu_Y)}{\sqrt{\dfrac{\tilde{\sigma}_X{}^2}{n_X} + \dfrac{\tilde{\sigma}_Y{}^2}{n_Y}}}$$

can be shown to possess an approximate Student t distribution. This is not surprising in view of the fact that Student's t is obtained by replacing the unknown variance by its unbiased sample estimate in the corresponding expression for a single variable. The number of degrees of freedom necessary to make (20) an approximate t variable is given by a rather elaborate formula, namely,

$$\nu = \frac{\left(\dfrac{\tilde{\sigma}_X{}^2}{n_X} + \dfrac{\tilde{\sigma}_Y{}^2}{n_Y}\right)^2}{\dfrac{\left(\dfrac{\tilde{\sigma}_X{}^2}{n_X}\right)^2}{n_X + 1} + \dfrac{\left(\dfrac{\tilde{\sigma}_Y{}^2}{n_Y}\right)^2}{n_Y + 1}} - 2\,.$$

Although ν is not likely to be an integer, it usually suffices to choose the nearest integer value in looking up critical values of t.

The foregoing problem is known as the Behrens-Fisher problem. There has been much controversy over how it should be solved, and the approximate solution here is but one version.

5.3 Confidence Limits for a Regression Coefficient

The problem to be considered in this section is that of determining whether the difference between the slopes of a sample and a theoretical regression line might reasonably be caused by sampling variation. Let $x_1, \cdots, x_n$ denote a set of selected X values and let $Y_1, \cdots, Y_n$ denote random variables corresponding to observations that are to be made of the random variable Y at those X values. The regression model introduced in section 3, Chapter 6, will be used here. It assumes that the Y_i are independently normally distributed with a common variance σ^2 and with means given by the formula

(21)
$$E[Y_i] = \alpha + \beta(x_i - \bar{x}), \qquad i = 1, \cdots, n\,.$$

The equation of the least squares estimate of this regression line, as derived in (7), Chapter 7, is given by

$$y' = a + b(x - \bar{x})$$

where

$$a = \bar{Y} \quad \text{and} \quad b = \frac{\sum\limits_{i=1}^{n}(x_i - \bar{x})Y_i}{\sum\limits_{i=1}^{n}(x_i - \bar{x})^2}.$$

In this regression model the x's are chosen in advance and then one observation on Y is made at each of those selected points. Repeated sampling experiments therefore imply that these same n values of X will be used each time. Although the x's fixed in this model, they can be obtained initially by choosing any desired set of values in advance of observing the Y's, or they can be obtained by taking a random sample of size n of the pair of random variables X, Y. But in the inference problems that follow it is assumed that the initial set of x's, however obtained, are employed in future experimentation and in the interpretation of those inferences.

For simplicity of notation let

(22)
$$w_i = \frac{x_i - \bar{x}}{\sum\limits_{j=1}^{n}(x_j - \bar{x})^2}.$$

Then

(23)
$$b = \sum_{i=1}^{n} w_i Y_i.$$

Since the x_i may be treated as constants with respect to the sampling, the w_i may also be so treated; hence b may be treated as a random variable that is a linear function of the random variables $Y_1, \cdots, Y_n$. Now Theorem 3, Chapter 5, shows that a linear combination of independent normal variables is also a normal variable; hence b is a normal variable. Since the mean and variance of b will be needed, consider their evaluations next.

From (23) and (21) it follows that

$$E[b] = \sum_{i=1}^{n} w_i E[Y_i] = \sum_{i=1}^{n} w_i[\alpha + \beta(x_i - \bar{x})]$$

$$= \alpha \sum_{i=1}^{n} w_i + \beta \sum_{i=1}^{n} w_i(x_i - \bar{x}).$$

Since $\sum_{i=1}^{n}(x_i - \bar{x}) = 0$, it follows from (22) that $\sum_{i=1}^{n} w_i = 0$. Furthermore, $\sum_{i=1}^{n} w_i(x_i - \bar{x}) = 1$; hence $E[b] = \beta$. This shows that b is an unbiased estimator of β.

From Theorem 3, Chapter 5, and (22) it follows that

$$\sigma_b^2 = \sum_{i=1}^{n} w_i^2 \sigma_{Y_i}^2 = \sigma^2 \sum_{i=1}^{n} w_i^2 = \frac{\sigma^2}{\sum_{i=1}^{n}(x_i - \bar{x})^2} .$$

Application of the preceding formulas shows that

$$U = \frac{b - \beta}{\sigma_b} = \frac{b - \beta}{\sigma} \sqrt{\sum_{i=1}^{n}(x_i - \bar{x})^2}$$

possesses the properties of the variable U in Theorem 3. In order to be able to apply Theorem 3 to this problem, it is necessary to find an independent χ^2 variable to serve as V^2. In the preceding applications of this theorem such a variable was obtained by recognizing that nS^2/σ^2 possesses a χ^2 distribution. Since σ^2 for this problem is the variance of the deviations of the Y_i from the true regression line, the quantity to use in place of nS^2, the sample estimate of $n\sigma^2$, is $\sum(Y_i - Y_i')^2$. With this choice, one would expect the variable

$$V^2 = \frac{\sum_{i=1}^{n}(Y_i - Y_i')^2}{\sigma^2}$$

to possess a χ^2 distribution. It can be shown with considerable difficulty that V^2 does possess a χ^2 distribution, but with $n - 2$, not $n - 1$, degrees of freedom, and that U and V^2 are independently distributed. These facts are assumed here. A direct application of Theorem 3 to the preceding U and V variables will then show that

$$(24) \qquad T = (b - \beta)\sqrt{\frac{(n - 2)\sum(x_i - \bar{x})^2}{\sum(Y_i - Y_i')^2}}, \qquad \nu = n - 2$$

possesses a Student's t distribution with $n - 2$ degrees of freedom. By means of (24) one can test hypothetical values of regression slopes and find confidence limits for them.

As an illustration of how (24) is applied, consider the data of Table 3 on the relationship between the thickness of coatings of galvanized zinc as measured by a standard stripping method Y and a magnetic method X.

TABLE 3

y_i	116	132	104	139	114	129	720	174	312	338	465
x_i	105	120	85	121	115	127	630	155	250	310	443

If the magnetic method were reliable for measuring the thickness of such coatings, it would be preferred to the standard stripping method because it does not destroy the sample being measured and the standard method does. Now suppose that the magnetic method yields the same mean thickness as the standard method for thicknesses in the normal range. Then the true regression line of Y on X will be the line $y' = x$. Thus, under this assumption of the consistency of the two methods, $\beta = 1$. If, contrary to the preceding supposition, the magnetic method were biased in giving, say, too small a reading for thin coatings, then the true regression line, provided that the regression curve is a straight line, would have a slope greater than 1.

In view of the preceding discussion, consider the problem of testing the consistency of the two methods for measuring the thicknesses of coatings. The problem may be treated as a problem of testing the hypothesis

$$H_0 : \beta = 1$$

against the alternative hypothesis

$$H_1 : \beta \neq 1 .$$

If it is assumed that the necessary conditions for applying (24) are satisfied, the data of Table 3 may be used to yield the information needed in (24). It will be found that the equation of the least squares line fitted to the data is

$$y' = -1.79 + 1.12x .$$

It will also be found that

$$\sum (x_i - \bar{x})^2 = 301{,}826$$

and

$$\sum (y_i - y_i')^2 = 2766 .$$

If these values are used in (24), then

$$t = .12 \sqrt{\frac{9(301{,}826)}{2766}} = 3.76 , \qquad \nu = 9 .$$

From Table IV the 5 percent critical value of t is 2.26; consequently this value is significant. It appears that there is a slight bias in the magnetic method of the type suggested earlier.

The preceding problem could have been solved by a slightly more efficient procedure because strictly speaking it does not conform to the regression model assumed in arriving at (24) in which both α and β are unknown parameters. The physical nature of this problem suggests that $\alpha = 0$, and therefore that a simpler model might have been appropriate. This refinement will not be studied here, however.

A confidence interval for β is readily obtained by means of formula (24) and the technique employed to obtain confidence limits for a normal mean.

This technique can be generalized to finding confidence limits for the regression coefficients in multiple and curvilinear regression. It can also be adapted to finding confidence limits for the ordinate of a regression curve corresponding to any fixed value of x. All of these problems give rise to the t distribution.

Thus far, Student's t distribution has been justified only on the grounds that it eliminates an inaccuracy of certain large sample methods. It is conceivable that there are other tests which overcome this inaccuracy and which at the same time are better tests than the t test in the sense of Chapter 8. It can be shown, however, that the tests using the t distribution that have been considered possess certain optimal properties from that point of view.

One of the striking advantages of the t distribution is that it is not affected much if the basic variable has a distribution that differs considerably from normality. This property, called robustness, was not true of the χ^2 variable of section 3.

6 THE F DISTRIBUTION

It will be recalled that it was necessary to assume that $\sigma_X = \sigma_Y$ in order to apply the t distribution to testing the difference between two means. For the purpose of checking on this assumption, a density function that can be used for testing the equality of two variances will be derived. It will be found that such a density function has many other uses as well.

Let U and V possess independent χ^2 distributions with ν_1 and ν_2 degrees of freedom, respectively. Then consider the problem of finding the density function of the variable

$$(25) \qquad F = \frac{U/\nu_1}{V/\nu_2}.$$

Formula (7) can be used to solve this problem in much the same manner as it was used to find the density function of Student's t variable. Since U possesses a χ^2 distribution with ν_1 degrees of freedom, the distribution of the numerator variable U/ν_1 in (25) can be found by using the change of variable technique (4). For this purpose let $x = u$ and $y = u/\nu_1$; then the change of variable is given by $y = x/\nu_1$, and formula (4), gives

$$g(y) = f(x)\nu_1 = \frac{\nu_1}{2^{\frac{\nu_1}{2}} \Gamma\left(\frac{\nu_1}{2}\right)} x^{\frac{\nu_1}{2}-1} e^{-\frac{x}{2}}$$

$$= a y^{\frac{\nu_1}{2}-1} e^{-\frac{1}{2}y\nu_1}$$

where the constant a is given by

$$a = \nu_1^{\frac{\nu_1}{2}}/2^{\frac{\nu_1}{2}}\Gamma\left(\frac{\nu_1}{2}\right).$$

The denominator variable V/ν_2 in (25) will possess a corresponding density function with a constant b that is obtained from the constant a by replacing ν_1 by ν_2.

It would be confusing to use f to denote a numerical value of the random variable F because f is being used to represent the density function of X; therefore F will be used here to represent both the random variable F and a possible numerical value of it. With this understanding formula (7) may now be applied, provided z is replaced by F, to give

$$q(F) = \int_0^\infty x \cdot bx^{\frac{\nu_2}{2}-1} e^{-\frac{1}{2}x\nu_2} \cdot a(xF)^{\frac{\nu_1}{2}-1} e^{-\frac{1}{2}xF\nu_1} \, dx$$

$$= abF^{\frac{\nu_1}{2}-1} \int_0^\infty x^{\frac{1}{2}(\nu_1+\nu_2-2)} e^{-\frac{x}{2}(\nu_2+\nu_1 F')} \, dx.$$

Let $w = x(\nu_2 + \nu_1 F)/2$; then $dx = 2dw/(\nu_2 + \nu_1 F)$ and

$$q(F) = \frac{abF^{\frac{\nu_1}{2}-1} 2^{\frac{1}{2}(\nu_1+\nu_2)}}{(\nu_2 + \nu_1 F)^{\frac{1}{2}(\nu_1+\nu_2)}} \int_0^\infty w^{\frac{1}{2}(\nu_1+\nu_2)-1} e^{-w} \, dw.$$

It will be observed that the value of this integral is $\Gamma[(\nu_1 + \nu_2)/2]$; consequently $q(F)$ reduces to

$$q(F) = c \frac{F^{\frac{\nu_1}{2}-1}}{(\nu_2 + \nu_1 F)^{\frac{1}{2}(\nu_1+\nu_2)}}$$

where

(26)
$$c = \frac{\nu_1^{\frac{\nu_1}{2}} \nu_2^{\frac{\nu_2}{2}} \Gamma\left(\frac{\nu_1 + \nu_2}{2}\right)}{\Gamma\left(\frac{\nu_1}{2}\right)\Gamma\left(\frac{\nu_2}{2}\right)}.$$

This derivation proves the following theorem.

THEOREM 4: *If U and V possess independent χ^2 distributions with ν_1 and ν_2 degrees of freedom, respectively, then*

$$F = \frac{U/\nu_1}{V/\nu_2}$$

has the F distribution with ν_1 and ν_2 degrees of freedom given by

$$f(F) = cF^{\frac{1}{2}(\nu_1 - 2)}(\nu_2 + \nu_1 F)^{-\frac{1}{2}(\nu_1 + \nu_2)}$$

where c is given by (26).

7 APPLICATIONS OF THE F DISTRIBUTION

Since the F distribution was derived partly in order to test the assumption of the equality of variances which is needed in the t test when that test is applied to testing the difference between two means, consider the problem of testing the hypothesis

$$H_0 : \sigma_X = \sigma_Y$$

against the alternative

$$H_1 : \sigma_X \neq \sigma_Y$$

under the assumption that X and Y are normally distributed.

Let S_X^2 and S_Y^2 be sample variances based on random samples of sizes n_X and n_Y, respectively, from these two populations. Then, since $n_X S_X^2 / \sigma_X^2$ and $n_Y S_Y^2 / \sigma_Y^2$ possess independent χ^2 distributions,

$$\frac{U}{\nu_1} = \frac{n_X S_X^2}{(n_X - 1)\sigma_X^2}$$

and

$$\frac{V}{\nu_2} = \frac{n_Y S_Y^2}{(n_Y - 1)\sigma_Y^2}$$

will satisfy the requirements for U/ν_1 and V/ν_2 in Theorem 4. Under the hypothesis H_0, $\sigma_X = \sigma_Y$; therefore, by Theorem 4,

$$F = \frac{n_X S_X^2/(n_X - 1)}{n_Y S_Y^2/(n_Y - 1)}$$

$$= \frac{\tilde{\sigma}_X^2}{\tilde{\sigma}_Y^2}$$

will possess an F distribution with $n_X - 1$ and $n_Y - 1$ degrees of freedom. Here $\tilde{\sigma}_X^2$ and $\tilde{\sigma}_Y^2$ denote the unbiased estimates of σ_X^2 and σ_Y^2. This notation is introduced to point out the fact that the value of F to use in testing $\sigma_X^2 = \sigma_Y^2$ is the ratio of the unbiased estimates of the two variances. This test, like the t test, possesses the desirable feature of being independent of population parameters.

As a numerical illustration, consider the problem that illustrated the application of the t distribution to the testing of the difference between two normal means. From Table 2 and immediately following it,

$$\tilde{\sigma}_X^2 = \frac{n_X S_X^2}{n_X - 1} = .071$$

and

$$\tilde{\sigma}_Y^2 = \frac{n_Y S_Y^2}{n_Y - 1} = .027 .$$

Therefore $F = 2.63$ with $\nu_1 = \nu_2 = 9$ degrees of freedom. It is necessary to consult tables of critical values of the F distribution in order to decide whether this value of F is unreasonably large or small. Such values are to be found in Table V in the appendix.

Since the F distribution depends on the two parameters ν_1 and ν_2, a three-way table would be needed to tabulate the values of F corresponding to different probabilities and values of ν_1 and ν_2. As a consequence, only the 5 and 1 percent right-tail area points are tabulated corresponding to various values of ν_1 and ν_2. The technique in the use of Table V is explained by means of the graph in Fig. 7, which illustrates the graph of $f(F)$ for a typical pair of values of ν_1 and ν_2. Let F_1 denote the value of F for which $P\{F < F_1\} = .025$ and F_2 the value for which $P\{F > F_2\} = .025$. If the sample value of F falls outside the interval (F_1, F_2), the hypothesis of a common σ^2 will be rejected. For convenience of notation, let $F' = 1/F$. Since $F = \tilde{\sigma}_X^2/\tilde{\sigma}_Y^2$ with ν_1 and ν_2 degrees of freedom, $F' = \tilde{\sigma}_Y^2/\tilde{\sigma}_X^2$ with ν_2 and ν_1 degrees of freedom. By means of the reciprocal function F', the probability of $F < F_1$ can be evaluated as follows:

$$.025 = P\{F < F_1\} = P\left\{\frac{1}{F} > \frac{1}{F_1}\right\} = P\left\{F' > \frac{1}{F_1}\right\} .$$

This result shows that the left critical value of the F distribution corresponds to the right critical value of the F' distribution. As a result, it is necessary to find only right critical values for F and F' to determine F_2 and F_1. The

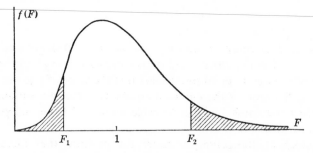

Fig. 7. A typical F distribution.

reciprocal of the right critical value for F' gives the left critical value for F. Because of this property of F, only right critical points for F are tabulated. Unfortunately, only the 5 and 1 percent critical points have been tabulated in Table V; consequently, it is necessary to interpolate between these two values in order to obtain an approximate $2\frac{1}{2}$ percent critical point.

In view of this reciprocal property, the procedure to be followed is always to place the larger of the two unbiased variance estimates in the numerator of F; consequently, $\tilde{\sigma}_X{}^2$ will always denote the larger of the two estimates. If the hypothesis of a common σ^2 is rejected whenever the sample value of this F exceeds its $2\frac{1}{2}$ percent point, the hypothesis will be rejected whenever the original F falls outside the interval (F_1, F_2), because, when $F > 1$, F_2 will serve as the critical value, and, when $F < 1$, F' will be used instead and F_2' will serve as the critical value. But, as demonstrated in the preceding paragraph, F_2' for F' corresponds to F_1 for F.

If this procedure is applied to the numerical problem being discussed, it will be found from Table V that the 5 percent critical value is, by interpolation,
$$F_2 = 4.5, \qquad v_1 = v_2 = 9 \,.$$

The sample value of $F = 2.63$ is therefore not significant. This result implies that the assumption of equal variances is a reasonable one and that the significant value of t obtained in connection with this problem when testing the hypothesis $\mu_X = \mu_Y$ may not be reasonably attributed to a lack of the assumption $\sigma_X = \sigma_Y$ being satisfied.

The preceding test is not a best test in the sense of Chapter 8, because it can be shown that there does not exist a best test for this problem; however, it is known that this is a good test from the type II error point of view.

Unfortunately, the preceding test is not reliable if X and Y do not possess normal distributions. Just as in the case of applying the χ^2 distribution to testing a single variance, the F distribution may be highly unreliable if X and Y possess distributions whose fourth moments are considerably larger than for a normal variable. Sampling experiments have shown that then there is a tendency for the critical region of the test to be considerably larger than it is when normality holds; consequently too many hypotheses of the form $H_0: \sigma_X = \sigma_Y$ will be rejected. If there is reason to believe that X and Y possess non-normal distributions of this particular type, the value of α could be changed from .05 to .01 to compensate for this non-robustness of the test. There are other tests available for treating this problem when the normality assumption is not justified, but they are not studied here.

Further applications of the F distribution are made in Chapter 11 on what is known as analysis of variance techniques. Because of the importance of such techniques in designing experiments, they have been incorporated in a separate chapter.

8   DISTRIBUTION OF THE RANGE

In certain fields of applied statistics the amount of routine computation becomes burdensome unless methods are chosen that involve only a small amount of it. In industrial quality-control work, for example, the repeated computation of standard deviations as measures of the variability of a product is undesirable. It is customary in such work to take the range as the measure of variability. Not only is the range easy to compute, but it is also simple to explain as a measure of variation to individuals without a statistical background. For small samples from a normal population it can be shown that the range is nearly as efficient for estimating σ as is the sample standard deviation; consequently for small samples the range is a highly useful statistic.

Consider a random sample, X_1', X_2', $\cdots$, X_n', drawn from the random variable X whose density function $f(x)$ is assumed to be continuous. Let these sample values be arranged in order of increasing magnitude and denote the ordered set by X_1, X_2, $\cdots$, X_n. Now, consider the problem of finding the joint density function of X_1 and X_n. The density function of the range $Z = X_n - X_1$ is easily found from the joint density of X_1 and X_n. Toward this end, let $U = X_1$ and $V = X_n$. Then

$$P\{U \geq u_0 \text{ and } V \leq v_0\} = P\{u_0 \leq X_1', \cdots, X_n' \leq v_0\}$$

$$= \left[\int_{u_0}^{v_0} f(x)\, dx \right]^n .$$

Let $g(u, v)$ be the desired joint density function of U and V; then since $U \leq V$ it follows from Fig. 8 that

$$P\{U \geq u_0 \text{ and } V \leq v_0\} = \int_{u_0}^{v_0} \int_u^{v_0} g(u, v)\, dv\, du .$$

Hence,

$$\int_{u_0}^{v_0} \int_u^{v_0} g(u, v)\, dv\, du = \left[\int_{u_0}^{v_0} f(x)\, dx \right]^n .$$

Rewrite this equality as follows:

$$-\int_{v_0}^{u_0} \int_u^{v_0} g(u, v)\, dv\, du = \left[-\int_{v_0}^{u_0} f(x)\, dx \right]^n .$$

Since u_0 and v_0 are arbitrary, hold v_0 fixed and differentiate both sides of this equality with respect to u_0. The standard calculus formula for differentiating

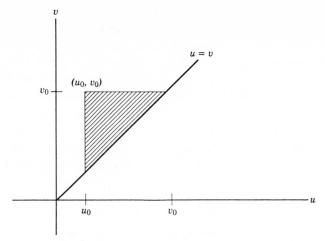

Fig. 8.   Sample space for U and V.

an integral with respect to its upper limit then gives

$$-\int_{u_0}^{v_0} g(u_0, v)\, dv = n\left[-\int_{v_0}^{u_0} f(x)\, dx\right]^{n-1} [-f(u_0)].$$

Now hold u_0 fixed and differentiate both sides with respect to v_0. This gives

$$-g(u_0, v_0) = -nf(u_0)(n - 1)\left[\int_{u_0}^{v_0} f(x)\, dx\right]^{n-2} f(v_0).$$

This is the negative of the density function of U and V at the point (u_0, v_0). Since u_0 and v_0 are arbitrary values, these results may be stated in the form of a theorem as follows:

THEOREM 5:   *If U and V denote the smallest and largest values, respectively, in a random sample of size n of a continuous random variable X that possesses a continuous density $f(x)$, then the joint density of U and V is given by*

$$g(u, v) = n(n - 1)f(u)f(v)\left[\int_{u}^{v} f(x)\, dx\right]^{n-2}.$$

The density function of the range can be obtained very easily from this result by means of formula (8). Since the range is given by $Z = V - U$, it is necessary to let $y = v$ and $x = u$ in formula (8). Then

$$q(z) = \int g(u, u + z)\, du$$

where the range of integration is over possible values of u when z is fixed.

If the variable X ranges over the interval (a, b), the range of u with z fixed will be from a to $b - z$. This upper limit arises from the fact that the smallest measurement, u, must be z units smaller than the largest measurement, v, and v cannot exceed the upper limit b for x. An expression for $q(z)$ may now be obtained by inserting the value of $g(u, v)$ given in Theorem 5 and using the limits of integration that were just found. The results of these operations are expressed in the form of a theorem.

THEOREM 6: *If the continuous variable X has the continuous density function $f(x)$ and if X assumes values in the interval (a, b) only, then the density function of the range Z for a random sample of size n is given by the formula*

$$q(z) = n(n - 1) \int_a^{b-z} f(u)f(u + z) \left[\int_u^{u+z} f(x)\, dx \right]^{n-2} du \, .$$

Unless the integral of $f(x)$ is quite simple, this expression is likely to be difficult to work with, even numerically. As an illustration of a simple problem, consider the range for a sample of size n from the rectangular distribution that is defined for $0 \leq x \leq 1$ by $f(x) = 1$. Here

$$\int_u^{u+z} f(x)\, dx = \int_u^{u+z} dx = z \, .$$

Therefore, by Theorem 6,

$$q(z) = n(n - 1) \int_0^{1-z} z^{n-2} \, du$$

$$= n(n - 1)z^{n-2}(1 - z) \, .$$

9 APPLICATIONS OF THE RANGE

In the introduction to the last section it was remarked that the range was useful as a substitute for the standard deviation as a measure of variability in certain routine operations. It should therefore be of interest to know what the relationship is between the range and the standard deviation for, say, a normal distribution. This relationship may be found by calculating the mean of Z. Since

$$E(Z) = \int_0^{b-a} zq(z) \, dz$$

it is clear from Theorem 6 that the evaluation of the desired relationship will give rise to a complicated double integral. Unfortunately, when $f(x)$ is a normal density function these integrations cannot be performed directly for general n; therefore numerical methods of integration are required. In spite of the complicated nature of the integral defining $E(Z)$, it can be shown that $E(Z)$ is a constant, depending on n, times σ. Tables are available for the normal variable case that expresses $E(Z) = \mu_Z$ in terms of σ_X for various values of n. Table 4 gives a few entires from a table to indicate the nature of the relationship.

As an illustration of the use of such tables, consider once more the technique of constructing a quality-control chart for $\bar{X}$ as given in section 6.1, Chapter 5. There a $3\sigma_{\bar{X}}$ band was constructed for controlling $\bar{X}$. If the range is taken as the measure of variability, $3\sigma_{\bar{X}} = 3\sigma_X/\sqrt{n}$ will be replaced by $3\mu_Z/d_n\sqrt{n}$, where d_n is the value obtained from the table, that is, the value of the ratio μ_Z/σ_X corresponding to the given value of n. Now the value of μ_Z

TABLE 4

n	2	3	4	5	10	40	100
$d_n = \dfrac{\mu_Z}{\sigma_X}$	1.128	1.693	2.059	2.326	3.078	4.498	5.015

can be estimated by using the sample mean of the Z values obtained for the various samples of n each. For such charts n is usually chosen to be an integer near 4 and a fairly large number of samples of this size is obtained before the chart is drawn; consequently, μ_Z is usually estimated quite accurately.

If n is chosen less than 10, the estimation of σ_X by means of the range rather than the standard deviation of a sample is quite efficient. Investigations have shown, for example, that the variance of the estimate of σ_X based on the range of a sample of size 6 is only about 15 percent larger than the variance of the sample standard deviation for a sample of size 6. From the point of view of Chapter 8, one can therefore conclude that the range is nearly as good as the standard deviation as an estimator for σ_X for small samples.

EXERCISES

Sec. 1

1. Given $f(x) = e^{-x}$, $x > 0$, find by the change of variable technique the density of the variable (a) $Y = 1/X$, (b) $Y = \log_e X$.

2. Given $f(x) = xe^{-x^2/2}$, $x > 0$, find the density of the variable (a) $Y = X + 1$, (b) $Y = X^2$, (c) $Y = \log_e X$.

3. Let X possess a uniform distribution over the interval $(-\pi/2, \pi/2)$. Show that $Y = \tan X$ possesses the Cauchy distribution

$$g(y) = \frac{1}{\pi(1 + y^2)}.$$

4. Given that θ is uniformly distributed over the interval $-\pi/2$ to $\pi/2$, find the density of $Z = A \sin \theta$, where A is a constant.

5. Show that if X possesses a rectangular distribution on $(0, 1)$, the variable $Y = -2 \log X$ will possess a chi-square distribution with two degrees of freedom.

6. Let X be a standard normal variable. Find the density of (a) $2X$, (b) $2X^2 + 1$.

7. If X possesses the density $f(x) = e^{-x/\theta}/\theta$, $x > 0$, find the density of $Z = (X - \mu)/\sigma$.

8. Given $f(x) = 2(1 - x)$, $0 < x < 1$, find the density of $Z = X^2$.

9. Given $f(x) = (x - 1)/2$, $1 < x < 3$, find an increasing function $y = h(x)$ such that Y will have a uniform distribution over $(0, 1)$.

10. Given that X has the continuous distribution function $F(x)$, find an expression for the distribution functions of the variables (a) $Y = e^X$, (b) $Y = \log_e X$, (c) $Y = F(X)$.

Sec. 2

11. Given $f(x) = 2x$, $0 < x < 1$, and a random sample of size two, X_1, X_2, calculate $P\{X_1 \leq 2X_2\}$.

12. Given that X has a uniform distribution over $(0, 1)$, find the density of $Z = X_1 X_2$, where X_1 and X_2 represent a random sample of size 2.

13. Given $f(x, y) = 2(1 + x + y)^{-3}$, $x > 0$, $y > 0$, find the density of $Z = X + Y$.

14. If $f(x, y) = e^{-(x+y)}$, $x > 0$, $y > 0$, find the density function of (a) $Z = X + Y$, (b) $Z = e^{-(X+Y)}$.

15. Let X_1 and X_2 denote a random sample of size two of a standard normal variable. Show that $Z = X_2/X_1$ possesses the Cauchy distribution

$$f(z) = \frac{1}{\pi(1 + z^2)}.$$

16. If $f(x, y) = 1$, $0 < x < 1$, $0 < y < 1$, find the density of (a) $Z = X^2$, (b) $Z = X + Y$, (c) $Z = Y/X$.

17. Let the random variable X possess the density $f(x) = \frac{1}{3}$, $x = -1$, 0, 1 and let $Y = X^2$. Find (a) $g(y)$ (b) $f(x, y)$ (c) Cov (X, Y).

18. If X and Y are independent standard normal variables, derive the density function of $Z = \sqrt{X^2 + Y^2}$. The variable Z represents radial error in problems for which X and Y represent independent coordinate axis errors with equal variability.

19. Use the general method for finding distribution functions to obtain a formula for the density of $Z = X^2/(X^2 + Y^2)$, given that X and Y are independent continuous variables with the same density function, f, and which can assume any real values.

Sec. 3

20. Show that the mean and variance of a χ^2 variable with ν degrees of freedom are given by ν and 2ν, respectively.

21. Given $f(x) = e^{-x}$, $x > 0$, find by moment generating function techniques the density function of $Z = 2n\bar{X}$.

22. Given that X is normal with mean μ and variance σ^2, show that the likelihood ratio test of the hypothesis $H_0: \sigma^2 = \sigma_0^2$ reduces to a χ^2 test.

23. Let X be a normal variable with mean 0 and variance σ^2 and let σ^2 be estimated by $Z = \sum_{i=1}^{n} X_i^2/n$. Show that $V(Z) = 2\sigma^4/n$. Use the results of problem 20.

24. Find the value of k that will minimize $E(kZ - \sigma^2)^2$, where Z is defined in problem 23. What does this imply about the unbiased estimate Z of σ^2 with respect to minimum mean square error estimates? Use the results of problem 23 here.

25. Determine what value of k will minimize $E[k \sum (X_i - \bar{X})^2 - \sigma^2]^2$ if X is a normal variable with mean μ and variance σ^2. Compare with the solution of problem 24 and comment concerning the two results and unbiased estimates.

26. Determine what non-zero values of c_1 and c_2 will make $Z = c_1 X + c_2 Y$ a chi-square variable if X and Y are independent chi-square variables with r and s degrees of freedom.

Sec. 3.2

27. Given that X is a normal variable and given the sample values $\bar{x} = 42$, $s = 6$, $n = 20$, (a) test the hypothesis that $\sigma = 9$, (b) find 96 percent confidence limits for σ^2.

28. Work problem 27 (b) using the normal approximation suggested in Table III in the appendix.

29. A sample of size 8 of a normal variable gave the values 10, 13, 10, 12, 8, 12, 11, 12. Find 90 percent confidence limits for σ^2.

30. Given the following sample values from a normal population, find 96 percent confidence limits for σ^2 based on combining these sample values properly. The sample variances are $s_1^2 = 25$, $s_2^2 = 36$, $s_3^2 = 16$ with $n_1 = 10$, $n_2 = 5$, $n_3 = 8$.

31. Find boundaries for a quality control chart for controlling variability if samples of size 6 are taken every hour and if it is known from past experience that $\sigma^2 = 10$. Use boundaries that will include 98 percent of the sample values of the variable $\sum_{1}^{5} (X_i - \bar{X})^2$.

Sec. 4

32. Show that $E[t] = 0$ for Student's t distribution.

33. Prove that the density function of the variable t approaches the density of the standard normal variable as the number of degrees of freedom ν becomes infinite. Assume the constant approaches $1/\sqrt{2\pi}$.

Sec. 5

34. For the data of problem 27, (a) test the hypothesis $H_0: \mu = 44$, (b) find 95 percent confidence limits for μ.

35. Given $\bar{x} = 20$, $s = 2$, $n = 10$, with X normally distributed, find 99 percent confidence limits for μ.

36. Compare the confidence limits obtained in problem 35 with those that would have been obtained if s had been treated as the true value of σ and normal curve large sample methods of Chapter 5 had been employed.

37. Work problem 29 for μ rather than σ.

38. The following data give the corrosion effects in various soils for coated and uncoated pipe. Taking differences of pairs of values test the hypothesis that the mean of such differences is 0.

Uncoated	42	37	61	74	55	57	44	55	37	70
Coated	39	43	43	52	52	59	40	45	47	62

Uncoated	52	55	60	48	52	44	56	44	38	47
Coated	40	27	50	33	56	36	54	32	39	40

39. Given two random samples of sizes 10 and 12 from two normal populations with $\bar{x}_1 = 20$, $\bar{x}_2 = 23$, $s_1 = 5$, $s_2 = 6$, (a) test the hypothesis $H_0: \mu_1 = \mu_2$, (b) find 90 percent confidence limits for $\mu_1 - \mu_2$ assuming that $\sigma_1 = \sigma_2$.

40. Treating the data of problem 38 as random sample values from two normal populations rather than as paired values, test the hypothesis $H_0: \mu_1 = \mu_2$. Explain why it is probably incorrect to apply this test to this problem.

41. The following data give the gains in weight of 20 rats, half of which received their protein from raw peanuts and half from roasted peanuts. Test to see whether roasting had any effect on their protein value with respect to weight.

Raw	62	60	56	63	56	63	59	56	44	61
Roasted	57	56	49	61	55	61	57	54	62	58

42. In an industrial experiment a job was performed by 30 workmen according to method I and by 40 workmen according to method II. The following data give the results of the experiment. Determine by means of a 95 percent confidence interval for $\mu_1 - \mu_2$ how much time on the average could be expected to be saved by using method I, where the variable is time to complete the job.

Time	50	51	52	53	54	55	56	57	58	59	60
I	1	3	5	4	7	5	3	1	1	0	0
II	0	1	2	5	8	9	6	3	3	1	2

43. In estimating the mean of a normal variable by means of a confidence interval, how large a sample is needed so that the length of a 90 percent confidence interval will be less than $\sigma/9$ if σ is known?

44. Prove that the likelihood ratio test of the hypothesis $\mu = \mu_0$ for a normal variable with unknown σ^2 is equivalent to student's t test for this hypothesis.

45. Find a likelihood ratio test for testing $H_0: \mu_1 = \mu_2$ for two normal variables with a common unknown variance σ^2. Assume equal size samples are taken of each variable.

46. For the data of problems 2 and 12, Chapter 7, find 95 percent confidence limits for the slope β of the theoretical regression line.

47. For a fixed value of x, find the distribution of the random variable $Y_x = a + b(x - \bar{x})$, where a and b are the least squares estimates.

48. Use the results in problem 47 to construct a Student's t variable for Y_x by means of which one can obtain confidence limits for $E[Y_x]$. Assume the independence of certain variables if necessary.

Sec. 6

49. Verify the .05 value of F for $\nu_1 = 2$ and $\nu_2 = 2$ by direct integration of the density function of F.

50. Show that $E[F] = \nu_2/(\nu_2 - 2)$ for the F distribution.

51. Prove that the variable t^2 with ν degrees of freedom is a special case of the variable F with $\nu_1 = 1$ and $\nu_2 = \nu$.

Sec. 7

52. Let X have an F distribution with degrees of freedom ν_1 and ν_2. Show that $1/X$ has an F distribution with degrees of freedom ν_2 and ν_1.

53. Samples of sizes 10 and 18 taken from two normal populations gave $s_1 = 14$ and $s_2 = 20$. Test the hypothesis $H_0: \sigma_1 = \sigma_2$.

54. The following table gives data on the hardness of wood stored outside and inside. Test to see whether the variability of hardness is affected by weathering.

	Outside	Inside
Sample size	40	100
Mean	117	132
Sum of squares about the mean	8,655	27,244

55. Derive a formula for obtaining confidence limits for σ_1^2/σ_2^2, where σ_1^2 and σ_2^2 are the variances of two normal variables, if samples of sizes n_1 and n_2, respectively, are taken of those variables.

56. Find 90 percent confidence limits for σ_x/σ_y if 21 samples are taken from each of two normal populations and if $s_x/s_y = 2$.

57. If one desires to have $\alpha = .05$ and $\beta = .05$ in testing the equality of two normal variances when actually one variance is twice the other, how large an equal size sample from each population should be taken if the right tail of the F distribution is chosen as critical region?

58. The time X between recordings of certain types of radiation activity is known to have the density $f(x) = \alpha e^{-\alpha x}$, $x > 0$. How would you proceed to construct a test of the hypothesis that the values of α for two different experiments are the same?

59. Given samples of sizes n_1 and n_2, respectively, from two normal populations with zero means and variances σ_1^2 and σ_2^2, construct a likelihood ratio test for testing $H_0: \sigma_1^2 = \sigma_2^2$ and show that it is equivalent to an F test for this problem.

60. The random samples $X_1, \cdots, X_{n_1}$ and $Y_1, \cdots, Y_{n_2}$ are taken from two standard normal variables. Find the density of the variable

$$Z = \sum_1^{n_1} X_i^2 \Big/ \left(\sum_1^{n_1} X_i^2 + \sum_1^{n_2} Y_i^2 \right).$$

Sec. 8

61. Find the density of the range Z if X has the density $f(x) = e^{-x}$, $x > 0$.

62. Find the probability that in a sample of size 10 from the uniform distribution $f(x) = 1$, $0 < x < 1$, the range will exceed .7.

63. Determine how large a sample must be taken from $f(x) = 1$, $0 < x < 1$, in order that the probability will exceed .90 that the range will exceed .80.

Sec. 9

64. For a control chart for the sample mean of a normal variable with $\mu = 40$, based on samples of size 5, find control boundaries in terms of the range.

65. For problem 14, Chapter 5, find control boundaries in terms of the mean of the sample ranges.

66. Suppose that samples of size 4 are taken from $f(x) = e^{-x}$, $x > 0$. (a) Calculate $E[Z]$. (b) Calculate σ and then compare the ratio $E[Z]/\sigma$ with that given by Table 4 for a normal variable. (c) Determine numbers Z_1 and Z_2 such that $P\{Z < Z_1\} = .025$ and $P\{Z > Z_2\} = .025$.

CHAPTER 11

Statistical Design in Experiments

It is a common occurrence for experimenters who are unacquainted with statistical principles to seek statistical assistance when their experiments fail to produce the results anticipated by them. In some experiments the data were obtained in such a manner as to exclude any valid conclusions of the type desired; in others, there is little that can be done to extract further information from the data because the experiment was not designed with a statistical analysis in mind. Only rarely are the experiments that give valid conclusions as sensitive as they would have been if a standard statistical design had been employed. Too many experimenters do not seem to appreciate the obvious injunction that the time to design an experiment is before the experiment is begun.

In this chapter, after a brief discussion of a few of the general principles involved in the design of experiments, some of the common techniques used in the design and analysis of experiments will be studied.

1 RANDOMIZATION, REPLICATION, AND SENSITIVITY

In most experiments there are several variables in addition to the one or more being investigated that need to be controlled if the experiment is to give valid conclusions. In some cases these interfering variables can be controlled by laboratory techniques; in others such control may be possible only by statistical design. As a simple illustration, consider an agricultural experiment in which two different seed varieties are to be tested on a piece of land. If the piece of land were divided into two equal pieces and one variety planted on each, the difference in yields could not be used as a valid estimate of the differential effect of the two seed varieties because of the possible difference in the soil fertility of the two pieces.

Experiments can often be made valid by applying the principles of *randomization* and *replication*. Thus, in the present illustration, if the piece of land were divided into a number of small plots of equal size and if one variety of seed were planted on half of these plots and the other variety on the remaining half, with the selection of the plots for each variety determined by a random process, then the varying fertility of the land would affect the two varieties approximately equally and therefore the difference in varietal yields would represent a valid estimate of the differential effects of the two seed varieties.

Randomization by itself is not necessarily sufficient to yield a valid experiment. For example, if one merely tossed a coin to determine which half of the original piece of land should be planted with one of the seed varieties, the selection would be random but it would not permit the two seed varieties to be equally affected by any varying soil fertility. If the two seed varieties were equally productive but the two halves of land were markedly different in fertility, then regardless of the seed variety selected for each half the conclusion would invariably be that the seed varieties differed in productivity. In order to insure validity, it would be necessary that the piece of land be divided into a sufficiently large number of similar plots so that the probability of having one of the seed varieties largely located on the more fertile plots would be very small. This repetition of an experiment or experimental unit is called replication. Thus, to insure validity in an experiment, randomization should be accompanied by sufficient replication.

Not only are randomization and replication useful techniques for assisting in the construction of valid experiments, but they are often essential to certain classes of experiments whose conclusions depend on the use of statistical techniques. Since the density functions of the various statistics considered in the preceding chapters were derived on the basis of random sampling, it follows that the methods employed in the preceding chapters are applicable to such samples only; consequently, any experiment whose conclusions depend on these methods requires randomization. Replication is also necessary for the application of any method that obtains its measure of variability directly from the data because at least two observations are needed to measure variation. For example, the illustrative experiment just discussed requires randomization and replication if the difference between mean yields is to be tested by means of Student's t distribution because the t distribution is based on random sampling and because sample variances are needed to evaluate t.

The requirement of random samples for the applicability of most statistical methods is not always easy to satisfy. For example, if the product of a machine is sampled every hour for several days, it may easily happen that the product of the machine changes during the day because of the operator's working pattern and also from day to day because of machine wear. For situations such as this in which observations are ordered with respect to time, one of the methods for testing randomness discussed in Chapter 12 should be applied before methods based on random samples are used.

In the preceding illustration the techniques of randomization and replication removed much of the danger of obtaining biased results; however, these techniques did not remove the effect of differences in soil fertility on the variability of yields. If the variation in fertility is increased, the variation in yield is thereby increased. As a consequence, if Student's t distribution for testing the difference between two means were applied, a considerably larger sample might be needed to produce a significant difference if large fertility differences existed between plots than if the plots were of uniform fertility because of the larger estimate of variance involved in the denominator of t. Such an experiment could therefore be made more sensitive by selecting plots of uniform fertility. Very often, however, it is not feasible to control the fertility in this manner. Now, by arranging the plots into small homogeneous groups and applying statistical design, it is often possible to eliminate statistically the greater share of the fertility variability effects in the t test and thereby make the experiment more sensitive.

2 ANALYSIS OF VARIANCE

One of the most useful techniques for increasing the sensitivity of an experiment is designing it in such a way that the total variation of the variable being studied can be separated into components that are of experimental interest or importance. Splitting up the total variation in this manner enables the experimenter to utilize statistical methods to eliminate the effects of certain interfering variables and thus to increase the sensitivity of his experiment. The analysis of variance is a technique for carrying out the analysis of an experiment designed from this point of view.

In designing an experiment, the experimenter usually has in mind the testing of a hypothesis or the estimation of some parameters. Although the analysis of variance technique enables the experimenter to design sensitive experiments for either of these basic problems, the explanation of the technique will be made largely from the point of view of testing hypotheses.

As an illustration of the type of problem for which the analysis of variance is useful, consider an experiment in which four different brands of machines are to be tested in a factory to see whether they are equally efficient. The experiment is to consist of having six machine operators, selected at random from the pool of such operators, use each of the four machines for a period of one week and recording the number of units of work produced by each operator on each brand of machine. The resulting scores are used to measure the relative efficiencies of the four brands. These scores may be arranged in a rectangular array containing six rows and four columns; however, for the purpose of considering other problems also, let the scores be displayed in a rectangular array containing a rows and b columns as shown in Table 1.

The entries in the margins of the table represent the means of the corresponding rows and columns. The location of the dot in the index shows whether the mean is a row mean or a column mean.

TABLE 1

x_{11}	x_{12}	$\cdots$	x_{1j}	$\cdots$	x_{1b}	$\bar{x}_{1.}$
x_{21}	x_{22}	$\cdots$	x_{2j}	$\cdots$	x_{2b}	$\bar{x}_{2.}$
						$\cdot$
x_{i1}	x_{i2}	$\cdots$	x_{ij}	$\cdots$	x_{ib}	$\bar{x}_{i.}$
						$\cdot$
x_{a1}	x_{a2}	$\cdots$	x_{aj}	$\cdots$	x_{ab}	$\bar{x}_{a.}$
$\bar{x}_{.1}$	$\bar{x}_{.2}$	$\cdots$	$\bar{x}_{.j}$	$\cdots$	$\bar{x}_{.b}$	$\bar{x}$

There are two well-known mathematical models that have been designed for experiments of the type being discussed. One of them is called the linear hypothesis model and the other is known as the components of variance model. The essential difference between the two models lies in the assumptions made concerning the population of experiments of which the given experiment is considered to be a random sample. Since the linear hypothesis model is the one most commonly used, it alone will be studied.

2.1 Linear Hypothesis Model

This model assumes that the random variable X_{ij} has a mean μ_{ij} that can be written in the form

$$(1) \qquad \mu_{ij} = \mu + \alpha_i + \beta_j$$

where μ denotes the expected value of $\bar{X}$, α_i denotes the expected value of $\bar{X}_{i.} - \bar{X}$, and β_j denotes the expected value of $\bar{X}_{.j} - \bar{X}$. Since $\bar{X}$ is the mean of both $\bar{X}_{i.}$ and $\bar{X}_{.j}$,

$$\sum_{i=1}^{a} (\bar{X}_{i.} - \bar{X}) = 0 \qquad \text{and} \qquad \sum_{j=1}^{b} (\bar{X}_{.j} - \bar{X}) = 0 .$$

Upon taking the expected value of each of these sums, it therefore follows that

$$(2) \qquad \sum_{i=1}^{a} \alpha_i = 0 \qquad \text{and} \qquad \sum_{j=1}^{b} \beta_j = 0 .$$

Assumption (1) essentially states that the mean of the variable X_{ij} is the sum of a general mean μ, a row effect α_i, and a column effect β_j. Thus, in the industrial experiment, if the ith operator were a superior workman, his mean score would be expected to exceed the mean score for all six operators by a positive amount, α_i, whereas if he were an inferior operator, α_i would be negative. Similarly, β_j is a number, positive or negative, that measures the superiority or inferiority of brand j with respect to the brands being tested. Assumption (1) is more restrictive than might appear at first glance, because in many practical problems it is unrealistic to assume that the two variables of classification have their effects additive in this simple fashion. For example, if the rows of Table 1 corresponded to different amounts of a chemical compound added to the soil, whereas the columns corresponded to different amounts of a second chemical compound added, one would not expect the effects of these compounds on crop productivity to operate independently in this manner.

In addition to assumption (1), the linear hypothesis model assumes that the variables X_{ij} are independently and normally distributed with the same variances σ^2.

Since the analysis of variance is being introduced as a technique for increasing the sensitivity of an experiment for testing hypotheses, consider the problem of testing the hypothesis that the theoretical column means of Table 1 are equal. For the illustration of operators and machine brands, this would mean testing the hypothesis that the four brands of machines are equally good, that is, that $\beta_1 = \beta_2 = \cdots = \beta_b$. This hypothesis is a generalization of the hypothesis $\mu_X = \mu_Y$ considered in Chapter 10. In terms of the notation introduced in (1), it follows from (2) that the hypothesis can be written in the form

$$(3) \qquad H_0 : \beta_1 = \beta_2 = \cdots = \beta_b = 0.$$

Under the foregoing assumptions and notation, the analysis of variance technique proceeds as follows. Write the total sum of squares of deviations of the observed values x_{ij} of Table 1 from their sample mean $\bar{x}$ in the following form:

$$\sum_{i=1}^{a} \sum_{j=1}^{b} (x_{ij} - \bar{x})^2 = \sum_{i=1}^{a} \sum_{j=1}^{b} [(\bar{x}_{i.} - \bar{x}) + (\bar{x}_{.j} - \bar{x}) + (x_{ij} - \bar{x}_{i.} - \bar{x}_{.j} + \bar{x})]^2.$$

If the trinomial on the right is squared and summed term by term, it will be found that the sums involving cross-product terms vanish; hence that

$$(4) \quad \sum_{i=1}^{a} \sum_{j=1}^{b} (x_{ij} - \bar{x})^2 = \sum_{i=1}^{a} \sum_{j=1}^{b} (\bar{x}_{i.} - \bar{x})^2 + \sum_{i=1}^{a} \sum_{j=1}^{b} (\bar{x}_{.j} - \bar{x})^2$$
$$+ \sum_{i=1}^{a} \sum_{j=1}^{b} (x_{ij} - \bar{x}_{i.} - \bar{x}_{.j} + \bar{x})^2.$$

As a partial verification of the fact that the cross-product terms do vanish, consider the evaluation of the second cross-product term. It is convenient to sum with respect to j first; thus

$$\sum_{i=1}^{a}\sum_{j=1}^{b}(\bar{x}_{i.} - \bar{x})(x_{ij} - \bar{x}_{i.} - \bar{x}_{.j} + \bar{x}) = \sum_{i=1}^{a}(\bar{x}_{i.} - \bar{x})\sum_{j=1}^{b}(x_{ij} - \bar{x}_{i.} - \bar{x}_{.j} + \bar{x}) \, .$$

But, summing term by term, it is clear from Table 1 that

$$\sum_{j=1}^{b}(x_{ij} - \bar{x}_{i.} - \bar{x}_{.j} + \bar{x}) = b\bar{x}_{i.} - b\bar{x}_{i.} - b\bar{x} + b\bar{x} = 0 \, .$$

Formula (4) shows that the total sum of squares can be broken down into three components, the first component measuring the variation of row means, the second component measuring the variation of column means, and the third component measuring the variation in the variables x_{ij} after the row and column effects have been eliminated.

The purpose of the breakdown in (4) was to separate the total variation into components that are of experimental interest and that can be used in a significance test using the F distribution of Chapter 10. It will turn out that the F value to use involves the ratio of two of the three sums of squares on the right side of (4). It is clear that the second sum of squares on the right should be used in the test because it measures the variation of column means and this variation is likely to be excessively large when H_0 is not true as compared to its value when H_0 is true. The last sum of squares should also be selected because it measures the variation of the x_{ij} after the variation due to row differences and column differences has been eliminated, and therefore it should prove useful as a basis for comparison for the second sum of squares. This technique of finding a measure of variation that has eliminated the effect of an interfering variable, such as operator skill, and using it as a basis for comparison with the variation of experimental interest, is a technique that often increases the sensitivity of the experiment remarkably. With the selection of these two sums of squares, the problem of testing H_0 is reduced to the problem of determining how to apply the F distribution to these two sums of squares. Consider, therefore, the method of converting these sums of squares into χ^2 variables.

The variable $\bar{X}_{.j}$ is a normal variable because it is a linear combination of the basic variables X_{ij} which are assumed to be normal. The mean of $\bar{X}_{.j}$, because of (1) and (2), is given by

$$E(\bar{X}_{.j}) = E\left(\frac{1}{a}\sum_{i=1}^{a} X_{ij}\right)$$

$$= \frac{1}{a}\sum_{i=1}^{a} E(X_{ij})$$

$$= \frac{1}{a}\sum_{i=1}^{a}(\alpha_i + \beta_j + \mu)$$

$$= \beta_j + \mu \, .$$

But when H_0 is true, this result together with (3) shows that $E(\bar{X}_{.j}) = \mu$. The variance of $\bar{X}_{.j}$ may be found by realizing that $\bar{X}_{.j}$ is the mean of a independent variables having the same variances σ^2. Thus the variance of $\bar{X}_{.j}$ is equal to σ^2/a. The variables $\bar{X}_{.j}$ are independent because the X_{ij} are independent; therefore, these results show that the variables $\bar{X}_{.j}$ are independently and normally distributed with the same means, μ, and the same variances, σ^2/a, when H_0 is true. By Theorem 2, Chapter 10, it therefore follows that

$$(5) \qquad \sum_{j=1}^{b} \frac{(\bar{X}_{.j} - \bar{X})^2}{\sigma^2/a} = \sum_{i=1}^{a} \sum_{j=1}^{b} \frac{(\bar{X}_{.j} - \bar{X})^2}{\sigma^2}$$

will possess a χ^2 distribution with $b - 1$ degrees of freedom. This proves that the second sum of squares on the right of (4), when divided by σ^2, and when expressed in terms of the random variables rather than their numerical values, will possess a χ^2 distribution with $b - 1$ degrees of freedom, provided that H_0 is true.

The demonstration that the last sum of squares of (4) when expressed in terms of random variables can be converted into a χ^2 variable is considerably more difficult than that just given for the second sum of squares. Because of the length and difficulty of the demonstration, the desired result is accepted without proof. Thus it is accepted that

$$(6) \qquad \frac{1}{\sigma^2} \sum_{i=1}^{a} \sum_{j=1}^{b} (X_{ij} - \bar{X}_{i.} - \bar{X}_{.j} + \bar{X})^2$$

possesses a χ^2 distribution with $(a - 1)(b - 1)$ degrees of freedom. The reason for this number of degrees of freedom is that in the derivation showing that (6) has a χ^2 distribution it is shown that the degrees of freedom on the left of (4) equals the sum of the degrees of freedom on the right. Since the left side of (4), when divided by σ^2, would possess a χ^2 distribution with $ab - 1$ degrees of freedom if the μ_{ij} were equal and since the first sum on the right has $a - 1$ degrees of freedom, it follows by subtraction that the last sum on the right must have

$$ab - 1 - [(a - 1) + (b - 1)] = (a - 1)(b - 1)$$

degrees of freedom.

Finally, in order to be able to apply Theorem 4, Chapter 10, to (5) and (6), it is necessary to know that (5) and (6) are independently distributed. The demonstration of this fact is quite difficult; hence the independence of (5) and (6) is also accepted without proof.

In view of the preceding discussion and Theorem 4, Chapter 10, if (5) is divided by $(b - 1)$ and (6) is divided by $(a - 1)(b - 1)$, the ratio of the resulting quantities will possess an F distribution. This result may be summarized in the following manner.

(7)   LINEAR HYPOTHESIS F TEST: *If the variables X_{ij} are independently and normally distributed with means $\mu_{ij} = \alpha_i + \beta_j + \mu$ and variances σ^2, the hypothesis $H_0: \beta_j = 0$ $(j = 1, \cdots, b)$ may be tested by using the right tail of the F distribution as critical region, where*

$$F = \frac{(a - 1) \sum\limits_{i=1}^{a} \sum\limits_{j=1}^{b} (\bar{X}_{\cdot j} - \bar{X})^2}{\sum\limits_{i=1}^{a} \sum\limits_{j=1}^{b} (X_{ij} - \bar{X}_{i\cdot} - \bar{X}_{\cdot j} + \bar{X})^2}$$

and where $v_1 = b - 1$ and $v_2 = (a - 1)(b - 1)$.

The right tail of the F distribution is selected as the critical region because the numerator of F is likely to be excessively large when H_0 is false. With this choice of critical region, the F test is known to be very good from the type II error point of view.

The equality of row means can be tested in a similar manner by using the first sum of squares on the right of (4) in the numerator of F and changing the degrees of freedom accordingly. Although the numerator in (7) can be written as a single sum, it is written as a double sum to remind one of the simple manner in which the value of F can be written down. All that one needs to do is to write out the fundamental identity (4), divide the sums of squares by their degrees of freedom, and take the proper ratio of two such quantities. The proper ratio depends on whether one is testing the equality of column means or the equality of row means.

2.2   Application of the Linear Hypothesis Model

For the purpose of illustrating the use of (7), consider the data of Table 2 on the yield of potatoes.   Four plots of land were divided into five subplots

TABLE 2

Treatment

		A	B	C	D	E
Plot	1	310	353	366	299	367
	2	284	293	335	264	314
	3	307	306	339	311	377
	4	267	312	312	266	342

each. For each plot, the five treatments were assigned at random to the five subplots. The problem here is to test whether the five treatments are equally effective with respect to mean yield.

The numerator sum of squares in (7) is readily computed directly from the means of the columns and the grand mean; however, the denominator sum of squares is most easily computed indirectly by computing the other sums of squares in (4) and then solving for this sum of squares. Calculations here yield the values

$$\sum_{j=1}^{5} (\bar{x}_{.j} - \bar{x})^2 = 3178$$

$$\sum_{i=1}^{4} (\bar{x}_{i.} - \bar{x})^2 = 1286$$

$$\sum_{i=1}^{4} \sum_{j=1}^{5} (x_{ij} - \bar{x})^2 = 21{,}530 .$$

Therefore, by formula (4), it follows that

$$\sum_{i=1}^{4} \sum_{j=1}^{5} (x_{ij} - \bar{x}_{i.} - \bar{x}_{.j} + \bar{x})^2 = 2388 .$$

As a result, the F value in (7) becomes

$$F = \frac{3 \cdot 4(3178)}{2388} = 16.0 , \qquad v_1 = 4 , \qquad v_2 = 12 .$$

From Table V it is clear that this result is significant; therefore, the five treatments undoubtedly differ in their affect on yield.

Since the preceding computations give the necessary sums of squares for testing the hypothesis that the row means are equal, that is, for testing the hypothesis

$$H_0 : \alpha_i = 0 , \qquad i = 1, \cdots, a ,$$

this hypothesis will also be tested. The value of F now becomes

$$F = \frac{4 \cdot 5(1286)}{2388} = 10.8 , \qquad v_1 = 3 , \qquad v_2 = 12 .$$

This result is also significant, which means that the four plots undoubtedly differ in fertility.

The computational results for analysis of variance problems are usually displayed in table form. Table 3 illustrates this type of summary for the problem just discussed.

The entries in the second column are the sums of squares in the fundamental identity (4). The third column lists the corresponding degrees of

TABLE 3

Source of variation	Sum of squares	d.f.	Mean square	F value
Columns	12,712	4	3178	16.0
Rows	6,430	3	2143	10.8
Remainder	2,388	12	199	
Totals	21,530	19		

freedom, and the fourth column gives each sum of squares divided by its degrees of freedom. These entries are the χ^2 values needed for the F ratios, which are displayed in the last column, and which were obtained by dividing each mean square by the remainder mean square.

In order to observe the increased sensitivity obtained by eliminating the variation due to differences in plot fertility when testing the hypothesis that the treatment means are equal, consider how the hypothesis would have been tested if the row classification were not available. This would be the situation, for example, if the five treatments had been assigned to the 20 subplots at random.

The fundamental identity (4) now reduces to

$$\sum_{i=1}^{a} \sum_{j=1}^{b} (x_{ij} - \bar{x})^2 = \sum_{i=1}^{a} \sum_{j=1}^{b} (\bar{x}_{\cdot j} - \bar{x})^2 + \sum_{i=1}^{a} \sum_{j=1}^{b} (x_{ij} - \bar{x}_{\cdot j})^2 .$$

It is easy to show that the second sum on the right, when divided by σ^2 and expressed in terms of the corresponding random variables, has a χ^2 distribution with $b(a - 1)$ degrees of freedom. Then accepting the fact that this χ^2 variable and the χ^2 variable given by (5) are independently distributed, it follows that the F distribution may be applied to the random variable

$$(8) \quad F = \frac{b(a - 1) \sum_{i=1}^{a} \sum_{j=1}^{b} (\bar{X}_{\cdot j} - \bar{X})^2}{(b - 1) \sum_{i=1}^{a} \sum_{j=1}^{b} (X_{ij} - \bar{X}_{\cdot j})^2}, \quad \nu_1 = b - 1, \quad \nu_2 = b(a - 1) .$$

The earlier calculations for Table 2 may be used to give the necessary values here. It will be found that

$$F = \frac{5 \cdot 3(12,712)}{4(8818)} = 5.4 , \quad \nu_1 = 4 , \quad \nu_2 = 15 .$$

The 5 percent and 1 percent critical values are 3.06 and 4.89; hence this F value is still significant at the 1 percent significance level but only barely so. A comparison of this result with the earlier result in which $F = 16.0$ shows that the segregation of plot differences in (4) gave rise to a much more sensitive experiment than that obtained by ignoring them.

The preceding illustration may give one the impression that the experimenter can choose either one of the two F tests applied there to determine whether a set of column means is equal. This is not strictly true, however, because the two models differ. The earlier problem concerning men and machines is a good one to illustrate the difference. The test based on (7) assumes that six men are selected and each performs four experiments, one with each brand of machine. The test based on (8) assumes that 24 men are selected and each performs one experiment with the brand of machine assigned him. In the first model the six men could have been selected at random, or otherwise, from a population of operators; however, it is assumed that in repetitions of the experiment the same six men are used. In the second model it is assumed that a fresh set of 24 men is selected at random from a population of operators every time the experiment is performed.

If H_0 were accepted in the first setup, the implication would be that there is no essential difference in the four machine brands as far as the six selected operators are concerned; however, in the second setup, the implication of no essential difference would apply to operators in general. Thus, the second model would be more desirable from an interpretation point of view, but because it does not eliminate the variability due to operator skill differences it might not be able to detect real differences in the machines, unless the experiment were very large. Since the analysis of variance technique was designed to increase the sensitivity of experiments and thereby decrease their costs, the first model is usually preferred in a situation such as this when large row variation is to be expected.

2.3 Analysis of Variance Estimation

Although the analysis of variance technique was presented as a technique for testing hypotheses, it is also a very useful tool for obtaining estimates of the various parameters, or functions of the parameters.

From (1) the parameters μ, α_i, and β_j were defined as the expected values of the corresponding random variables $\bar{X}$, $\bar{X}_{i.} - \bar{X}$, and $\bar{X}_{.j} - \bar{X}$; hence these random variables yield unbiased estimates of those parameters. Thus, using a wave over a parameter to denote an unbiased estimate of it,

$$\tilde{\mu} = \bar{X}, \quad \tilde{\alpha}_i = \bar{X}_{i.} - \bar{X}, \quad \tilde{\beta}_j = \bar{X}_{.j} - \bar{X}.$$

An unbiased estimate of σ^2 is given by

$$(9) \qquad \tilde{\sigma}^2 = \frac{\sum_{i=1}^{a} \sum_{j=1}^{b} (X_{ij} - \bar{X}_{i.} - \bar{X}_{.j} + \bar{X})^2}{(a-1)(b-1)}.$$

A demonstration of the fact that this estimator is unbiased can be based on (6). It was accepted earlier that (6) possesses a χ^2 distribution with $(a - 1)(b - 1)$ degrees of freedom. In Chapter 3 it was shown that the expected value of a χ^2 variable is equal to its degree of freedom; hence the expected value of (9) must be σ^2.

The preceding estimators may be used to give estimates of interesting functions of the parameters. For example, an experimenter interested in estimating the difference between two treatment effects corresponding to, say, the first and second columns, could use the estimator $\overline{X}_{.1} - \overline{X}_{.2}$. They may also be used to obtain confidence intervals for α_i or β_j by means of the technique employed in section 5.1, Chapter 10, based on the t distribution.

2.4 Generalizations

The linear hypothesis model can be generalized to cover situations in which there are more than two variables of classification. Thus, if there were three variables of classification, the fundamental identity (4) would assume the form

$$
\begin{aligned}
(10) \quad \sum \sum \sum (x_{ijk} - \overline{x})^2 &= \sum \sum \sum (\overline{x}_{i..} - \overline{x})^2 + \sum \sum \sum (\overline{x}_{.j.} - \overline{x})^2 \\
&+ \sum \sum \sum (\overline{x}_{..k} - \overline{x})^2 + \sum \sum \sum (\overline{x}_{ij.} - \overline{x}_{i..} - \overline{x}_{.j.} + \overline{x})^2 \\
&+ \sum \sum \sum (\overline{x}_{i \cdot k} - \overline{x}_{i..} - \overline{x}_{..k} + \overline{x})^2 + \sum \sum \sum (\overline{x}_{.jk} - \overline{x}_{.j.} - \overline{x}_{..k} + \overline{x})^2 \\
&+ \sum \sum \sum (x_{ijk} - \overline{x}_{ij.} - \overline{x}_{i \cdot k} - \overline{x}_{.jk} + \overline{x}_{i..} + \overline{x}_{.j.} + \overline{x}_{..k} - \overline{x})^2 .
\end{aligned}
$$

The general theory for the linear hypothesis model shows that one proceeds in the same manner as before. Thus it is merely necessary to divide each of the preceding sums of squares by its degrees of freedom and then take the proper ratio, depending on the hypothesis to be tested, to obtain an F ratio. For example, if the equality of row means were being tested, one would choose the first and last sums of squares on the right of (10) to form the F ratio. The new feature in (10), not found in (4), is that now there are sums of squares that measure the *interaction* between two variables. Thus the fourth sum of squares on the right measures the extent to which the first and second variables interact on each other. If, for example, different amounts of two different chemicals were applied to experimental plots of ground, it might happen that increased amounts of each chemical alone would increase yield but that when both chemicals were applied equally no appreciable increase in yield would result.

For the purpose of seeing how the fourth sum of squares is capable of measuring the interaction of the first two variables, consider the expression

$$(x_{ij} - \overline{x}_{i.}) - (\overline{x}_{.j} - \overline{x}) = y_{ij} - \overline{y}_{.j} .$$

This quantity is the typical term, before squaring and with the third variable dots omitted, in the sum of squares being discussed. If the row effects were strictly additive, then subtracting the row mean $\bar{x}_i$. from every cell entry, x_{ij}, in a two-way table would yield a set of observational values, y_{ij}, that are random except for column effects. The sample variance of these adjusted values in any given column would therefore yield an estimate of the basic variance σ^2. Since $\bar{y}._j$ is the column mean of the y_{ij}, it follows that $\sum_{i=1}^{a} (y_{ij} - \bar{y}._j)^2$, when divided by the appropriate number of degrees of freedom, would be expected to be a valid estimate of σ^2 and therefore that the double sum when divided by the proper number would yield such an estimate also.

Now suppose that the two variables do not act independently of each other in this additive fashion and that, say, the first row variable is beneficial in conjunction with the first column variable but harmful otherwise. Then the value of $y_{11} = x_{11} - \bar{x}_1$. would be expected to be larger than under independence, whereas the values of the y_{1j} for $j = 2, \cdots, b$ would be expected to be smaller (larger negatively) than under independence. If the other row variables also interact in various ways with the column variables, the net effect will be to produce sets of y_{ij} in the various columns that are more variable than under independence. As a result, the sum of squares being discussed will tend to be larger when interaction is present than when it is absent. The general theory of analysis of variance shows that a valid test of the hypothesis that there is no interaction between the first and second variables is given by applying the F test to the fourth and last sums of squares in the usual manner.

An analysis of variance design in which there are two variables of classification but in which there are k observations in each cell can be treated as a special case of a three-variable problem. Since the index k on x_{ijk} corresponds to a kth replication rather than to a third variable, the fundamental identity (10) would be rewritten to eliminate terms involving a segregation for the third variable. Thus the third, fifth, and sixth terms on the right would be combined with the seventh term to give a new remainder term. The breakdown would then become

$$(11) \quad \sum\sum\sum (x_{ijk} - \bar{x})^2 = \sum\sum\sum (\bar{x}_i.. - \bar{x})^2 + \sum\sum\sum (\bar{x}._j. - \bar{x})^2$$
$$+ \sum\sum\sum (\bar{x}_{ij}. - \bar{x}_i.. - \bar{x}._j. + \bar{x})^2 + \sum\sum\sum (x_{ijk} - \bar{x}_{ij}.)^2 .$$

The analysis then proceeds as usual. It is now possible to test for interaction between the two variables, whereas when there was but one observation per cell, as in (4), this was not possible.

The material presented here on the analysis of variance is only an introduction to this important topic. Books on experimental design discuss many other models and generalizations and give detailed discussions of applications.

3 STRATIFIED SAMPLING

The technique of breaking down the variation of a variable into useful components in order to decrease the experimental variation, as done in the analysis of variance, can also be used to advantage in designing experiments for estimating means of populations. It turns out that a more accurate estimate of the mean can often be obtained by taking restricted random samples than by using completely random samples. For example, suppose that an accurate estimate of the mean weight of fifth grade pupils is desired by a school system. By taking the proper size random samples in the various age groups, or in the various schools of the system, a more accurate estimate of the population mean can usually be obtained than by taking the same total sample at random in the system. In order to determine the proper size subsamples, consider the following general problem.

Let a population be divided into k distinct subpopulations. Further, let the mean and variance of this population be μ and σ^2 and of the ith subpopulation, μ_i and σ_i^2. Then consider as estimates of μ the quantities $\bar{X}$ and $\bar{X}_R$, where $\bar{X}$ is the mean of a random sample of size n and where

$$(12) \qquad \bar{X}_R = \frac{n_1}{n}\,\bar{X}_1 + \cdots + \frac{n_k}{n}\,\bar{X}_k$$

in which $\bar{X}_i$ is the mean of a random sample of size n_i drawn from the ith subpopulation and $\sum_1^k n_i = n$. This restricted type of random sampling is called *stratified sampling*.

For the purpose of comparing the relative precision of these two estimates of μ, consider their respective variances. The variance of $\bar{X}$ is given by $\sigma_{\bar{X}}^2 = \sigma^2/n$. Since the $\bar{X}_i$ are independent, the variance of (12) is given by

$$(13) \qquad \sigma_{\bar{X}_R} = \sum_1^k \left(\frac{n_i}{n}\right)^2 \sigma_{\bar{X}_i}^2 = \sum_1^k \left(\frac{n_i}{n}\right)^2 \frac{\sigma^2}{n_i} = \sum_1^k \frac{n_i \sigma^2}{n^2}.$$

In order to express the variance of $\bar{X}$ in terms of the σ_i^2, it is necessary to express the density function of the population in terms of those of the subpopulations. This may be done by applying the two basic rules of probability to the problem of determining the probability that X will assume a value within any specified interval. If p_i denotes the probability that X, which is assumed to be a continuous random variable, will come from the ith subpopulation and $f_i(x)$ denotes the density function for this subpopulation, then

$$p_i \int_\alpha^\beta f_i(x)\,dx$$

gives the probability that X will come from the ith subpopulation and will assume a value between α and β. Since these subpopulations are mutually exclusive, the probability that X will assume a value between α and β is the sum of all such probabilities; hence

$$\int_\alpha^\beta f(x)\,dx = p_1 \int_\alpha^\beta f_1(x)\,dx + \cdots + p_k \int_\alpha^\beta f_k(x)\,dx .$$

But α and β are arbitrary; consequently, under the assumption that these densities are continuous functions, it follows by the same reasoning as in (2), Chapter 6, that

$$f(x) = p_1 f_1(x) + \cdots + p_k f_k(x) .$$

Now

$$(14) \qquad \mu = \int_a^b x f(x)\,dx = p_1 \int_a^b x f_1(x)\,dx + \cdots + p_k \int_a^b x f_k(x)\,dx$$

$$= p_1 \mu_1 + \cdots + p_k \mu_k .$$

Furthermore

$$\sigma^2 + \mu^2 = \int_a^b x^2 f(x)\,dx$$

$$= p_1 \int_a^b x^2 f_1(x)\,dx + \cdots + p_k \int_a^b x^2 f_k(x)\,dx$$

$$= p_1(\sigma_1^2 + \mu_1^2) + \cdots + p_k(\sigma_k^2 + \mu_k^2) .$$

If the value of μ^2 is eliminated by means of (14) and the fact that $\sum_1^k p_i = 1$ is used, this will reduce to

$$\sigma^2 = \sum_1^k p_i[\sigma_i^2 + (\mu_i - \mu)^2] .$$

From this result it follows that the variance of $\bar{X}$ can be written in the form

$$(15) \qquad \sigma_{\bar{X}}^2 = \frac{1}{n}\sum_1^k p_i[\sigma_i^2 + (\mu_i - \mu)^2] .$$

Now consider a special type of sampling called *representative sampling* in which the subpopulation sample sizes n_i are chosen so that $n_i/n = p_i$. For a finite population this means that the relative sizes of the subpopulation samples are chosen equal to the relative sizes of the subpopulations. For representative sampling, (15) may be reduced by means of (13) to the form

$$(16) \qquad \sigma_{\bar{X}}^2 = \sigma_{\bar{X}_R}^2 + \sum_1^k \frac{n_i}{n^2}(\mu_i - \mu)^2 .$$

This shows that $\sigma_{\bar{X}}^2 > \sigma_{\bar{X}_R}^2$, unless the subpopulations have equal means. Representative sampling is of particular advantage for populations whose subpopulations have widely differing means. It follows from taking the expected value of (12) and using (14) that $\bar{X}_R$ is an unbiased estimate of μ; therefore a comparison of the variances of $\bar{X}_R$ and $\bar{X}$ is a comparison of their mean square errors of estimation.

Public opinion polls are familiar examples of representative sampling. For such polls it is customary to stratify the population in several ways. For example, it may be divided into several income groups, into several vocational groups, etc. Then, within strata, random samples are taken proportional to the relative sizes of those strata.

Various other types of restricted random sampling are available, most of which have been developed by governmental and industrial agencies for their particular needs.

As an illustration of the increased precision of estimating μ by the use of representative sampling, suppose for the sake of simplicity that a district is made up of 45 percent Democrats and 55 percent Republicans and that 70 percent of the Democrats will vote for a certain "nonpartisan" candidate in a primary election but only 20 percent of the Republicans will do so. Now suppose that a sample of size 200 is taken by each method. Although experience indicates that the precision of poll percentages is not so great as that given by binomial theory, the precisions here will be compared on a theoretical basis; consequently

$$\sigma_{\bar{X}}^2 = \sigma_{\hat{p}}^2 = \frac{pq}{n} = \frac{(.425)(.575)}{200} = .00122$$

and

$$\sum_1^k \frac{n_i}{n^2}(\mu_i - \mu)^2 = \frac{90}{(200)^2}(.70 - .425)^2 + \frac{110}{(200)^2}(.20 - .425)^2.$$

$$= .00031.$$

Therefore, from (16)

$$\sigma_{\bar{X}_R}^2 = .00091.$$

Since $\sigma_{\bar{X}_R}^2/\sigma_{\bar{X}}^2 = .75$ here, a considerable increase in precision would result from using representative sampling in preference to pure random sampling. Formulas (15) and (16) are valid for discrete variables; therefore the preceding application is a proper use of those formulas.

4 SAMPLING INSPECTION

The discussion thus far in this chapter has been concerned with techniques for designing valid experiments and for increasing the sensitivity of such experiments. In the remainder of this chapter the emphasis will be on a method that directly tries to minimize the amount of sampling needed to attain a desired sensitivity in the experiment. Since a smaller size analysis of variance experiment usually suffices to attain the same sensitivity as a corresponding more elementary design, the analysis of variance can be considered indirectly as a technique for decreasing the amount of sampling needed to attain a desired sensitivity. Thus all the techniques of this chapter can be thought of as techniques for decreasing the amount of sampling needed to attain the desired objective.

One of the very useful applications of the design of experiments to minimize the amount of sampling occurs in industrial sampling inspection. If a certain type of sampling procedure is agreed on, the notion of the two types of error can be used to advantage to design a good inspection procedure.

It is a common practice in industry to accept or reject lots of merchandise on the basis of a sample drawn from the lot. This practice arises from the fact that it is often more economical to tolerate a small percentage of defectives than to bear the cost of 100 percent inspection. The basis for accepting a lot of merchandise usually consists in specifying the maximum number of defective pieces that will be tolerated in a random sample of a given size. By means of such samples and specifications the purchaser is protected against receiving bad lots of merchandise.

Sampling inspection is quite different from quality control. It is a method for protecting the purchaser against poor quality after the product has been manufactured rather than a method for finding and correcting flaws in the manufacturing process, as in quality-control methods. When sampling inspection methods are applied to continuous manufacturing processes, however, they are often useful in helping to control the quality of the product.

From the consumer's point of view, there is a maximum percentage of defectives that he will tolerate. This percentage when expressed as a decimal is known as the *lot tolerance fraction defective* and is denoted by p_t. Without almost 100 percent inspection, it may be impossible to be certain that the quality is better than p_t; however, it is possible to set up a sampling procedure that will insure this quality with a certain probability. To this end consider a lot of N pieces from which a random sample of n pieces is to be selected. Let c denote the maximum number of defective pieces that will be tolerated in the sample if the lot is to be accepted.

Although numerous sampling schemes are available, only one common type of sampling procedure, known as *single sampling*, is considered here. This scheme proceeds as follows:

(17) (i) Inspect a sample of n pieces.

(ii) If the number of defective pieces does not exceed c, accept the lot; otherwise, inspect the entire lot.

(iii) Replace all defective pieces found by nondefective pieces.

Now consider the computation of the probability that the consumer will receive a bad lot under this sampling procedure, where bad is defined to be quality worse than p_t. Suppose that the lot being submitted for inspection is of p fraction defective so that it contains exactly Np defective and $N - Np$ nondefective items. Then the probability of obtaining exactly x defectives in a sample of size n is given by the ratio of the number of ways of choosing x things from Np things and $n - x$ things from $N - Np$ things to the number of ways of choosing n things from N things. Using the hypergeometric distribution given in (19), Chapter 3, this probability, $P\{x\}$, may be expressed in the form

(18)
$$P\{x\} = \frac{\binom{Np}{x}\binom{N - Np}{n - x}}{\binom{N}{n}}.$$

Under the sampling scheme (17), the consumer will accept a bad lot provided that $p > p_t$ and that $x \le c$. Assuming, therefore, that $p > p_t$, the probability that the consumer will accept a bad lot is given by summing the probabilities, $P\{x\}$, given by (18), for $x = 0, 1, \cdots, c$. Now it can be shown that $\sum_{x=0}^{c} P\{x\}$ decreases as p increases; consequently the probability that the consumer will accept a bad lot will be a maximum when p is as small as possible. But p cannot be less than p_t if the lot is to be judged as bad; hence the probability that the consumer will accept a bad lot cannot exceed the value

(19)
$$P_c = \frac{\sum_{x=0}^{c} \binom{Np_t}{x}\binom{N - Np_t}{n - x}}{\binom{N}{n}}.$$

This value is given a special name.

(20) DEFINITION: *The probability P_c that a consumer will accept a lot of fraction defective p_t, where p_t is his lot tolerance fraction defective, is called the consumer's risk.*

By demanding a small value of P_c, the consumer is protected against poor quality. If the actual fraction defective p is smaller than p_t, the consumer will be satisfied with the lot, whereas if it is larger than p_t he will wish to reject the lot, and the probability that he will fail to do so will not exceed P_c. Thus P_c is a conservative estimate of the probability that he will accept a bad lot.

From the producer's point of view any sampling scheme for deciding on the quality of a lot possesses the disadvantage of occasionally rejecting a lot of satisfactory quality. Most producers, however, are concerned principally with the percentage of the lots that are likely to be rejected, whether those lots deserve to be rejected or not. Thus the producer is interested in knowing what the probability is that a lot of N selected at random from his production line will be rejected. In order to calculate this probability, assume that the manufacturing process is under control with a process fraction defective p. This means that individual items coming off the production line may be considered to be random samples from a binomial population for which the probability that an item will be defective is p. From this point of view selecting a random sample of size N from the production process and then selecting a subsample of size n from the sample of N already selected is equivalent to selecting a random sample of size n directly from the production process. Since a lot will be rejected only if $x > c$, it follows that the probability that a lot will be rejected is equivalent to the probability of getting more than c successes in n trials of an experiment for which p is the probability of success in a single trial. The desired probability is therefore given by

$$(21) \qquad P_p = \sum_{x=c+1}^{n} \frac{n!}{x!\,(n-x)!}\, p^x q^{n-x}.$$

This value is also given a special name.

(22) DEFINITION: *The probability P_p that a producer will have a lot rejected when his production process is under control is called the producer's risk.*

The producer's risk is sometimes defined with respect to a single lot of size N, similar to the consumer's risk, in which case the binomial term being summed in (21) would be replaced by (18). The definition given by (22) is more realistic from the producer's point of view because he is interested in the long-run percentage of lots that will be rejected rather than in a particular lot of given quality. It should be observed that the consumer's risk does not depend on the process fraction defective p, whereas the producer's risk does.

4.1 Minimum Single Sampling

Thus far nothing has been said concerning the method of selecting values of n and c. The consumer's requirements fix the values of p_t and P_c in (19). Since N is specified, (19) places a single restriction on n and c. Now from the producer's point of view one desirable method of approach is to select that pair of values which minimizes the amount of inspection. Since a sample of size n is always inspected and the remainder of the lot is inspected with a probability given by (21), the mean number of pieces inspected per lot under the sampling scheme (17) will be given by

$$(23) \qquad I = n + (N-n)P_p.$$

In order to satisfy the consumer's demands and also minimize the amount of inspection, it is necessary to find that pair of values of n and c which satisfies (19) and minimizes (23). These quantities are difficult to manipulate; consequently the minimizing solution is obtained numerically for different values of the parameters involved. Extensive tables are available for the minimizing values of n and c for various values of the parameters and for P_c chosen equal to .10.

As an illustration, consider a lot of 1000 pieces for which the fraction defective is $p = .01$ and for which the consumer is willing to assume a risk of $P_c = .10$ of accepting a lot with a fraction defective of $p_t = .05$. By allowing c to assume small integer values and working numerically by trial and error methods, it will be found that the minimum amount of inspection will occur if a sample of 130 is taken and if the maximum allowable number of defectives is 3. With these values of n and c, it will also be found that the mean number of pieces inspected will be 164 as long as production remains in control. These results are easily obtained by consulting the Dodge and Romig tables referred to in exercise 17 at the end of this chapter.

4.2   Average Outgoing Quality Limit

A somewhat different approach to the problem of protecting the consumer from an inferior product is to attempt to guarantee him a certain quality level of the product after inspection, regardless of what quality level is being maintained by the producer. Toward this end, consider the problem of determining the mean value of the fraction defective after inspection if the producer's fraction defective is p.

First, it is necessary to derive a formula for calculating the expected value of a variable Y in terms of conditional expected values of Y when the population is split into two groups by means of a related variable X. For example, suppose the mean grade-point average of students in a given college is desired. It might be interesting to obtain the mean grade-point averages of students whose intelligence-quotient scores are less than 110 and of those whose scores are at least 110 and then combine the two means properly to give the desired mean.

Suppose that each member of a finite population is measured with respect to two variables X and Y. Let the population be split into two parts by the criterion that $X \leq c$ or $X > c$ and let $y_1, y_2, \cdots, y_k$ denote the possible values of y. The mean value of Y is given by

$$(24) \qquad\qquad E[Y] = \sum_{i=1}^{k} y_i P\{y_i\} .$$

Now, the probability $P\{y_i\}$ that Y will assume the value y_i can be obtained by considering the two mutually exclusive ways in which this event can occur. Either the individual selected at random will belong to the first group ($X \leq c$) and possess this value or he will belong to the second group ($X > c$) and possess this value. The sum of the probabilities of these two possibilities will yield $P\{y_i\}$; hence

$$P\{y_i\} = P\{X \leq c\}P\{y_i \mid X \leq c\} + P\{X > c\}P\{y_i \mid X > c\}.$$

If this formula is substituted in (24), it will follow that

$$(25) \quad E[Y] = P\{X \leq c\} \sum_{i=1}^{k} y_i P\{y_i \mid X \leq c\} + P\{X > c\} \sum_{i=1}^{k} y_i P\{y_i \mid X > c\}.$$

But each of the two sums on the right, when compared with (24), is seen to be a conditional expected value of Y. As a result, (25) may be written in the form

$$(26) \quad E[Y] = P\{X \leq c\}E[Y \mid X \leq c] + P\{X > c\}E[Y \mid X > c].$$

In order to use this formula to determine the mean value of the fraction defective after inspection, let Y be the number of defectives left in a lot when the procedure given in (17) is followed. When $X > c$, the entire lot will be inspected and all defective pieces will be replaced by nondefective pieces; hence the value of Y will be 0 and therefore the value of $E[Y \mid X > c]$ will be 0. When $X \leq c$, there will be $N - n$ uninspected pieces; hence the value of Y can range from 0 to $N - n$. But, since these $N - n$ pieces constitute a random sample of this size from a binomial population with probability p, the mean number of defectives will be $(N - n)p$; hence this is the value of $E[Y \mid X \leq c]$. If these values are inserted in (26), it will reduce to

$$(27) \quad E[Y] = P\{X \leq c\}(N - n)p.$$

In order to obtain the mean fraction defective after inspection, rather than the mean number of defectives as given by (27), it is merely necessary to divide both sides of (27) by N. If $\tilde{p}$ denotes the mean fraction defective after inspection, it therefore follows that

$$(28) \quad \tilde{p} = E\left[\frac{Y}{N}\right] = P\{X \leq c\}\left(1 - \frac{n}{N}\right)p.$$

Since X is the binomial variable discussed in connection with (21), $P\{X \leq c\}$ may be expressed as the sum of the corresponding binomial probabilities; hence (28) can be written

$$(29) \quad \tilde{p} = \left(1 - \frac{n}{N}\right)p \sum_{x=0}^{c} \frac{n!}{x!\,(n - x)!} p^x q^{n-x}.$$

This formula gives the mean fraction defective after inspection when following the inspection procedure (17); however, it is more commonly called the *average outgoing quality*.

When the sampling procedure given by (17) has been specified, the values of N, n, and c may be treated as given. Although (29) was calculated on the assumption that the producer's fraction defective is p and is known, the consumer is not likely to accept the producer's claim that p is the actual process fraction defective; hence p may not be treated as given. If $\tilde{p}$ is considered as a function of p only, it will be found that $\tilde{p}$ ordinarily possesses a maximum value that is assumed for a single value of p. This maximum value of $\tilde{p}$, denoted by $\tilde{p}_L$, is given the following special name.

(30) DEFINITION: *The maximum value, $\tilde{p}_L$, of the mean fraction defective after inspection, as a function of p, is called the average outgoing quality limit.*

The average outgoing quality limit is a number such that, regardless of what the producer's fraction defective may be, the mean fraction after inspection will not exceed it. This does not prevent a particular lot from containing worse quality, but in the long run the average value of the fraction defective in the inspected lots will not exceed $\tilde{p}_L$. Since these calculations are based on the assumption that the binomial distribution may be applied to defective parts coming off the production line, it is necessary that the production process be under control at some quality level p, even though the particular value of p being maintained is irrelevant.

The average outgoing quality limit has a certain appeal to many consumers that is not possessed by the protection afforded by a specified consumer's risk. For this reason, it is widely used as a basis for consumer protection.

It is usually possible to select several pairs of values of c and n that will yield functions, $\tilde{p}$, having approximately the same value of $\tilde{p}_L$. Figure 1 illustrates a typical situation. Since it is immaterial to the consumer which pair of values of c and n is chosen for a specified value of $\tilde{p}_L$, the producer is at

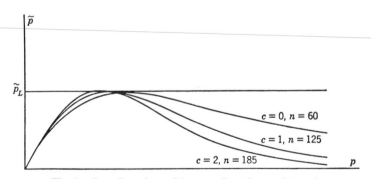

Fig. 1. Sampling plans with approximately equal p_L values.

liberty to choose them to his advantage. From his point of view it would be highly desirable to select that pair of values which minimizes the amount of inspection given by (23). As in the minimum single sampling scheme of the preceding section, the minimizing pair of values of c and n is obtained numerically. Tables are also available for determining these minimizing values corresponding to useful ranges of values of N, $\tilde{p}_L$, and p. It should be noted that the value of the process fraction defective p is required in order to minimize I, just as it was in the case of minimum single sampling.

As an illustration of the preceding ideas, consider the problem that was used as an illustration for minimum single sampling. There $N = 1000$ and $p = .01$. If the consumer requests an average outgoing quality limit of, say, $\tilde{p}_L = .03$, the Dodge and Romig tables referred to previously will give $c = 2$ and $n = 44$ as the values that will minimize the amount of inspection.

EXERCISES

Sec. 2.2

1. Given the following analysis of variance breakdown and the corresponding numerical values of the sums of squares, test the hypothesis that the column means are equal

$$\sum_{i=1}^{4}\sum_{j=1}^{6}(x_{ij} - \bar{x})^2 = \sum_{i=1}^{4}\sum_{j=1}^{6}(\bar{x}_{.j} - \bar{x})^2 + \sum_{i=1}^{4}\sum_{j=1}^{6}(x_{ij} - \bar{x}_{.j})^2$$

$$\qquad\quad 400 \qquad\qquad\qquad 170 \qquad\qquad\qquad 230$$

2. Suppose that the last sum of squares in problem 1 had been further analyzed to measure the row variability and had yielded the following indicated numerical values. Now test the hypothesis that the column means are equal.

$$\sum_{i=1}^{4}\sum_{j=1}^{6}(x_{ij} - \bar{x}_{.j})^2 = \sum_{i=1}^{4}\sum_{j=1}^{6}(\bar{x}_{i.} - \bar{x})^2 + \sum_{i=1}^{4}\sum_{j=1}^{6}(x_{ij} - \bar{x}_{i.} - \bar{x}_{.j} + \bar{x})^2$$

$$\qquad\quad 230 \qquad\qquad\qquad 110 \qquad\qquad\qquad 120$$

3. The following table gives the gains of four different types of hogs fed three different rations. Test to see whether the rations or the hog types differ in their effect on mean weight.

Type

Ration	I	II	III	IV
A	7	16	10.5	13.5
B	14	15.5	15	21
C	8.5	16.5	9.5	13.5

4. The following data represent the number of units of production per day turned out by 5 workmen using 4 types of machines. (a) Test to see whether the machines differ in mean productivity. (b) Test to see whether the workmen differ with respect to mean productivity.

Machine Type

		1	2	3	4
	1	44	38	47	36
	2	46	40	52	43
Workmen	3	34	36	44	32
	4	43	38	46	33
	5	38	42	49	39

5. Suppose an analysis of variance experiment involving 10 rows and 4 columns gave a significant result when testing the hypothesis of equal column means and that the first two column means were larger than the others. The error sum of squares in the denominator of F based on 36 degrees of freedom was equal to 180. If you now wished to test $H_0: \mu_1 = \mu_2$ against $H_1: \mu_1 > \mu_2$, approximately how large an equal size sample should you expect to take from each if you wanted to be certain with a probability of .90 of detecting a difference of $\mu_1 - \mu_2 = 1$?

6. Describe an analysis of variance experiment involving two variables for which it is clear that the linear model is a natural model for testing row or column variability.

Sec. 2.3

7. Assuming that the row means of problem 1 are equal, use the last sum of squares in that problem to obtain an unbiased estimate of σ^2.

8. Use the last sum of squares in problem 2 to find 96 percent confidence limits for σ^2.

Sec. 3

9. Suppose a population consists of two subpopulations whose means are 26 and 20 and whose variances are 8 and 6. If the two subpopulations are equally probable in random sampling, calculate the advantage in estimating the mean in taking two samples of 50 each from the two subpopulations over taking a random sample of 100 from the entire population.

10. Let p denote a cost factor that represents what a single sample costs when it is necessary to sample from one of the subpopulations in problem 9 as contrasted to one unit of cost for a single random sample from the entire population. Thus, if $p = 1.2$, it follows that 100 samples from a subpopulation will cost as much as 120 samples from the entire population. Determine the largest value of p such that the two methods of problem 9 will cost the same and yield the same precision in estimating μ.

11. A sample of size n is to be taken from a population made up of k strata consisting of N_i ($i = 1, \cdots, k$) members. If n_i denotes the size sample to be taken from the ith stratum and c_i the cost per sample from that stratum, and if a total of $c = \sum_{i=1}^{k} c_i n_i$ dollars can be spent for the sample, show that the variance of the estimate of the population mean will be a minimum if n_i is chosen proportional to $N_i \sigma_i / \sqrt{c_i}$, where σ_i^2 is the variance for the ith stratum population.

12. A population is made up of k subpopulations. The probabilities of success for an experiment for these subpopulations are $p_1, \cdots, p_k$, respectively. A set of n experiments is conducted by first drawing one of the k subpopulations at random and then performing the n experiments with it. If X denotes the number of successes obtained, show that the variance of X is given by

$$V(X) = n\mu_p(1 - \mu_p) + n(n - 1)V(p),$$

where μ_p is the mean of the p's and $V(p)$ is their variance. Explain what this implies concerning sampling from a population made up of highly different subpopulations.

13. If infected plants tend to occur in groups that are randomly distributed over an area and if p is the proportion of sampling areas that contain at least one group of infected plants, show that an estimate of the plant density of infected plants is $-\mu \log (1 - p)$, where μ is the mean number of infected plants per group.

Sec. 4

14. Using the Poisson approximation

$$\sum_{x=0}^{c} \frac{e^{-np}(np)^x}{x!}$$

with p equal to .05 and .01, respectively, to obtain the values of P_c and P_p given by (19) and (21), respectively, verify that the values of c and n given in the illustration on minimum single sampling are approximately correct to yield $P_c = .10$ and $I = 164$.

15. Using the Poisson approximation of problem 14, determine by numerical methods the values of c and n that minimize the amount of inspection for $N = 400$, $P_c = .10$, $p_t = .05$, and $p = .02$. Proceed by assigning c a value, beginning with 2, then determining the value of n to satisfy (19), and finally by selecting that pair of values which make (23) a minimum.

16. Explain why you should prefer to calculate the producer's risk from formula (21) than from the corresponding formula based on the hypergeometric distribution used in (18).

17. Consult the tables found in Dodge and Romig, *Sampling Inspection Tables* to verify that your results in problem 15 are approximately correct. Use these same tables to determine the average out-going quality limit for that problem.

18. Show that $P\{x \mid p\} \geq P\{x \mid p'\}$, where $Np' = Np + 1$, if $x < n(Np + 1)/(N + 1)$. This shows that $\sum_{x=0}^{c} P\{x \mid p\} \geq \sum_{x=0}^{c} P\{x \mid p'\}$, provided that $c < np$, and that $\sum_{x=0}^{c} P\{x \mid p\}$ is a decreasing function of p as assumed in the discussion following formula (18) in the text.

CHAPTER 12

Nonparametric Methods

Most of the statistical methods that have been considered so far have possessed two features in common. They have assumed that the functional form of the basic density function is known and they have been concerned with testing hypotheses about parameters of this density function or with estimating its parameters. For example, all the small sample methods developed in Chapter 10, with the exception of the material on the range, require that the basic variable be normally distributed and are concerned with testing or estimating means and variances of those variables. The χ^2 distribution of Chapter 9, however, was not restricted to problems of this type and is a striking exception to the general pattern of methods found in other chapters.

For situations in which very little is known about the distribution of the basic variable or for which it is known that the distribution is not of the required type, it is necessary to develop methods that do not depend on the particular form of the basic density function. A number of methods of this type have been designed. The only assumption that is needed for most of these methods is that the density function be continuous. A few of them, however, require that the density function possess low order moments.

Since the methods being described are not concerned with testing or estimating the parameters of a density function of a given type, they are usually called *nonparametric methods*. Such methods are also called *distribution-free methods* because they do not require a knowledge of how the basic variables are distributed. Since neither name is strictly correct for all of the methods usually listed under these names, the first name is used here because of tradition. Although a large number of nonparametric techniques are available to solve certain types of problems, only a few of the more important ones that are fairly easy to discuss are considered in this chapter. The first two are concerned with testing the location of a density function or the difference of the locations of two density functions. The third is designed to test the randomness of a sequence of observations taken over time. The last two are concerned with estimation and are designed to estimate the variability of a random variable and the distribution function of a random variable.

1 SIGN TEST

For nonparametric problems related to continuous variables the median is a more natural measure of location for a distribution than the mean. The median has the desirable property that the probability is $\frac{1}{2}$ that a sample value will exceed the population median regardless of the nature of the distribution. As a result, it is possible to design tests for testing hypothetical values of the median without knowing the form of the underlying distribution. The simplest of these is the *sign test*. For the purpose of describing this test assume that $f(x)$ is a continuous density function and assume that its median, denoted by ξ, is uniquely defined by

$$\int_{\xi}^{\infty} f(x)\, dx = \tfrac{1}{2}.$$

Let $X_1, X_2, \cdots, X_n$ be a random sample of X and consider testing the hypothesis

$$H_0: \xi = \xi_0 \quad \text{against} \quad H_1: \xi > \xi_0.$$

From the definition of the median it follows that when H_0 is true $P\{X - \xi_0 \geq 0\} = \frac{1}{2}$ and therefore that $P\{X_i - \xi_0 \geq 0\} = \frac{1}{2}$, $i = 1, \cdots, n$. Let

$$Z_i = \begin{cases} 1, & \text{if } X_i - \xi_0 \geq 0 \\ 0, & \text{if } X_i - \xi_0 < 0 \end{cases}.$$

Then the variable Z_i is a binomial variable corresponding to a single trial of an experiment for which $p = \frac{1}{2}$. Since the Z_i are independent, their sum $U = \sum_{i=1}^{n} Z_i$ will be a binomial variable corresponding to n independent trials of an experiment for which $p = \frac{1}{2}$.

Under H_1 the X_i will tend to be larger than ξ_0 and therefore the variable U will tend to exceed the value to be expected when H_0 is true. As a result, the right tail of the binomial distribution should be chosen as the critical region of the test. When H_0 is true,

$$E[U] = \sum_{i=1}^{n} E[Z_i] = \sum_{i=1}^{n} \frac{1}{2} = \frac{n}{2}.$$

For very small values of n it is necessary to calculate the right tail probabilities until a total probability of approximately α has been obtained to obtain the critical region for the test. However, for $p = \frac{1}{2}$ the binomial distribution is approximated well by its normal approximation for fairly small values of n; therefore it usually suffices to use the normal approximation on these problems. If the normal approximation is used, it is best to employ the small sample correction factor of $\frac{1}{2}$ based on the geometry of the problem and therefore treat τ as a standard normal variable where

$$(1) \qquad \tau = \frac{U \pm \frac{1}{2} - \frac{n}{2}}{\sqrt{\frac{n}{4}}}.$$

The correction $+\frac{1}{2}$ is used for a left tail critical region and $-\frac{1}{2}$ for a right tail region.

As an illustration, consider the following data obtained from testing the breaking strength of ceramic tile manufactured by a new cheaper process: 20, 42, 18, 21, 22, 35, 19, 18, 26, 20, 21, 32, 22, 20, 24. Suppose that experience with the old process produced a median of 25. Then a natural hypothesis to test here is

$$H_0: \xi = 25 \quad \text{against} \quad H_1: \xi < 25$$

because there is reason to believe that the cheaper process may have lowered quality.

Subtracting 25 from each of the observed values will yield the following Z values: 0, 1, 0, 0, 0, 1, 0, 0, 1, 0, 0, 1, 0, 0, 0, and the value $U = 4$. It is customary in carrying out the test to record only the signs of the numbers $X_i - \xi$ rather than the Z values, in which case one would record the sequence

$$- + - - - + - - + - - + - - - .$$

Then the total number of positive signs gives the value of U. It is for this reason that the test is called the sign test. Since $n = 15$, application of formula (1) then gives

$$\tau = \frac{4 + \frac{1}{2} - \frac{15}{2}}{\sqrt{\frac{15}{4}}} = -1.55 .$$

From Table II it will be found that $P\{\tau \leq -1.55\} = .06$; therefore H_0 is accepted if $\alpha = .05$.

Calculations by means of the binomial density with $n = 15$ and $p = \frac{1}{2}$ will show that

$$P\{0\} = .000, \quad P\{1\} = .000, \quad P\{2\} = .003, \quad P\{3\} = .014, \quad P\{4\} = .042 .$$

Hence it follows that $P\{U \leq 4\} = .059$, which demonstrates that the normal approximation with the $\frac{1}{2}$ correction is excellent here.

If testing a hypothetical value of the median is the nonparametric analogue of testing a hypothetical value of the mean for a normal distribution, testing the difference of two medians should be the analogue of testing the difference of two normal means. The sign test can be adapted to solving that problem also.

In applying Student's t distribution to testing the difference of the means of two normal distributions it was necessary to assume that the variances of the two distributions are the same. As a result, that test can be thought of as a *slippage test* in the sense that geometrically the graph of the density for Y can be obtained from the graph of the density for X by sliding the latter normal curve along the axis until it is centered over μ_y. The nonparametric test that is about to be discussed is the nonparametric analogue of this normal distribution slippage test.

In applying the sign test to the slippage problem it is necessary to assume that equal size samples are taken from the two populations. In the next section a different type of test will be introduced that is not subject to this restriction. The sign test however, is particularly useful when observations come in related pairs and it will therefore be described from that point of view.

Let f_1 and f_2 be the two continuous density functions under discussion and let $(X_1, Y_1), (X_2, Y_2), \cdots, (X_n, Y_n)$ denote n paired sample values to be drawn from the two populations. Consider the hypothesis

$$(2) \qquad\qquad H_0 : f_1(x) = f_2(x) .$$

For the purpose of testing this hypothesis, it is convenient to consider the differences $X_i - Y_i$, $(i = 1, 2, \cdots, n)$. When H_0 is true, X_i and Y_i constitute a random sample of size two from the same population. Since the probability that the first of two sample values will exceed the second is the same as the probability that the second will exceed the first and since, theoretically, the probability of a tie is zero, it follows that the probability that $X_i - Y_i$ will be positive is $\frac{1}{2}$. Thus, if only the signs of the differences are considered, a nonparametric test for H_0 can be constructed. Toward this end, let

$$Z_i = \begin{cases} 1, & \text{if } X_i - Y_i \geq 0 \\ 0, & \text{if } X_i - Y_i < 0 \end{cases}.$$

Then the variable Z_i is a binomial variable corresponding to a single trial of an experiment for which $p = \frac{1}{2}$. Since the Z_i are independent, their sum $U = \sum_{i=1}^{n} Z_i$ will be a binomial variable corresponding to n independent trials of an experiment for which $p = \frac{1}{2}$.

In order to use this last result for testing H_0, consider as an alternative to H_0 the hypothesis

$$(3) \qquad\qquad H_1 : f_1(x) = f_2(x - c) .$$

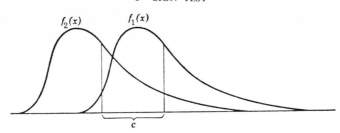

Fig. 1. A density function and its translation.

where c is some positive constant. H_1 states that the second density function is merely the first density function shifted to the left a distance of c units. Figure 1 illustrates the relationship between $f_1(x)$ and $f_2(x)$.

Under H_1, the X_i will tend to be larger than the Y_i and the variable U will tend to exceed its expected value of $n/2$. One would therefore choose as critical region the right tail of the binomial distribution. If c had been negative, the left tail would have been chosen; however, if H_1 were the alternative that a translation of unknown direction had occurred, both tails would be used.

A useful feature of the sign test is that it is applicable to situations in which the density functions f_1 and f_2, although identical under H_0 for each pair of samples, change from sample pair to sample pair. For example, in determining whether two types of coating for soil pipe are equally resistant to corrosion, the experimenter would ordinarily subject each type of coating to the same set of soil types and thus obtain a pair of experimental values for each soil type. In such an experiment it might happen that for one type of soil there is very little corrosion for either coating, whereas for another type there is a great deal of corrosion for both coatings. One would expect the variation in the amount of corrosion for the first soil type to be considerably smaller than the variation for the second. Such differences, however, do not affect the sign test because the distribution of each Z_i is binomial regardless of the nature of the density for each i.

As an illustration of how the sign test is applied, consider the data found in Table 1. These data were obtained by taking random samples of 30 from two normal populations with means 14 and 16 and standard deviations 2 and 2, respectively. The reason for choosing normal data is that it is interesting to compare this and other nonparametric methods with standard methods based on normality.

If the differences $x_i - y_i$ are taken, it will be found that there are 10 positive and 20 negative differences; hence $U = 10$. Since the normal approximation to a binomial distribution with $p = \frac{1}{2}$ is excellent for n as large as 30, U may be treated as a normal variable here.

TABLE 1

x	13.3	14.6	13.6	17.2	14.1	10.6	15.9	14.7	14.2	14
y	14.1	15.1	9.9	14.5	17.9	16.1	16.8	15.1	13.2	18

x	17.4	15.6	8.2	13.8	15.4	16.3	17.7	15	13.4	13.4
y	16.3	13.3	15.8	18	20.4	15.7	21.5	14.5	16.7	13.7

x	16	13.3	14.9	12.9	14	16.2	11.5	10.4	12.6	18.1
y	13.6	17	15.7	16.8	18.8	18.8	16	14.6	12.3	17.7

Because it is known that the alternative given by (3) holds here with $c = -2$, let the hypothesis to be tested be that given by (2) and let the alternative be that given by (3), where $c < 0$. The left tail of the U distribution will therefore be chosen as the critical region for the test. In order to test the hypothesis, it suffices to calculate the probability that $U \leq 10.5$. Thus

$$\tau = \frac{10.5 - 15}{\sqrt{30 \cdot \frac{1}{2} \cdot \frac{1}{2}}} = -1.643$$

and hence $P\{U \leq 10.5\} \doteq .050$. Since this probability is the borderline value for making a decision, one might toss a coin or compute P more accurately; however, the hypothesis is known to be false here.

If this problem is worked on the assumption that the differences of the paired values can be treated as random sample values of a normal variable and if Student's t is calculated for the differences, it will be found that

$$t = -3.16 \,.$$

For 29 degrees of freedom, the .005 point for the t distribution is 2.76; hence the probability of obtaining a value less than -3.16 is much smaller than .005. A comparison of this result with that above using the sign test shows that the t test was able to demonstrate the real difference existing in the two populations with greater assurance than the sign test. This, of course, is to be expected when the normality assumption holds because the t test is known to possess optimal properties for this type of problem.

Although the sign test may appear to be rather weak, it can be shown that it possesses certain optimal properties for testing a hypothetical median when nothing is known about the underlying distribution. It is only when one compares the sign test with a test based on a given distribution, such as the normal distribution, that the sign test suffers.

2 RANK SUM TEST

If the data for testing the hypothesis $f_1(x) = f_2(x)$ do not consist of matched pairs, the sign test is not the natural test to apply. This is particularly true if the sizes of the samples from the two populations differ considerably because then the sign test will waste some of the data. A simple nonparametric test for this more general situation can be obtained by studying the possible ordered arrangements of the combined sample values.

Let x_1', x_2', $\cdots$, x_{n_1}' and y_1', y_2', $\cdots$, y_{n_2}' denote random samples of sizes n_1 and n_2 taken from the populations with the continuous density functions f_1 and f_2, respectively. Let these two sets of sample values be arranged in order of increasing magnitude and denote the ordered sets by x_1, x_2, $\cdots$, x_{n_1} and y_1, y_2, $\cdots$, y_{n_2}. If the two ordered sets are combined into a single ordered set, a typical arrangement such as

$$(4) \qquad y_1, y_2, x_1, y_3, x_2, y_4, y_5, x_3, \cdots$$

will be obtained. When $f_1(x) = f_2(x)$ the x_i' and y_j' represent random sample values from the same population. The combined set x_1', $\cdots$, x_{n_1}', y_1', $\cdots$, y_{n_2}' therefore represents a random sample of size $n_1 + n_2$ from this common population. Since the sampling is random, any particular order of these sample values should have the same probability of occurring as any other order. For example, the first sample value to be drawn, x_1', has the same probability of being, say, the largest value obtained as the second sample value x_2'. Thus each of the $(n_1 + n_2)!$ possible permutations of x_1', $\cdots$, x_{n_1}', y_1', $\cdots$, y_{n_2}' has the same probability of being the ordered set of values. In calculating the probability that a particular type of ordered set will be obtained, it is therefore necessary to count the number of the $(n_1 + n_2)!$ permutations that give rise to the desired order type and divide this number by $(n_1 + n_2)!$.

Although it makes no difference which variable is labeled X and which Y, it is convenient when consulting the tables that have been constructed for the test to be presented here to have $n_1 \leq n_2$ and therefore to label the variables accordingly.

After this joint ordering, as in (4), has been performed, write down the ranks of the x values and let T denote the sum of those ranks. For example, in (4) one would write down beneath the consecutive x values displayed there the numbers 3, 5, 8, $\cdots$ because those are the ranks of the x values in the combined set. The value of T will then be the sum of those numbers.

Now the sampling distribution of T under the assumption that $f_1(x) = f_2(x)$, and therefore based on the resulting assumption that all $(n_1 + n_2)!$ permutations of the combined set of values have the same probability of being the ordered set, has been worked out by combinatorial methods. The distribution of T depends, of course, on the sizes, n_1 and n_2, of the two samples. Table VII in the appendix gives the necessary critical values corresponding to various sample sizes for $n_2 \leq 10$. For small values of n_1 and n_2 this computation can be done by listing the various permutations that give rise to fixed values of T. For example, suppose $n_1 = 3$ and $n_2 = 4$ and the value of $P\{T = 10\}$ is desired. Since the possible ranks for the x's range from 1 to 7, the only possible sets of ranks that will yield a total of 10 are $(1, 2, 7)$, $(1, 3, 6)$, $(1, 4, 5)$, and $(2, 3, 5)$. The x's can be permuted in their rank positions in 3! ways and the y's in their positions in 4! ways; therefore there are $4 \cdot 3! \, 4!$ permutations that give rise to $T = 10$. As a result

$$P\{T = 10\} = \frac{4 \cdot 3! \, 4!}{7!} = \frac{4}{35}.$$

For larger sample sizes the distribution of T can be approximated satisfactorily by the proper normal distribution. This is the normal distribution with mean and variance given by the formulas

(5)
$$E(T) = \frac{n_1(n_1 + n_2 + 1)}{2}$$

$$\sigma_T^{\,2} = \frac{n_1 n_2(n_1 + n_2 + 1)}{12}.$$

The problem used to illustrate the sign test is also used to illustrate the rank sum test. If the 30 values of x and the 30 values of y from Table 1 are ordered in magnitude, the following combined ordering will be obtained. The y values are the italicized values. Tied ranks may be averaged or alternated in the ordering. Here they were alternated.

(6)
8.2, *9.9*, 10.4, 10.6, 11.5, *12.3*, 12.6, 12.9, *13.2*, 13.3,

13.3, 13.3, 13.4, 13.4, *13.6*, 13.6, *13.7*, 13.8, 14.0, 14.0,

14.1, *14.1*, 14.2, *14.5*, *14.5*, *14.6*, 14.6, 14.7, 14.9, 15.0,

15.1, *15.1*, 15.4, 15.6, *15.7*, *15.7*, *15.8*, 15.9, *16.0*, 16.0,

16.1, 16.2, 16.3, *16.3*, *16.7*, *16.8*, *16.8*, 17.0, 17.2, 17.4,

17.7, 17.7, *17.9*, 18.0, *18.0*, 18.1, *18.8*, *18.8*, 20.4, 21.5 .

In order to calculate the value of T, it is necessary to write down the ranks of the x values in (6).

$$1, \ 3, \ 4, \ 5, \ 7, \ 8, 10, 12, 13, 14, 16, 18, 19, 20, 21, 23,$$
$$27, 28, 29, 30, 33, 34, 38, 40, 42, 43, 49, 50, 52, 56 \ .$$

The sum of these ranks is $T = 745$. Since $n_1 = n_2 = 30$, application of formulas (5) will show that

$$E(T) = 915, \quad \sigma_T{}^2 = 4575 \ .$$

Consequently,

$$\tau = \frac{T - E(T)}{\sigma_T} = \frac{745 - 915}{67.6} = -2.51 \ .$$

Since τ is an approximate standard normal variable, it follows from Table II that $P\{\tau \leq -2.51\} \doteq .006$. A comparison of this result with that of the sign test shows that the rank sum test is superior to the sign test for this problem. The value of $t = -3.16$ for the Student's t test shows, however, that the t test is still somewhat better than either of the nonparametric tests for this problem, as was to be expected.

When testing the hypothesis $f_1(x) = f_2(x)$ against the alternative hypothesis that the first population is situated to the left of the second population, the left tail of the T distribution should obviously be chosen as the critical region. One should, of course, use the right tail if the shift has been to the right. The rank sum test is known to be an excellent test for the slippage problem being considered here. It can be shown, for example, that even when the variables are normal, the rank sum test based on 100 observations is approximately as good as Student's t test based on 95 observations.

The rank sum test is an example of a test constructed on the *randomization principle*. In such a test one studies the distribution of a statistic, such as T, under all possible permutations of the sample values rather than under repeated random sampling. For the purpose of explaining how this philosophy of testing differs from that of traditional testing, suppose a random sample of size n is taken of some random variable X and that it yielded the values $x_1, x_2, \cdots, x_n$. In traditional testing, a statistic, such as $\bar{X}$, is chosen and then its sampling distribution under repeated sampling is found in order to determine a critical region for the test. A test based on the randomization principle is constructed in much the same manner except that in determining the critical region one considers the distribution of the statistic, such as $\bar{X}$, under all possible permutations of the values $x_1, x_2, \cdots, x_n$. In a sense, a new random sample under the traditional approach corresponds to a new permutation of the original sample values rather than a new set of values. This randomization principle can be used to construct nonparametric versions of the standard tests based on the t and F statistics, for example; however, it is very difficult to find the distribution of such statistics under randomization for even fairly small samples. The rank sum test is one such test where the computations are feasible and for which a good large sample approximation exists.

3 SERIAL CORRELATION

Most of the statistical methods that have been considered in the preceding chapters were designed to be applied to data for which no useful information was gained by preserving the time order of the observations. It was assumed that the observations constitute a random sample from a fixed population, in which case the time order can be ignored. If there is reason to believe that the observations may not behave like a random set when they are taken over some time interval, then it is necessary to test the randomness of the sequence before the usual statistical methods based on randomness can be applied.

Although a few of the statistical methods in the preceding chapters have been based on the order of the observations, they assume that the distribution of the basic variable is known. For example, the quality control chart technique was applied to such variables as normal and binomial variables. In this section a nonparametric method based on the randomization principle is discussed for testing the randomness of sequences of observations.

For data obtained from consecutive daily observations of some economic variable one might expect to find these daily values correlated. This would certainly be true for a variable such as a stock quotation. In studying the number of daily automobile accidents occurring in a city over a period of time, one might expect to find a positive correlation between values that are seven days apart because certain days of the week have more traffic and hence more accidents than others. Prices of many seasonal products vary regularly over the year; therefore they would show a cyclical pattern and give rise to a positive correlation if monthly prices twelve months apart were correlated.

In view of examples such as these it is to be expected that a test for randomness based on correlation should prove effective in detecting a lack of randomness in many sets of observations taken over time.

Let $x_1, x_2, \cdots, x_n$ denote the sequence to be tested for randomness and consider the ordinary correlation coefficient calculated for this sequence when y_i is chosen as x_{i+1} for $i = 1, 2, \cdots, n - 1$. With this choice for y, the corresponding values of x and y are those indicated in Table 2.

The correlation coefficient of the values given in Table 2 is called the *serial correlation coefficient* with lag 1. If y_i had been chosen equal to x_{i+k}, the correlation coefficient of the resulting x and y values would have

TABLE 2

x	x_1	x_2	$\cdots$	x_i	$\cdots$	x_{n-1}
y	x_2	x_3	$\cdots$	x_{i+1}	$\cdots$	x_n

been called the serial correlation coefficient with lag k. Such a correlation would be effective in detecting a cycle of length k.

If the sequence $x_1, x_2, \cdots, x_n$ could be treated as a random sample of size n from a normal population, one would expect to be able to apply some normal distribution theory such as the regression theory of Chapter 10 to solve the present problem. However, since the y_i no longer constitute a set of random sample values for a fixed set of x's, nor do the pairs of values (x_i, y_i) constitute a set of random sample values from a joint distribution, the ordinary regression and correlation theory is not applicable here. Furthermore, since this chapter is concerned with methods that do not require a normality or similar assumption, the methods of Chapter 10 would not be available for this reason also.

A nonparametric method based on serial correlation can be devised on the randomization principle if it is assumed that all permutations of the sequence being considered are equally probable. For each such permutation, one can calculate the value of the serial correlation coefficient. Since there are $n!$ permutations possible, there are $n!$ values of the serial correlation coefficient to be computed. The ordered set of values obtained, together with the frequencies of those values that are obtained more than once, give the distribution of the serial correlation coefficient with respect to the set of $n!$ permutations of the sequence. For most sequences one would expect the distribution to be fairly symmetrical and to be centered near the origin. If the sequence being tested yielded a large positive or negative value of the serial correlation coefficient, its randomness would be questioned. In order to obtain a critical region for testing randomness, it would be necessary to find two values of the serial correlation coefficient, one for each tail of the distribution, such that, say, 5 percent of the $n!$ values of the serial correlation coefficient lay outside the interval determined by the two values.

It is obvious from the preceding discussion that the computational difficulties of the proposed test become prohibitive for n at all large; hence it is necessary to find an approximation for the distribution of the serial correlation coefficient when n is large. This problem is considered next.

For ease of discussion, let y_n be defined to be x_1. Then Table 2 will contain n pairs of values rather than $n - 1$ pairs. The resulting correlation coefficient is called the *circular form* of the serial correlation coefficient. For the extended table the serial correlation coefficient may be expressed in the form

$$r = \frac{\sum\limits_{i=1}^{n} X_i X_{i+1} - n\overline{X}\,\overline{Y}}{n s_X s_Y}.$$

Since all n values of the sequence occur in both rows of the extended Table 2 and the statistics $\overline{X}$, $\overline{Y}$, s_X, and s_Y are independent of the order of the sample values, it follows that $\overline{X}$, $\overline{Y}$, s_X, and s_Y are unchanged under permutations of the sequence. Now the only quantity in r that is affected by permutations of the sequence is the sum $\sum_{i=1}^{n} X_i X_{i+1}$; therefore it suffices to study the distribution of this sum rather than the distribution of r itself. Furthermore, as n becomes large, any differences in the distribution of r, or of this sum, for the standard definition and the circular definition of the serial correlation coefficient disappear; therefore it suffices to consider the statistic

$$R = \sum_{i=1}^{n} X_i X_{i+1}.$$

If it is assumed that the values of the sequence being tested constitute a random sample from a population that possesses low order moments, then it can be shown that the random variable R has an approximate normal distribution for large n. In order to test the hypothesis of zero serial correlation, one must know the mean and variance of R. The necessary values are given by the formulas

$$E(R) = \frac{S_1^{\,2} - S_2}{n - 1}$$

and

$$\sigma_R^{\,2} = \frac{S_2^{\,2} - S_4}{n - 1} + \frac{S_1^{\,4} - 4S_1^{\,2}S_2 + 4S_1 S_3 + S_2^{\,2} - 2S_4}{(n - 1)(n - 2)} - E^2(R)$$

where

$$S_k = x_1^{\,k} + x_2^{\,k} + \cdots + x_n^{\,k}.$$

Unfortunately, the computations involved in evaluating the mean and variance are somewhat lengthy if n is at all large. Since the test is not affected by adding the same constant to each member of the sequence or multiplying each member by the same constant, these computations can be simplified considerably by replacing the observed values by reduced values.

As an illustration of how the modified serial correlation test is applied, consider the following sequence of values that was obtained from taking daily samples from an industrial production line: .220, .213, .221, .222, .219, .214, .222, .216, .212, .221, .223, .214, .221, .216, .217, .215. If .218 is subtracted from each value and the resulting values are multiplied by 1000, the following sequence will be obtained:

$$2, -5, 3, 4, 1, -4, 4, -2, -6, 3, 5, -4, 3, -2, -1, -3 .$$

Computations yield the values

$$S_1 = -2, \qquad S_2 = 200, \qquad S_3 = -170, \qquad S_4 = 3944 .$$

If these values are substituted in the formulas for $E(R)$ and $\sigma_R{}^2$, it will be found that

$$E(R) = -13.1 \qquad \text{and} \qquad \sigma_R = 48.7 .$$

It will also be found that $R = -67$, hence that

$$Z = \frac{R - E(R)}{\sigma_R} = -1.11 .$$

Assuming that the normal approximation is satisfactory, one would choose equal tails of the approximating normal curve as the critical region. Since, for a normal variable,

$$P\{|Z| > 1.11\} = .27$$

it follows that the hypothesis of randomness is accepted.

It should be pointed out that the preceding theory was discussed from the point of view of the serial correlation with lag 1; however, it is applicable to other lags also. One calculates the corresponding value of R and performs the test as usual.

The preceding section has been concerned with a nonparametric test for deciding whether a sequence in time is random. The problem of discovering a lack of randomness in time data and the nature of it is a very important and difficult problem in statistics. The technique presented here is one of the simplest available to describe, and is intended only as a mild introduction to one feature of the analysis of time series.

4 TOLERANCE LIMITS

The preceding nonparametric methods have been concerned with testing hypotheses. In this and the next section interest will center on nonparametric estimation. Since the sample median is an estimate of the population median ξ and probability statements concerning its accuracy can be made by means of the sign test techniques, it is not necessary to discuss further the estimation of the location of an unknown distribution. What is lacking, however, is a nonparametric method for estimating the variability of a random variable. It will not suffice to calculate the sample variance and merely state that it is an estimate of the population variance if it is necessary to know the accuracy of the estimate because that can be determined only if one knows the distribution of the random variable. In this section a method will be presented for measuring the variability of a random variable by means of samples and for which probability statements concerning accuracy can be made without knowing its density function.

Let the continuous random variable X possess the continuous density function $f(x)$ and let a random sample of size n be taken of it. Then from Theorem 5, Chapter 10, the density function of the smallest, U, and largest, V, values in such a sample is given by

$$(7) \qquad f(u, v) = n(n - 1)f(u)f(v)\left[\int_u^v f(x)\, dx\right]^{n-2}.$$

The integral within the brackets is the proportion of the distribution lying between the extreme values of the sample; therefore it is a measure of the variability of the random variable X. If this integral has a value near 1 and u and v are close together, that would imply that there is very little variability in X because a high percentage of the population is confined to a narrow interval of values. The nonparametric random variable that will be used to measure variability is therefore chosen to be the variable

$$Z = \int_U^V f(x)\, dx\,.$$

For the purpose of deriving the density function of Z, let the variable u be held fixed in (7) and consider the change of variable from v to z where

$$(8) \qquad z = \int_u^v f(x)\, dx\,.$$

From formula (6), Chapter 10, it follows that the joint density function of U and Z will be given by

$$h(u, z) = \frac{f(u, v)}{\left|\dfrac{\partial z}{\partial v}\right|}\,.$$

But from the familiar calculus formula for differentiating an integral with respect to its upper limit, it follows from (8) that

$$\frac{\partial z}{\partial v} = f(v)\,.$$

Hence, from (7),

$$(9) \qquad h(u, z) = \frac{f(u, v)}{f(v)} = n(n - 1)f(u)z^{n-2}\,.$$

Now hold z fixed and consider the change of variable from u to w, where

(10) $$w = \int_a^u f(x)\, dx \, .$$

Since $dw/du = f(u)$, it follows from formula (6), Chapter 10, and (9) that the joint density of W and Z is given by

$$k(w, z) = \frac{h(u, z)}{\left| \dfrac{dw}{du} \right|} = n(n - 1)z^{n-2} \, .$$

Finally, to obtain the density of Z it is merely necessary to integrate $k(w, z)$ with respect to w over its range of values for z fixed. From (8) and (10) it will be observed that $w + z$ equals the probability that X will be inside the interval (a, v); therefore $w + z \leq 1$. Since z is being held fixed, w can assume values from 0 to $1 - z$ only; hence the density of Z is given by

$$g(z) = \int_0^{1-z} k(w, z)\, dw$$

$$= \int_0^{1-z} n(n - 1)z^{n-2}\, dw$$

$$= n(n - 1)z^{n-2}(1 - z) \, .$$

This result may be expressed in the following form.

THEOREM: *If the random variable X possesses the continuous density function $f(x)$ and if Z denotes the proportion of the distribution that lies between the extreme values of a random sample of size n, then the density function of Z is given by*

$$g(z) = n(n - 1)z^{n-2}(1 - z), \qquad 0 \leq z \leq 1 \, .$$

As an illustration, consider the problem of determining how large a sample must be taken in order to be certain with a probability of .90 that at least 95 percent of the distribution will be between the extreme values of the sample. The solution is given by determining the value of n which satisfies the equation

$$\int_{.95}^{1.00} g(z)\, dz = .90 \, .$$

If the value of $g(z)$ is inserted and the integration is performed, this equation becomes

$$n(n - 1)\left[\frac{1}{n - 1} - \frac{1}{n} - \frac{(.95)^{n-1}}{n - 1} + \frac{(.95)^n}{n} \right] = .90$$

which simplifies to

$$(.95)^{n-1} = \frac{2}{n + 19}.$$

It will be found by trial and error that the integer that most nearly satisfies this equation is $n = 76$; consequently a sample of this size is required to obtain the desired coverage. The extreme values U and V are called tolerance limits in connection with the variability problem. They are special cases of other types of tolerance limits used in studying the variability problem. Similar formulas can be derived for the coverage between, say, the rth largest and rth smallest sample values.

5 KOLMOGOROV-SMIRNOV STATISTIC

In this section a technique for finding a confidence band for the distribution function of a continuous variable is presented. By modifying this technique slightly, it can also be used to test hypotheses of the type treated in sections 1 and 2.

Let x_1', x_2', $\cdots$, x_n' denote a random sample from a population with distribution function $F(x)$ and let x_1, x_2, $\cdots$, x_n denote the ordered sample. The problem now is to use this ordered sample to obtain a confidence band for $F(x)$. It should be noted that it is the distribution function $F(x)$ and not the density function $f(x)$ that is being considered here.

The desired method, which was first presented by the two Russian mathematicians whose names are attached to it, consists in using the ordered sample to construct an upper and lower step function such that $F(x)$ will be contained between them with a specified probability. Toward this end, consider the sample distribution function, which is a step function, given by the formula

$$S_n(x) = \begin{cases} 0, & x < x_1 \\ \dfrac{k}{n}, & x_k \le x < x_{k+1} \\ 1, & x \ge x_n \end{cases}.$$

The graph of this function for a typical sample, together with the graph of a typical $F(x)$, is shown in Fig. 2.

Now suppose that $F(x)$ is a known continuous function. Then it would be possible to calculate the value of $|F(x) - S_n(x)|$ for any desired value of x. It is clear from inspecting Fig. 2 that it would suffice to look at the left and right end points of an interval (x_i, x_{i+1}) to determine how large the function $|F(x) - S_n(x)|$ can become in that interval. Since $S_n(x)$ is constant over each

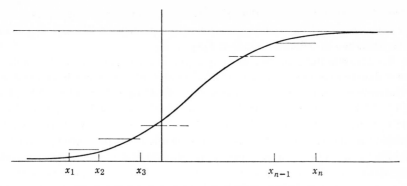

Fig. 2. A sample and theoretical distribution function.

interval of the form $x_i \leq x < x_{i+1}$, and $F(x)$ is a nondecreasing continuous function of x, it follows that it is possible for $|F(x) - S_n(x)|$ to assume its maximum value at the left end point of such an interval but not at the right end point because the right end point is not included in the interval. Hence it is necessary to replace maximum by least upper bound in studying how large this function can become and to study the function

$$\sup_x |F(x) - S_n(x)| \, .$$

This function gives the maximum vertical distance possible between the graphs of $F(x)$ and $S_n(x)$ over the range of possible x values. It can be shown that the distribution of this maximum possible distance does not depend upon $F(x)$. As a consequence, this quantity, which is denoted by D_n, can be used as a nonparametric variable for constructing a confidence band for $F(x)$.

Since $S_n(x)$ varies from sample to sample D_n is obviously a random variable. In order to use it as a tool for finding a confidence band for $F(x)$ it is necessary to find its distribution. This distribution can be worked out numerically for any particular value of n by using combinatorial methods that are too lengthy and involved to be presented here. Certain critical values of this distribution, however, are given in Table VIII in the appendix. Let $D_n{}^\alpha$ denote such a critical value that satisfies the relation

(11) $$P\{D_n \leq D_n{}^\alpha\} = 1 - \alpha \, .$$

In view of the definition of D_n and (11), the following successive equalities can be written down:

$$1 - \alpha = P\{\sup_x |F(x) - S_n(x)| \leq D_n{}^\alpha\}$$

$$= P\{|F(x) - S_n(x)| \leq D_n{}^\alpha \text{ for all } x\}$$

$$= P\{S_n(x) - D_n{}^\alpha \leq F(x) \leq S_n(x) + D_n{}^\alpha \text{ for all } x\} \, .$$

This last equality shows that the two step functions, $S_n(x) + D_n^{\alpha}$ and $S_n(x) - D_n^{\alpha}$, yield a confidence band with confidence coefficient $1 - \alpha$ for the unknown distribution function $F(x)$.

To illustrate the preceding technique, the sample values for the variable X in Table 1 are employed to construct a 95 percent confidence band for $F(x)$. The ordered values of this sample are 8.2, 10.4, 10.6, 11.5, 12.6, 12.9, 13.3, 13.3, 13.4, 13.4, 13.6, 13.8, 14.0, 14.0, 14.1, 14.2, 14.6, 14.7, 14.9, 15.0, 15.4, 15.6, 15.9, 16.0, 16.2, 16.3, 17.2, 17.4, 17.7, 18.1. From Table VIII it will be found that the value of $D_{30}^{.05}$ is .24; consequently this is the value that must be added to and subtracted from the sample distribution function $S_{30}(x)$ to yield the desired confidence band. Since the step function $S_{30}(x)$ increases by the amount $\frac{1}{30}$ at each distinct sample point, it is easily constructed. The graph of $S_{30}(x)$, together with the graphs of the step functions that determine the desired confidence band, are shown in Fig. 3. Vertical lines have been added to the confidence band step function graphs for better delineation.

The statistic D_n can also be used to test the hypothesis that a random sample came from a population with a specified distribution function. This is accomplished by calculating the maximum difference between the hypothetical distribution function, say $F_0(x)$, and the sample distribution function $S_n(x)$, and then determining whether this difference exceeds the critical value given in Table VIII. This use of the D_n statistic yields another method for solving the "goodness of fit" problem that was treated in Chapter 9 by means of the χ^2 test. The D_n statistic possesses the advantage that it is an exact method, whereas the χ^2 method is valid only for fairly large samples. There is no such restriction here as requiring all cell frequencies to exceed 5 as in the case of the χ^2 test because there is no necessity to classify the observations in carrying out the present test. As a matter of fact, the test then is no longer an exact one because the maximum difference for classified and unclassified data may

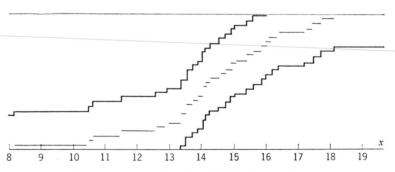

Fig. 3.   Confidence band for a distribution function.

not be the same; however, the discrepancy is usually slight if the classification is not too coarse.

The problem of testing "goodness of fit" is ordinarily a parametric type problem, and therefore strictly speaking it does not belong in this chapter; however, it is included here because it arises naturally in a discussion of confidence band methods. Furthermore, the test based on D_n possesses such striking advantages over the χ^2 test in certain respects that it is important to make this test available to students.

As an illustration of how D_n is used to test a hypothetical distribution, consider the problem of testing whether the following data came from sampling a normal distribution with mean $\frac{1}{2}$ and variance 1. The data, which were obtained from taking a sample of 20 from a normal distribution with mean 0 and variance 1, are the following: .36, .92, −.56, 1.86, 1.74, .56, −.95, .24, −.15, −.74, .32, .82, .70, −.10, −1.06, .15, .55, −.48, −.49, −1.26.

To obtain $S_n(x)$ it suffices to order these values and then add .05 to $S_n(x_i)$ to obtain $S_n(x_{i+1})$. The results of the computations are shown in Table 3.

The values of $F(x_i)$ are obtained from Table II by means of the following relationship:

$$F(x) = \int_{-\infty}^{x} \frac{e^{-\frac{1}{2}(t-\frac{1}{2})^2}}{\sqrt{2\pi}}\, dt = \int_{-\infty}^{x-\frac{1}{2}} \frac{e^{-\frac{z^2}{2}}}{\sqrt{2\pi}}\, dz.$$

The values of $F(x_i)$ obtained in this manner are given in Table 3. For the purpose of determining the value of D_n it is instructive to inspect the graphs of $F(x)$ and $S_n(x)$ as shown in Fig. 4. It will be observed from this figure that in finding the maximum distance between the graphs of $F(x)$ and $S_n(x)$ it is necessary to consider both $|F(x_i) - S_n(x_i)|$ and $|F(x_i) - S_n(x_{i-1})|$ because at the point x_i one is permitted to choose either the top or bottom of the step. An inspection of Table 3 will show that $D_n = .24$, which occurs at $x_i = .92$.

From Table VIII, $D_{20}^{.05} = .29$. Since $D_n < D_{20}^{.05}$, the hypothesis is accepted. Although the hypothesis is false here, the sample was not quite large enough to reject it by this goodness of fit test.

The nonparametric methods that have been presented in this chapter were constructed on an intuitive basis. The critical region for each test was chosen by analogy with the critical region selected in a similar parametric problem. A satisfactory theory of "best tests" has not yet been developed for nonparametric methods; therefore it is necessary to rely heavily on intuition, and attempt to show that the nonparametric test selected is superior to other available tests of this type for the problem being considered.

TABLE 3

x_i	-1.26	-1.06	-.95	-.74	-.56	-.49	-.48	-.15	-.10	.15	.24	.32
$S_n(x_i)$	.05	.10	.15	.20	.25	.30	.35	.40	.45	.50	.55	.60
$F(x_i)$	.04	.06	.07	.11	.14	.16	.16	.26	.27	.36	.40	.43

.36	.55	.56	.70	.82	.92	1.74	1.86
.65	.70	.75	.80	.85	.90	.95	1.00
.44	.52	.52	.58	.63	.66	.89	.91

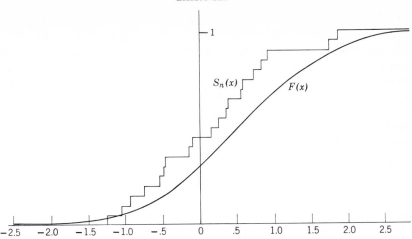

Fig. 4. The graphs of $F(x)$ and $S_n(x)$ for the data of Table 3.

EXERCISES

Sec. 1

1. The following data represent the grammar scores made by an entering group of students at a private school. Past experience has shown that fifty percent of such students will score 35 or more points on a test of this type. Use the sign test to determine whether this group is inferior with respect to grammar, assuming that the test is not more difficult than usual.

44, 22, 31, 12, 21, 36, 33, 34, 29, 25, 11, 51, 18, 46, 23,
38, 29, 32, 18, 41, 26, 17, 30, 32.

2. In an elementary school 17 pairs of first grade children were formed on the basis of similarity of intelligence and background. One child of each pair was taught to read by method I and the other child by method II. After a period of training, the children were given a reading test with the following results.

Method I	65	68	70	63	64	62	73	75	72	78	64	73	79	80	67	74	82
Method II	63	68	68	60	65	60	72	75	73	70	66	70	77	78	63	74	78

Using the sign test and ignoring ties, test to see whether the methods are equally effective.

3. Use the sign test to work problem 38, Chapter 10.

Sec. 2

4. Work problem 2 by means of the rank sum test. Why is the rank sum test not strictly applicable to a problem such as this?

5. Compare the results in problems 2 and 4 with those obtained by applying the *t* test to the differences of paired values.

6. Take random samples of size 10 each from the two horizontal distributions given by $f_1(x) = 1$, $0 < x < 1$ and $f_2(x) = 2$, $0 < x < \frac{1}{2}$ by choosing the proper sets of numbers from the table of random numbers. Test the hypothesis $H_0 : f_1(x) = f_2(x)$ by (a) pairing values and applying the sign test (b) applying the rank sum test.

7. The following two sets of observed values were obtained from sampling two populations. Using the rank sum test, test the hypothesis that the two populations possess the same density function.

I	25	30	28	28	24	25	13	32	24	30	31	35		
II	44	34	22	8	47	31	40	30	32	35	18	21	35	29	22

Sec. 3

8. Given the sequence of numbers 1, 1, 3, 3, 1, 1, 3, 3, 1, 1, 3, 3, 1, 1, 3, 3, 1, 1, 3, 3, name two lags that would be particularly effective for the serial correlation test to show up this lack of randomness.

9. Alternate the two sets of values obtained in problem 6 to obtain a sequence of 20 values. Test for randomness by means of serial correlation with lag 1.

10. Obtain the annual rainfall records for your community for the last 50 years and test for randomness by the serial correlation test.

11. Test the following set of measurements for a trend by means of serial correlation, applying the formulas to the measurements after they have been reduced by the subtraction of 28. The measurements are 28, 32, 37, 25, 31, 29, 33, 28, 27, 28, 23, 22, 18, 17.

12. Prove that the serial correlation test is unaffected by subtracting the same constant from each observed value.

13. Prove that the serial correlation test is unaffected by multiplying each observed value by the same constant.

Sec. 4

14. Find the probability that (a) the larger of 2 observations taken from a continuous distribution will exceed the true median and (b) the smaller of 2 observations will exceed the median.

15. For a sample of 40, calculate the probability that at least 90 percent of a distribution will be included between the extreme values of the sample.

16. Using the data of problem 1, determine the probability that at least 80 percent of the distribution from which this sample came will lie between the two extreme values of the sample.

17. How large a sample is necessary in order that the probability will be .80 that at least 90 percent of the distribution will be expected to be between the extreme values of the sample?

18. Using the density function for the smallest and largest observations in a sample of size n, derive the density function for the smallest observation.

19. Using the formula derived in problem 18, find the expected value of the proportion of the distribution to the left of the smallest observed value. Comment on the result.

Sec. 5

20. Draw a sample of 20 two-digit random numbers from Table VI and apply the Kolmogorov-Smirov test to test the hypothesis that they came from a normal distribution with mean 50 and standard deviation 20.

21. Find an 80 percent confidence band for the distribution function corresponding to the y values in Table 1.

22. Find a 95 percent confidence band for the distribution function corresponding to the data on problem 12, Chapter 9, ignoring the fact that the data have been grouped.

CHAPTER 13

Other Methods

The hypothesis testing methods that have been presented thus far possess two restrictive characteristics. They are based on the assumption that a sample of fixed size is to be taken and that a choice is to be made in favor of one of two possible decisions.

If samples can be taken one at a time and the information from them accumulated, one would expect to be in a better position to make decisions than if no attempt were made to look at the data until a sample of fixed size had been taken. There are methods available, known as *sequential methods*, that operate on this accumulation-of-information basis and that require considerably less sampling on the average than the fixed-size sample methods.

The restriction that only two choices for decision making are possible can be bothersome in a problem in which there are several natural choices available. Thus, in the study of blood types there are four natural categories, and it would be unrealistic to reduce them to two because one's statistical techniques were designed for only two possibilities. There are methods, known as *multiple decision methods*, for treating such more general problems.

The material in this chapter is devoted principally to explaining some of the basic but elementary ideas in these two new decision-making methods, and also in looking at a different approach to estimation and hypothesis testing.

1 SEQUENTIAL ANALYSIS

Sequential methods possess striking advantages for testing hypotheses; therefore they are discussed here from that point of view.

In testing a hypothesis, the sequential method gives a rule of procedure for making one of the following three decisions at each stage of the experiment: (1) accept the hypothesis, (2) reject the hypothesis, or (3) continue the experiment by taking an additional observation.

For the purpose of describing a sequential test, consider a single continuous variable X whose density function $f(x; \theta)$ depends on the single parameter θ. Although the sequential test about to be described may be applied to either discrete or continuous variables, the description is given for a continuous variable.

Let the hypothesis to be tested be

$$H_0 : \theta = \theta_0$$

and let the alternative hypothesis be

$$H_1 : \theta = \theta_1 .$$

Since a simple hypothesis is being tested against a simple alternative, the Neyman-Pearson lemma given in section 3.2, Chapter 8, would suggest using the likelihood ratio

$$\frac{\prod\limits_{i=1}^{n} f(x_i; \theta_1)}{\prod\limits_{i=1}^{n} f(x_i; \theta_0)}$$

as a basis for deciding between H_0 and H_1. For a fixed-size sample of size n, the Neyman-Pearson method chooses as critical region those sample points for which this likelihood ratio is larger than a certain constant k. The region in which this ratio is smaller than k would then constitute the region for accepting H_0. A sequential test can be constructed by extending this fixed-size sample method slightly to include a region for continuing the sampling.

In discussing sequential methods, it is convenient to use the letter m in place of n to denote the size of a sample in order to distinguish it from the fixed n situation. The letter n is reserved for the size sample that is required to reach a final decision. As a consequence, n is a random variable in sequential methods. Another convenient symbol is p_{im} to denote the likelihood function when H_i is true and a sample of size m is taken. Now consider the likelihood ratio

(1)
$$\frac{p_{1m}}{p_{0m}} = \frac{\prod\limits_{i=1}^{m} f(x_i; \theta_1)}{\prod\limits_{i=1}^{m} f(x_i; \theta_0)}, \qquad m = 1, 2, \cdots .$$

By analogy with the fixed-size sample test, one would choose as the region for accepting H_0 those sample points for which (1) is small and as the region for accepting H_1 those sample points for which (1) is large. The new idea in sequential testing is to use part of the sample space for a third region such that if the sample point falls in this region the decision to accept H_0 or H_1

will be postponed. From the preceding remarks, this postponement region should consist of those points for which (1) is neither small nor large. Thus in the sequential test being described, two numbers c_1 and c_2 are chosen and successive observations are taken, $m = 1, 2, \cdots$, as long as

$$c_1 < \frac{p_{1m}}{p_{0m}} < c_2 .$$

However, whenever $p_{1m}/p_{0m} \leq c_1$, sampling ceases and the decision is made to accept H_0, and whenever $p_{1m}/p_{0m} \geq c_2$, sampling ceases and the decision is made to accept H_1.

Now it can be shown that if c_1 and c_2 are chosen properly this sequential test will have prescribed values, α and β, for the two types of error. The exact values of c_1 and c_2 are not available; however, excellent approximations are given by choosing

$$(2) \qquad c_1 = \frac{\beta}{1 - \alpha} \qquad \text{and} \qquad c_2 = \frac{1 - \beta}{\alpha} .$$

A justification for these approximations is given in section 1.1. With these choices for c_1 and c_2, the test is now complete. The name given to this test and the technique for carrying it out may be expressed as follows.

(3) SEQUENTIAL PROBABILITY RATIO TEST: *To test the hypothesis $H_0 : \theta = \theta_0$ against the alternative $H_1 : \theta = \theta_1$, calculate the likelihood ratio p_{1m}/p_{0m} and proceed as follows:*

(i) *if* $\dfrac{p_{1m}}{p_{0m}} \leq \dfrac{\beta}{1 - \alpha}$, *accept H_0*

(ii) *if* $\dfrac{p_{1m}}{p_{0m}} \geq \dfrac{1 - \beta}{\alpha}$, *accept H_1*

(iii) *if* $\dfrac{\beta}{1 - \alpha} < \dfrac{p_{1m}}{p_{0m}} < \dfrac{1 - \beta}{\alpha}$, *take an additional observation .*

One of the striking features of this test is that it is not necessary to derive the density function of a statistic such as t or F in order to carry out the test. Furthermore, one can decide in advance what size type I and type II errors to tolerate rather than fix the type I error size and then be forced to calculate the type II error size as in fixed-size sample tests. On the other hand, one never knows how large a sample will be required to arrive at a decision, because n, the size sample needed, is now a random variable. A general formula exists for calculating the mean value of n, so that one can determine in advance how large n is likely to be.

As an illustration of how a sequential test is constructed, consider the problem of determining whether the mean of a normal variable with variance

1 has the mean θ_0 or the mean θ_1. Here

$$f(x; \theta) = \frac{e^{-\frac{1}{2}(x-\theta)^2}}{\sqrt{2\pi}} \ ;$$

hence (1) becomes

$$\frac{p_{1m}}{p_{0m}} = \frac{\prod_{i=1}^{m} e^{-\frac{1}{2}(x_i-\theta_1)^2}}{\prod_{i=1}^{m} e^{-\frac{1}{2}(x_i-\theta_0)^2}} = \frac{e^{-\frac{1}{2}\sum_{i=1}^{m}(x_i-\theta_1)^2}}{e^{-\frac{1}{2}\sum_{i=1}^{m}(x_i-\theta_0)^2}}.$$

$$= e^{(\theta_1-\theta_0)\sum_{i=1}^{m} x_i + \frac{m}{2}(\theta_0^2-\theta_1^2)}.$$

Now (iii) of (3) is equivalent to

$$\log \frac{\beta}{1-\alpha} < \log \frac{p_{1m}}{p_{0m}} < \log \frac{1-\beta}{\alpha}.$$

For this problem, these inequalities become

$$\log \frac{\beta}{1-\alpha} + \frac{m}{2}(\theta_1^2 - \theta_0^2) < (\theta_1 - \theta_0) \sum_{i=1}^{m} x_i$$

$$< \log \frac{1-\beta}{\alpha} + \frac{m}{2}(\theta_1^2 - \theta_0^2).$$

If $\theta_1 > \theta_0$, this is equivalent to

$$(4) \quad \frac{1}{\theta_1 - \theta_0} \log \frac{\beta}{1-\alpha} + \frac{m}{2}(\theta_0 + \theta_1) < \sum_{i=1}^{m} x_i$$

$$< \frac{1}{\theta_1 - \theta_0} \log \frac{1-\beta}{\alpha} + \frac{m}{2}(\theta_0 + \theta_1).$$

For $\theta_1 < \theta_0$, these inequalities would be reversed.

As a numerical illustration, suppose that $\alpha = .05$, $\beta = .10$, $\theta_0 = 9.5$, and $\theta_1 = 10$. Then (4) becomes

$$-4.50 + 9.75m < \sum_{i=1}^{m} x_i < 5.78 + 9.75m \ .$$

Following (3), the test now proceeds as follows:

(i) if $\sum_{i=1}^{m} x_i \leq -4.50 + 9.75m$, accept $\theta = 9.5$

(ii) if $\sum_{i=1}^{m} x_i \geq 5.78 + 9.75m$, accept $\theta = 10$

(iii) if neither inequality is satisfied, take another observation .

An experiment was performed by taking successive samples from a normal population with mean $\theta = 10$ and variance 1 until a decision was reached. The decision to accept H_1, which is the correct decision here, occurred at the twelfth observation. The values of Σx_i obtained in the experiment, together with the values of the decision boundaries, are displayed in Table 1.

As a second illustration, consider the problem of determining whether $p = p_0$ or $p = p_1$ for a binomial distribution. If one chooses $X = 1$ for success and $X = 0$ for failure, $f(x; \theta)$ will be given by $f(1; p) = p$ and $f(0; p) = q$. Now, suppose that the first m trials of the event produced d_m successes. Then the likelihood function $\prod_{i=1}^{m} f(x_i; \theta)$ will consist of the product of p's and q's, a p occurring as a factor whenever a success occurred and a q otherwise. The likelihood ratio (1) then becomes

$$\frac{p_{1m}}{p_{0m}} = \frac{p_1{}^{d_m} q_1{}^{m-d_m}}{p_0{}^{d_m} q_0{}^{m-d_m}}.$$

If this expression is substituted in (3) and the desired numerical values are assigned to p_0, p_1, α, and β, the test procedure will be determined.

As a numerical illustration, let $p_0 = .5$, $p_1 = .7$, $\alpha = .10$, and $\beta = .20$. These values may be thought of as those that might be used to test the honesty of a coin when that coin is suspected of giving too many heads. Here $\beta/(1-\alpha) = \frac{2}{9}$, $(1-\beta)/\alpha = 8$, and

$$\frac{p_{1m}}{p_{0m}} = \frac{(.7)^{d_m}(.3)^{m-d_m}}{(.5)^{d_m}(.5)^{m-d_m}} = \left(\frac{3}{5}\right)^m \left(\frac{7}{3}\right)^{d_m}.$$

The first inequality in (3),

$$\left(\frac{3}{5}\right)^m \left(\frac{7}{3}\right)^{d_m} \leq \frac{2}{9}$$

can be written more conveniently in the form

$$d_m \leq \frac{\log \frac{2}{9}}{\log \frac{7}{3}} + m \frac{\log \frac{5}{3}}{\log \frac{7}{3}}.$$

In a similar manner the second inequality becomes

$$d_m \geq \frac{\log 8}{\log \frac{7}{3}} + m \frac{\log \frac{5}{3}}{\log \frac{7}{3}}.$$

If these logarithms are evaluated, the test will proceed as follows:

(i) if $d_m \leq -1.78 + .603m$, accept $p = .5$
(ii) if $d_m \geq 2.45 + .603m$, accept $p = .7$
(iii) if neither inequality is satisfied, take another trial.

TABLE 1

$5.78 + 9.75m$	15.53	25.28	35.03	44.78	54.53	64.28	74.03	83.78	93.53	103.28	113.03	122.78
$\sum x_i$	10.47	20.98	30.76	42.93	52.88	64.10	73.41	83.16	91.72	101.24	112.89	123.24
$-4.50 + 9.75m$	5.25	15.00	24.75	34.50	44.25	52.00	61.75	71.50	81.25	91.00	100.75	110.50

TABLE 2

m	1	2	3	4	5	6	7	8	9	10	11	12	13	14	15
x_m	0	0	1	1	1	0	1	1	0	1	0	0	1	0	0
d_m	0	0	1	2	3	3	4	5	5	6	6	6	7	7	7

Tosses of a coin gave the results shown in Table 2. For the purpose of determining when one of the inequalities is satisfied, it is convenient to represent these inequalities and the results of the successive trials graphically. If m and d_m are treated as the coordinates of a point, the straight lines $d_m = -1.78 + .603m$ and $d_m = 2.45 + .603m$ will serve to divide the m, d_m plane into three regions corresponding to the three possible decisions at each trial. The graph corresponding to this problem is given in Fig. 1. From this graph it will be observed that the experiment terminated after 15 trials because inequality (i) was then satisfied. In accepting the hypothesis that $p = .5$ the experimenter does so in preference to accepting the hypothesis that $p = .7$.

As stated earlier, sequential methods often reduce considerably the size sample needed to arrive at a reliable decision in testing hypotheses. For example, in the preceding illustration it is not difficult to show that a

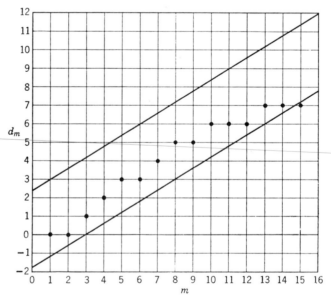

Fig. 1. Sequential test for testing $p = .5$ against $p = .7$.

fixed-size sample of approximately 26 will suffice to test $H_0:p = .5$ against $H_1:p = .7$ with $\alpha = .10$ and yield a type II error of $\beta = .20$. However, in the theory of sequential analysis it can be shown that the average size sample needed to arrive at a decision in this illustration with $\alpha = .10$ and $\beta = .20$ is approximately 13. The experimental result of 15 is in good agreement with expectation. For problems of this type, on the average one saves 40 to 50 percent in sampling by using sequential methods. This follows from a formula in section 1.1 which gives a good approximation to the value of $E[n]$.

The reason for the advantage of the sequential approach over the fixed-size-sample approach lies in the ability of the sequential method to reach an early decision for samples that are extremely favorable to either H_0 or H_1. Thus, if a good coin were tossed a number of times and gave rise to a fairly high percentage of tails, it would be clear rather early that $p = .5$ should be accepted in preference to $p = .7$. Conversely, if the coin were biased toward heads and if a high percentage of heads occurred in the early stages one would accept $p = .7$ without continuing the experiment further.

The savings that can be realized by the sequential approach may be even greater than theory would indicate because in real-life experiments the actual value of p, for example, might heavily favor either H_0 or H_1. Thus in the preceding illustration, if it were true that $p = .4$, the sequential test would quickly accept H_0 in preference to H_1. This ability to arrive at an early decision can be very useful in such fields as sampling inspection, where it is not uncommon for lots to be very bad when they are bad or very good when they are good.

The sequential probability ratio test given by (3) possesses the disadvantage that, strictly speaking, it applies only to testing a simple hypothesis $\theta = \theta_0$ against a simple alternative $\theta = \theta_1$. This disadvantage can often be circumvented by properly rephrasing the problem to be solved. For example, suppose that a consumer wishes to determine whether a producer's fraction defective is actually p_0 as claimed by the producer. He may fear that the true fraction defective is larger than p_0; consequently he would be interested in testing $p = p_0$ against $p > p_0$. But he may be willing to state a value of $p > p_0$, say $p = p_1$, such that it would begin to become a serious matter if p exceeded p_1; whereas, if p were less than p_1, even though $p > p_0$, no serious harm would result. By this device of deciding on an upper limit p_1 for p, the problem can be reduced to the ordinary sequential test of testing $H_0:p = p_0$ against $H_1:p = p_1$. Similar devices can sometimes be used to arrive at satisfactory tests for composite hypotheses as well.

1.1 Approximations for c_1, c_2, and $E[n]$

For each value of $m = 1, 2, 3, \cdots$, let the sample space be divided into the three regions R_{0m}, R_{1m}, and R_{cm} corresponding to the three possible decisions of accepting H_0, accepting H_1, or continuing to sample, respectively. These regions are in the m-dimensional space determined by the variables $X_1, X_2, \cdots, X_m$. When $m = n$, the sample point must have fallen in R_{cm} for all values of $m < n$ because n denotes the sample size for which the decision to accept H_0 or H_1 is first made. In terms of this notation, consider the problem of calculating the value of $1 - \beta$. In this calculation the sample point is denoted by X, regardless of the dimension of the sample space. Thus

$$1 - \beta = P\{\text{accept } H_1 \mid H_1\} = \sum_{n=1}^{\infty} P\{X \in R_{1n} \mid H_1\}$$

$$= \sum_{n=1}^{\infty} \int_{R_{1n}} \cdots \int p_{1n} \, dx_1 \cdots dx_n .$$

The symbol $P\{X \in R_{1n} \mid H_1\}$ denotes the probability that the sample point X will fall in the region R_{1n}, given that H_1 holds. But the sample point X can be in R_{1n} only if it satisfies the inequality $p_{1n}/p_{0n} \geq c_2$; consequently this inequality holds for all points in R_{1n}. As a result,

$$1 - \beta \geq \sum_{n=1}^{\infty} \int_{R_{1n}} \cdots \int c_2 p_{0n} \, dx_1 \cdots dx_n$$

$$= c_2 \sum_{n=1}^{\infty} \int_{R_{1n}} \cdots \int p_{0n} \, dx_1 \cdots dx_n$$

$$= c_2 \sum_{n=1}^{\infty} P\{X \in R_{1n} \mid H_0\}$$

$$= c_2 P\{\text{accept } H_1 \mid H_0\} = c_2 \alpha .$$

This shows that the number c_2 satisfies the inequality

(5) $$c_2 \leq \frac{1 - \beta}{\alpha} .$$

The same type of calculations when applied to finding the value of β will lead to the inequality

(6) $$c_1 \geq \frac{\beta}{1 - \alpha} .$$

It is illuminating to look at these inequalities from a geometrical point of view. For this purpose the two lines in the α, β plane whose equations are

(7) $$c_2 \alpha = 1 - \beta \qquad \text{and} \qquad c_1(1 - \alpha) = \beta .$$

have been graphed in Fig. 2 for a typical choice of c_1 and c_2. These are merely the relations given by the formulas in (2). The shaded area represents all pairs of α, β that satisfy the inequalities (5) and (6).

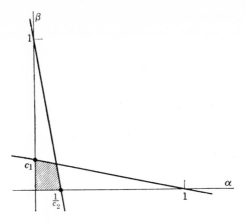

Fig. 2. Inequality region for α, β.

Suppose α and β are temporarily chosen equal to .10 and that c_1 and c_2 are chosen to be the corresponding values given by (2). Then

$$c_1 = \tfrac{1}{9} \quad \text{and} \quad c_2 = 9 \,.$$

The values of c_1 and $1/c_2$ shown in Fig. 2 are somewhat larger than these values; consequently the shaded region for these choices would be smaller and more nearly a square region than that shown in Fig. 2. The point of intersection of the two lines would, of course, be given by $\alpha = .10$ and $\beta = .10$. Since the shaded region is very nearly square, it follows that when c_1 and c_2 are chosen in this manner, the actual value of α (and similarly of β) can possibly exceed .10 by only a small amount. Furthermore, if one of them should exceed .10, then the shaded area shows that the other would need to be less than .10. The greater the excess over .10 for one of them, the smaller in value the other must be.

In view of the preceding discussion, if c_1 and c_2 are chosen to be the values given by (2), the true values of α and β will probably differ very little from those that were selected to be used in (2), particularly when c_1 and $1/c_2$ are very small. As a consequence, if a sequential test is constructed with values of c_1 and c_2 given by (2), it will for all practical purposes possess sizes of the two types of error that do not exceed the selected values of α and β. Thus one can decide in advance what protection against error is desirable and then by choosing c_1 and c_2 as in (2) be assured that at least this much protection will be realized in carrying out the test.

The derivation of the formula alluded to earlier for approximating the value of $E[n]$ is quite lengthy; consequently, only the result of the derivation is given here. In this connection, let $P(\theta)$ denote the probability of rejecting H_0 if θ is the true parameter value. Thus $P(\theta_0)$ is equal to α and $1 - P(\theta_1)$

is equal to β when testing $H_0: \theta = \theta_0$ against $H_1: \theta = \theta_1$. Further, let $Z = \log f(X; \theta_1) - \log f(X; \theta_0)$. Then the desired formula may be expressed as follows:

(8) $$E[n] \doteq \frac{P(\theta) \log c_2 + [1 - P(\theta)] \log c_1}{E[Z]}.$$

In applying this formula, one uses the approximations for c_1 and c_2 given in (2).

For the purpose of illustrating how to use this formula, consider the binomial distribution problem that was discussed in the preceding section. For that problem,

$$f(x; \theta) = \begin{cases} p, & \text{if } x = 1 \\ q, & \text{if } x = 0 \end{cases}.$$

Since the value of $E[n]$ depends on what value is assigned to θ, it is necessary to specify the value of p. The hypothesis H_0 was actually true here; consequently in making comparisons the value of $E_0[n]$ will be used. When H_0 is true

$$E_0[Z] = E_0\left[\log\frac{f(X; \theta_1)}{f(X; \theta_0)}\right]$$

$$= q_0 \log \frac{q_1}{q_0} + p_0 \log \frac{p_1}{p_0}.$$

In view of the numerical values $p_0 = .5$, $p_1 = .7$, $\alpha = .10$, and $\beta = .20$ and the fact that $P(\theta_0) = \alpha$, it follows from (8) and calculations that

$$E_0[n] \doteq \frac{.10 \log 8 + .90 \log \frac{2}{9}}{.5 \log \frac{3}{5} + .5 \log \frac{7}{5}} \doteq 13.$$

This is the value that was claimed for $E[n]$ in the earlier illustration. Similar calculations will show that if H_1 is true then $E_1[n] \doteq 17$.

2   MULTIPLE CLASSIFICATION TECHNIQUE

In all the problems of testing hypotheses that have been considered thus far it was necessary either to accept or reject some hypothesis. This is true for sequential methods also, even though a final decision may be postponed for some time. There are many problems, however, that cannot be treated in this simple manner because they involve more than just two possible choices. For example, a botanist may wish to classify a mixed set of plants involving three different varieties into their proper variety. This is a three-decision problem and it is unnatural to attempt to solve it, say, by successive two-decision problem methods. In this section a multiple-decision technique for such classification problems is presented as an introduction to general multiple-decision methods.

For simplicity of exposition, the following discussion is limited to the case in which there are three possible categories of classification; however, the method presented can obviously be extended to any number of categories.

Let X_1, X_2, $\cdots$, X_k denote k random variables corresponding to k different measurements that are to be taken on an individual of some population. Thus, for a population of flowers, the X's might represent such characteristics as petal length, petal width, and stamen length. Let the population to be sampled consist of the three subpopulations P_1, P_2, and P_3, and let these subpopulations constitute the proportions p_1, p_2, and p_3 of this population.

Since an individual will be determined by his values of X_1, X_2, $\cdots$, X_k, the problem of classification is the problem of dividing the k-dimensional sample space into three parts, say, S_1, S_2, and S_3, corresponding to the three subpopulations, and agreeing to classify an individual as belonging to subpopulation P_i if his sample point lies in S_i. In this connection, it seems reasonable to choose as a criterion of optimality a division of the sample space that will maximize the probability of classifying an individual correctly.

From the theory in section 3.2, Chapter 8, on how best tests are constructed and from the theory of sequential analysis, one would guess that the ratios of likelihood functions will undoubtedly play a leading role in the determination of a best set of S's. For example, a sample point at which the probability density under P_1 is considerably larger than under either P_2 or P_3 should certainly be placed in the region S_1. The feature that makes this problem somewhat different from earlier problems, in addition to that of having more possibilities for decisions, is the introduction of the proportions p_1, p_2, and p_3 for the relative frequency of occurrence of samples from the various subpopulations. If p_1, for example, were close to zero, then for all practical purposes the problem would reduce to a two-decision problem and earlier methods could be used to solve it. Thus it is clear that the p's must also enter in the determination of a best set of S's. As in the theory of best tests, a theorem will be stated, and then proved, that yields the desired optimal solution. In this theorem the letter X is used to denote the vector variable $X = (X_1, X_2, \cdots, X_k)$ and $f_i(x)$ will denote the density function of the vector variable X_1, X_2, $\cdots$, X_k in the subpopulation p_i $(i = 1, 2, 3)$.

THEOREM 1: *The regions* R_i $(i = 1, 2, 3)$ *into which a k-dimensional sample space should be divided to maximize the probability of correctly classifying an individual selected at random from the population composed of the subpopulations* P_i $(i = 1, 2, 3)$, *which constitute the proportions*

	R_1	R_2	R_3
S_3	S_{31}	S_{32}	S_{33}
S_2	S_{21}	S_{22}	S_{23}
S_1	S_{11}	S_{12}	S_{13}

Fig. 3.   Two classifications of a sample space.

p_i $(i = 1, 2, 3)$ *of that population and which possess the density functions* $f_i(x)$ $(i = 1, 2, 3)$, *are determined by the points x that satisfy the inequalities*

$$R_i : p_i f_i(x) \geq p_j f_j(x), \quad \text{for all } j \neq i .$$

Proof: Let S_1, S_2, S_3 be any other division of the sample space. Now any one of these regions S_i $(i = 1, 2, 3)$ can be divided into three subregions S_{i1}, S_{i2}, and S_{i3} such that S_{ij} $(j = 1, 2, 3)$ contains only points found in R_j. Thus S_{ij} contains all the points common to the two regions S_i and R_j.

The geometry of these relations is shown in Fig. 3 for $k = 2$. The regions R_1, R_2, and R_3 are those determined by the two vertical lines and the regions S_1, S_2, and S_3 are those determined by the two horizontal lines. The entire sample space here consists of the points in the rectangle. From Fig. 3 it is clear that this subdivision of S_i can be expressed by means of the formula

$$(9) \qquad S_i = S_{i1} + S_{i2} + S_{i3}, \qquad i = 1, 2, 3 ,$$

and that in terms of this notation, it is also true that

$$(10) \qquad R_j = S_{1j} + S_{2j} + S_{3j}, \qquad j = 1, 2, 3 .$$

The probability of correctly classifying an individual selected at random from the population when using the division of the sample space determined by the S's is given by

$$P_S = \sum_{i=1}^{3} p_i P\{X \in S_i \mid P_i\}$$

$$= \sum_{i=1}^{3} p_i \int_{S_i} f_i(x)\, dx .$$

The preceding integrals are k-dimensional multiple integrals with the arguments $x_1, x_2, \cdots, x_k$; however, they are written symbolically as single integrals with respect to x. This is the same convenient notational device that was used in section 3.2, Chapter 8. Now, by means of formula (9),

P_S can be written in the form

(11) $$P_S = \sum_{i=1}^{3} \left[\int_{S_{i1}} p_i f_i(x) \, dx + \int_{S_{i2}} p_i f_i(x) \, dx + \int_{S_{i3}} p_i f_i(x) \, dx \right].$$

Since all the points lying in S_{i1} are in R_1, it follows from the definition of R_1, given in the theorem, that they must satisfy the inequality

(12) $$p_1 f_1(x) \geq p_j f_j(x), \qquad j = 2, 3 .$$

Similarly, all points in S_{i2} must satisfy

(13) $$p_2 f_2(x) \geq p_j f_j(x), \qquad j = 1, 3 ,$$

and all points in S_{i3} must satisfy

(14) $$p_3 f_3(x) \geq p_j f_j(x), \qquad j = 1, 2 .$$

In view of (12), if $p_i f_i(x)$ is replaced by $p_1 f_1(x)$ in the first of the three integrals in (11), that integral will become at least as large as it was before, for all three values of i. For $i = 1$ there is, of course, no change; however, for $i = 2$ and $i = 3$ the value of the integral will be increased unless inequality (12) is an equality for the points of S_{i1}. Similarly, because of (13), if $p_i f_i(x)$ is replaced by $p_2 f_2(x)$ in the second of the three integrals in (11), that integral will become at least as large as before. Finally, the same conclusion will hold for the third integral if $p_i f_i(x)$ is replaced by $p_3 f_3(x)$. Thus, it follows that

(15) $$P_S \leq \sum_{i=1}^{3} \left[\int_{S_{i1}} p_1 f_1(x) \, dx + \int_{S_{i2}} p_2 f_2(x) \, dx + \int_{S_{i3}} p_3 f_3(x) \, dx \right].$$

But from (10) it follows that $S_{11} + S_{21} + S_{31} = R_1$ and therefore that the sum with respect to i of the first integral in (15) must yield the integral of $p_1 f_1(x)$ over the region R_1. Similar reasoning may be applied to the other two sums of integrals to yield the result

(16) $$P_S \leq \int_{R_1} p_1 f_1(x) \, dx + \int_{R_2} p_2 f_2(x) \, dx + \int_{R_3} p_3 f_3(x) \, dx .$$

If the p's are factored out and the right side is expressed in probability language, (16) will yield the desired result, namely

$$P_S \leq \sum_{i=1}^{3} p_i P\{X \in R_i \mid P_i\} = P_R .$$

Since the set of regions S_1, S_2, S_3 was any set other than R_1, R_2, R_3 defined in the theorem, this inequality proves that R_1, R_2, R_3 is an optimal set in the sense of maximizing the probability of a correct classification.

Points that satisfy the equality part of the inequalities defining the regions R_1, R_2, R_3 may be placed in any one of the regions for which the equality holds. In applications the boundaries of the regions are usually surfaces ($k \geq 3$) so that there is seldom any problem on this score.

The preceding method of proof can obviously be applied to the case in which there are more than three categories of classification.

The difficulty in applying the preceding theorem arises from the fact that one seldom knows the values of the p's and even the values of the parameters determining the density functions. Then it is necessary to estimate such parameters by means of a random sample from the total population, and in the process be able to classify each individual into its proper subpopulation. The resulting decision regions will be estimates of the optimal decision regions given by the theorem.

As an illustration of how the theorem is applied when no estimation is required, consider the following information. A population consists of three subpopulations in the proportions $\frac{1}{4}$, $\frac{1}{4}$, and $\frac{1}{2}$. Each subpopulation is a two-variable normal population with independently distributed variables possessing unit variances and means $(-1, 0)$, $(0, 1)$, and $(1, 0)$, respectively.

In the notation of the theorem $p_1 = \frac{1}{4}$, $p_2 = \frac{1}{4}$, $p_3 = \frac{1}{2}$, and

$$f_1(x) = \frac{e^{-\frac{1}{2}[(x_1+1)^2+x_2^2]}}{2\pi}$$

$$f_2(x) = \frac{e^{-\frac{1}{2}[x_1^2+(x_2-1)^2]}}{2\pi}$$

$$f_3(x) = \frac{e^{-\frac{1}{2}[(x_1-1)^2+x_2^2]}}{2\pi}.$$

The region R_1 is therefore the region determined by the two inequalities

$$\frac{1}{4}\frac{e^{-\frac{1}{2}[(x_1+1)^2+x_2^2]}}{2\pi} \geq \frac{1}{4}\frac{e^{-\frac{1}{2}[x_1^2+(x_2-1)^2]}}{2\pi}$$

and

$$\frac{1}{4}\frac{e^{-\frac{1}{2}[(x_1+1)^2+x_2^2]}}{2\pi} \geq \frac{1}{2}\frac{e^{-\frac{1}{2}[(x_1-1)^2+x_2^2]}}{2\pi}.$$

These inequalities are easily shown to reduce to the inequalities

$$x_1 + x_2 \leq 0$$

and

$$x_1 \leq -\tfrac{1}{2}\log_e 2.$$

Similar calculations will show that R_2 is determined by the inequalities

$$x_1 + x_2 \geq 0$$

and

$$x_2 - x_1 \geq \log_e 2.$$

The region R_3 is clearly the remaining part of the x_1, x_2 plane not occupied by R_1 and R_2; however, one can calculate it directly from definition and show

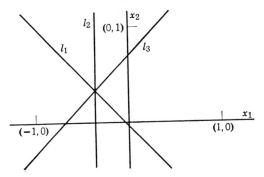

Fig. 4. Boundaries for classification regions.

that it is determined by the inequalities

$$x_1 \geq -\tfrac{1}{2} \log_e 2$$

and

$$x_2 - x_1 \leq \log_e 2 .$$

In order to determine the regions R_1, R_2, and R_3, it suffices to graph the lines whose equations are

$$l_1 : x_1 + x_2 = 0$$
$$l_2 : x_1 = -\tfrac{1}{2} \log_e 2$$
$$l_3 : x_2 - x_1 = \log_e 2 .$$

The graphs of these lines are shown in Fig. 4. From the inequalities defining R_1, R_2, and R_3, it is clear that R_1 is the region below line l_1 and to the left of line l_2, that R_2 is the region above l_1 and above l_3, and that R_3 is the remaining part of the x_1, x_2 plane. These regions are shown more clearly in Fig. 5.

If the p's had all been equal, log 2 would have been replaced by log 1 = 0 in the preceding inequalities and then the regions would have been those obtained by shifting the three-ray boundary configuration of Fig. 5 parallel

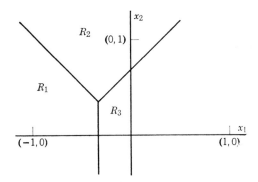

Fig. 5. Optimal classification regions.

to itself until the vertex was at the origin. This illustrates the affect the p's have on the classification regions.

3 BAYES TECHNIQUES

A somewhat different approach to decision making can be formulated in terms of economic losses or gains rather than in terms of the probability of making the correct decision. It may well be, for example, that making one of two possible incorrect decisions is not nearly so serious economically as making the other incorrect decision. In such situations it would be desirable to weight the relative importance of the various errors that can be made. Decision-making procedures based on such notions can be constructed that are capable of treating both estimation and hypothesis testing problems.

In constructing decision making procedures of the foregoing type, it is necessary to introduce the concept of a density, called the prior density, of the parameter that is being estimated or tested. In some statistical problems this is a natural thing to do; in others it is unnatural but serves a useful mathematical purpose. For example, in the quality control of a manufactured product, a buyer may wish to decide whether to accept or reject a shipment of the product on the basis of a sample taken from the shipment. If X represents a quality characteristic of the product, the decision could be made by calculating $\bar{X}$ and determining whether it is compatible with the hypothesis that, say, $\mu > \mu_0$. However, if the buyer has been purchasing shipments of this product regularly and has a record of his sample means, he will be in a position to postulate a distribution for μ, where μ is the true shipment mean and varies from shipment to shipment. It would be inefficient not to use this type of information concerning μ in conjunction with $\bar{X}$ to decide whether $\mu > \mu_0$. Statistical methods that are based on treating the parameter to be estimated or tested as a random variable and which postulate a density function for it are called *Bayesian methods*.

For the purpose of describing these methods, consider first the problem of estimation and how it is formulated. Let X be a continuous random variable with density function $f(x; \theta)$ and let $t = t(X_1, X_2, \cdots, X_n)$ be any estimate of θ based on a random sample of size n. Furthermore, let $W(\theta, t)$ be a weight function that measures the economic loss in claiming (estimating) that the true value of the parameter is t when it is actually θ. For example, one might choose W as the function $W(\theta, t) = c(t - \theta)^2$ if large errors of estimation are very serious, or as the function $W(\theta, t) = c |t - \theta|$ if they are not quite so serious.

As a criterion for determining whether $t = t(X_1, X_2, \cdots, X_n)$ is a good estimator, one can use the expected value of the weight function. Thus one considers the quantity

$$(17) \quad R(\theta, t) = E[W(\theta, t)] = \int \cdots \int W(\theta, t) \prod_{i=1}^{n} f(x_i; \theta) \, dx_1 \cdots dx_n \, .$$

The weight function $W(\theta, t)$ is usually called the *loss function* and the quantity $R(\theta, t)$ is called the *risk function*. An estimate that makes the risk function small in some sense would be considered a desirable estimate. The difficulty with (17) as a basis for judgment is that the result of the integration is usually a function of θ, and an estimate t seldom minimizes (17) for all possible values of θ. It is therefore necessary to introduce some further criteria before (17) can be used effectively to determine whether an estimate is a good one.

One approach to a solution is to study the behavior of the risk function and use some property of it as a basis for judgment. Since, as indicated in the preceding paragraph, it is unlikely that the graph of $R(\theta, t)$ for one estimate will lie below the graphs of all other estimates, it is customary to look only at the maximum value on the graph and then hunt for an estimate t that has the smallest such maximum value. If there exists an estimate with this property, it is called the *minimax* estimate. Figure 6 illustrates this criterion as it applies to only two possible estimates. For this situation, estimate t_2 is the minimax estimate.

A second approach to a solution is to introduce a density function for the parameter θ and then calculate the expected value of $R(\theta, t)$ with respect to this density function. Since the result will be a number rather than a function of θ, the comparison of estimates becomes a simple matter. The principal difficulty with this approach is that one seldom has any precise knowledge of what density function for θ is realistic or whether it is realistic to introduce such a function at all. There are problems, however, for which one can postulate a realistic distribution. For example, familiarity with manufacturing processes suggests that the assumption of a normal distribution for p, the probability of getting a head in tossing a randomly selected penny, is a

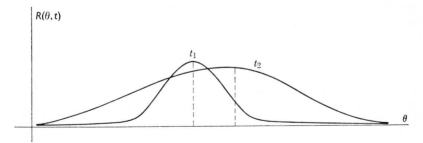

Fig. 6. Graphs of expected losses for two estimates.

reasonable one. One would probably choose the mean to be $\frac{1}{2}$ but would need to perform some experiments on pennies to obtain a variance estimate. If, however, one were interested in estimating the mean brain weight of a certain race of prehistoric people by means of skulls found in an archaeological excavation, it is unnatural to assume that this mean possesses some probability distribution.

For the purpose of developing this approach further, let $\pi(\theta)$ denote the density function that has been selected for the parameter θ. Then the expected value of $R(\theta, t)$ with respect to this density function is given by

$$r(\pi, t) = \int R(\theta, t)\pi(\theta)\, d\theta$$

$$= \int \cdots \int W(\theta, t) \prod_{i=1}^{n} f(x_i; \theta)\pi(\theta)\, dx_1 \cdots dx_n\, d\theta\,.$$

This expected value is usually called the *mean risk*. It depends on the estimating function $t(x_1, x_2, \cdots, x_n)$ selected and also on the density function $\pi(\theta)$ chosen. For a given $\pi(\theta)$, one tries to find an estimating function $t(x_1, x_2, \cdots, x_n)$ that minimizes the mean risk. If such a function exists, it is called the *Bayes solution* to the problem corresponding to the prior density function $\pi(\theta)$. Different choices for $\pi(\theta)$ may well give rise to different minimizing estimating functions; therefore one does not speak of a Bayes solution without specifying the function $\pi(\theta)$.

Bayesian methods is a name given to statistical methods that introduce distributions of parameters at some stage of their development. Although a Bayes approach to estimation has been explained here as an alternative to the minimax approach, it can be shown that the Bayes technique is often a useful one for obtaining a minimax solution; consequently, Bayes methods are useful even in situations in which it seems unrealistic to introduce a distribution for a parameter.

When the loss function in estimation is chosen as squared error, there is often a simple technique for determining a Bayes solution. For the present assume that only a single observation of X is to be made. Under these assumptions the mean risk reduces to

$$r(\pi, t) = \iint (t - \theta)^2 f(x; \theta)\pi(\theta)\, dx\, d\theta\,.$$

When X and θ are both treated as random variables, $f(x; \theta)$ is the conditional density function of X with θ held fixed. Consequently, if $g(x, \theta)$ denotes the joint density function of X and θ, it follows that

$$g(x, \theta) = \pi(\theta)f(x; \theta)\,.$$

But $g(x, \theta)$ can also be written in the form

$$(18) \qquad\qquad g(x, \theta) = h(x)g(\theta \mid x)$$

where $h(x)$ is the marginal density function of X and $g(\theta \mid x)$ is the conditional density function of θ with X held fixed. The equivalence of these two ways of expressing $g(x, \theta)$ enables the mean risk to be written as

$$r(\pi, t) = \iint (t - \theta)^2 h(x)g(\theta \mid x)\, dx\, d\theta$$

or, if the order of integration is interchanged, as

$$r(\pi, t) = \int h(x)\left[\int (\theta - t)^2 g(\theta \mid x)\, d\theta \right] dx .$$

Now the inner integral is merely the second moment about the point t of the variable θ, for X fixed. Since the second moment of a variable is a minimum when it is taken about the mean of the variable, it follows that this integral is minimized for each value of x if t is chosen as the mean of the conditional distribution of θ for X fixed. The double integral is therefore also minimized by this choice; consequently

$$t(x) = E(\theta \mid x)$$

yields the desired solution. The results of this derivation may be expressed in the following form.

THEOREM 3: *If θ possesses the prior density $\pi(\theta)$, X possesses the conditional density $f(x; \theta)$, and if the loss function is chosen to be squared error, then the Bayes estimate of θ is given by $E(\theta \mid x)$, where x is the observed value of X.*

The problem of finding a Bayes solution is now seen to reduce to the problem of finding the conditional expected value of the parameter θ when the variable X is held fixed. This solution is based on the assumption that only a single value of X is to be observed. It is considerably more general than this, however, because X may be chosen to be a statistic, such as a maximum likelihood estimate, from which an estimate t is to be constructed, or it may be a vector variable. Thus, in estimating the mean of a normal distribution, X might well represent $\bar{X}$, or the vector variable $X_1, X_2, \cdots, X_n$.

As an illustration of how a Bayes solution is obtained when the loss function is chosen to be squared error, let

$$f(x; \mu) = \frac{e^{-\frac{1}{2}\left(\frac{x-\mu}{\alpha}\right)^2}}{\sqrt{2\pi}\,\alpha}$$

and

$$\pi(\mu) = \frac{e^{-\frac{1}{2}\left(\frac{\mu-\mu_0}{\beta}\right)^2}}{\sqrt{2\pi}\,\beta}\,.$$

Then

(19) $$g(x, \mu) = \pi(\mu)f(x; \mu) = \frac{e^{-\frac{1}{2}\left[\left(\frac{x-\mu}{\alpha}\right)^2+\left(\frac{\mu-\mu_0}{\beta}\right)^2\right]}}{2\pi\alpha\beta}\,.$$

The marginal density function $h(x)$ is obtained by integrating the joint density function $g(x, \mu)$ with respect to μ; hence

$$h(x) = \frac{1}{2\pi\alpha\beta}\int_{-\infty}^{\infty} e^{-\frac{1}{2}\left[\left(\frac{\mu-x}{\alpha}\right)^2+\left(\frac{\mu-\mu_0}{\beta}\right)^2\right]}\,d\mu\,.$$

This integral can be evaluated by squaring out the two binomials, collecting terms in μ^2 and μ, completing the square in μ, and then recognizing the value of the resulting integral. The result of such manipulations is

$$h(x) = \frac{e^{-\frac{1}{2}\frac{(x-\mu_0)^2}{\alpha^2+\beta^2}}}{\sqrt{2\pi}\sqrt{\alpha^2+\beta^2}}\,.$$

From (18) and (19), it then follows that

$$g(\mu\mid x) = \frac{\sqrt{\alpha^2+\beta^2}}{\sqrt{2\pi}\,\alpha\beta}\,e^{-\frac{1}{2}\left[\left(\frac{x-\mu}{\alpha}\right)^2+\left(\frac{\mu-\mu_0}{\beta}\right)^2-\frac{(x-\mu_0)^2}{\alpha^2+\beta^2}\right]}\,.$$

But the calculations that produced $h(x)$ show at once that the expression for $g(\mu\mid x)$ must reduce to

$$g(\mu\mid x) = \frac{\sqrt{\alpha^2+\beta^2}}{\sqrt{2\pi}\,\alpha\beta}\,e^{-\frac{1}{2}\frac{\alpha^2+\beta^2}{\alpha^2\beta^2}\left(\mu-\frac{x\beta^2+\mu_0\alpha^2}{\alpha^2+\beta^2}\right)^2}\,.$$

This result demonstrates that the conditional distribution of μ for X fixed is normal with the mean

$$E(\mu\mid x) = \frac{x\beta^2+\mu_0\alpha^2}{\alpha^2+\beta^2}\,.$$

The Bayes solution for this estimation problem and for this particular choice of density function $\pi(\mu)$ is therefore given by

$$t(X) = \frac{X\beta^2+\mu_0\alpha^2}{\alpha^2+\beta^2}\,.$$

It is convenient to write this result in the form

$$t(X) = \frac{X + \mu_0 \delta}{1 + \delta}$$

where $\delta = \alpha^2/\beta^2$ is the ratio of the two variances.

Suppose experience has shown that a quality characteristic, such as breaking strength of a manufactured product, is a normal variable and that if μ denotes the mean of this normal variable for a shipment of this product then μ may also be treated as a normal variable, corresponding to successive shipments. Let μ_0 denote the grand mean, that is the mean of the various shipment means, and assume that experience has yielded a grand mean of $\mu_0 = 200$ and a standard deviation of $\beta = 5$ for such shipment means. Now suppose a fresh shipment comes in and the mean μ of this shipment is to be estimated by means of a sample of 25 items selected at random from the shipment. Assume that experience has yielded a standard deviation of 50 for the random variable X that represents the quality characteristic of a single item in a shipment. If X is now chosen as $\overline{X}$ in the preceding theory, the value of α becomes $\alpha = 50/\sqrt{25} = 10$. Calculations with these values yield the estimate

$$t(\overline{X}) = \frac{\overline{X}}{5} + \frac{4}{5} 200 .$$

This estimate gives four times as much weight to past experience as to the sample estimate $\overline{X}$. Additional calculations will show that the mean risk for this estimate, that is the value of $r(\pi, t)$ is $\alpha^2\beta^2/(\alpha^2 + \beta^2) = 20$. Similar calculations will show that the mean risk for the estimate $\overline{X}$, which is based on the sample only, is $\alpha^2 = 100$. Thus there is a very large reduction in the mean risk when past experience is incorporated into the design of the estimate.

The difficulty in applying these methods is that there is seldom experience of the kind assumed here. Furthermore, even when there is experience available, that experience often shows that successive shipments, for example, cannot be treated as random samples from a population of shipments. There is often strong correlation over time between the quality of successive shipments of a manufactured product. Thus, although the Bayes approach may yield considerably better results in the sense of mean risk, the results may not be altogether trustworthy.

Now turn to the problem of how Bayesian methods are used to test hypotheses. In particular consider the problem that gave rise to the Neyman-Pearson lemma of testing a simple hypothesis against a simple alternative. Let X be a continuous random variable with density $f(x; \theta)$ and let the two alternative hypotheses be $H_0 : \theta = \theta_0$ and $H_1 : \theta = \theta_1$. Assume that the test is to be based on a single observed value of X.

The simplest loss function that is used for testing, and which is the one customarily used for this purpose, is the function

(20)
$$W(\theta, t) = \begin{cases} 0, & \text{if the correct decision is made} \\ 1, & \text{if an incorrect decision is made .} \end{cases}$$

One could just as well have picked out two other numbers to measure the losses provided the larger number replaces 1, but there is no loss of generality in using 0 and 1. As before, let $\pi(\theta)$ denote the prior density function of θ. Since there are but two values of θ involved in testing H_0 against H_1, it is necessary to know the values of $\pi(\theta_0) = \pi_0$ and $\pi(\theta_1) = \pi_1$ only. Although $\pi_1 = 1 - \pi_0$ in this problem because it is assumed that θ_0 and θ_1 are the only conceivable values of θ, it is convenient to retain π_1 rather than use the relation $1 - \pi_0$ in the following computations.

For the purpose of finding a Bayes solution to the testing problem, it is first necessary to calculate the risk function corresponding to θ_0 and θ_1. Toward this objective, let A_0 and A_1 be those parts of sample space that correspond to accepting the respective hypotheses H_0 and H_1. The union of A_0 and A_1 will be the entire sample space, but this fact is not needed now. In terms of this notation it follows from the choice of loss function given by (20) that

$$R(\theta_0, t) = EW(\theta_0, t) = 0 \cdot P\{X \in A_0 \mid \theta_0\} + 1 \cdot P\{X \in A_1 \mid \theta_0\}$$
$$= P\{X \in A_1 \mid \theta_0\} .$$

Similarly,

$$R(\theta_1, t) = EW(\theta_1, t) = 0 \cdot P\{X \in A_1 \mid \theta_1\} + 1 \cdot P\{X \in A_0 \mid \theta_1\}$$
$$= P\{X \in A_0 \mid \theta_1\} .$$

There are but two values of θ to consider; therefore the mean risk is now easily calculated. It is given by

$$r(\pi, t) = ER(\theta, t) = \pi_0 R(\theta_0, t) + \pi_1 R(\theta_1, t)$$
$$= \pi_0 P\{X \in A_1 \mid \theta_0\} + \pi_1 P\{X \in A_0 \mid \theta_1\} .$$

Since $P\{X \in A_1 \mid \theta_0)$ is the probability of making the wrong decision when H_0 is true and $P\{X \in A_0 \mid \theta_1\}$ is the probability of making the wrong decision when H_1 is true, it follows that $r(\pi, t)$ gives the probability of making an incorrect decision when choosing A_0 and A_1 as the acceptance regions for H_0 and H_1, respectively. Thus, a Bayes solution to the testing problem can be obtained by choosing A_0 and A_1 to minimize the probability of making an incorrect decision. But the solution to this problem is available in Theorem 1 for $k = 2$ because that theorem explains how the A's should be chosen to maximize the probability of correctly classifying an individual, and hence minimizing the probability of incorrectly doing so. The proof of Theorem 1 treated X as the vector random variable $X = (X_1, \cdots, X_k)$, where the k random variables correspond to k different measurements taken on an individual of some population. By letting these k measurements represent a random sample of size k of the basic random variable involved in testing H_0, correctly classifying an individual by means of his X vector value is equivalent to making the correct decision between H_0 and H_1 by means of a random sample of size k. Since $f_i(x)$ of Theorem 1 is the density function of the vector variable $X = (X_1, \cdots, X_k)$ corresponding to subpopulation i, it follows that

$$f_1(x) = \prod_{i=1}^{k} f(x_i; \theta_0)$$

and

$$f_2(x) = \prod_{i=1}^{k} f(x_i; \theta_1) .$$

In terms of this notation and for a random sample of size n, Theorem 1 then yields the following result.

THEOREM 3: *Given the random variable X whose density function is $f(x; \theta)$, a Bayes solution to testing $H_0: \theta = \theta_0$ against $H_1: \theta = \theta_1$ on the basis of a random sample of size n and the prior probabilities π_0 and π_1 for the values θ_0 and θ_1 is given by choosing as critical region those sample points where*

$$\frac{\prod\limits_{i=1}^{n} f(x_i; \theta_1)}{\prod\limits_{i=1}^{n} f(x_i; \theta_0)} \geq \frac{\pi_0}{\pi_1} .$$

A comparison of this result with that given by the Neyman-Pearson lemma of Chapter 8 will show that here π_0/π_1 takes the place of the constant k in that lemma. In applying the Neyman-Pearson lemma, the value of k was chosen to produce a critical region of any desired size. Here there is no such choice and the values of α and β are determined by the value of π_0/π_1.

As an illustration of how the Bayes solution compares with the non-Bayesian solution based on the Neyman-Pearson lemma, consider the problem of testing $H_0: \mu = \mu_0$ against $H_1: \mu = \mu_1 > \mu_0$ for a normal variable with known variance σ^2. Since the likelihood ratio of Theorem 3 was calculated previously in section 3.2, Chapter 8, for this problem, a direct application of Theorem 3 and those earlier computations will yield the inequality

$$\exp\left[\frac{\mu - \mu_0}{\sigma^2} \sum x_i + \frac{n(\mu_0^2 - \mu_1^2)}{2\sigma^2}\right] \geq \frac{\pi_0}{\pi_1} .$$

Taking logarithms and performing some algebra will yield the equivalent inequality

(21)
$$\bar{x} \geq \frac{\mu_0 + \mu_1}{2} + \frac{\sigma^2}{n} \frac{\log(\pi_0/\pi_1)}{\mu_1 - \mu_0}.$$

From this result it will be observed that when $\pi_0 = \pi_1$ the borderline for deciding in favor of $H_0 : \mu = \mu_0$ against $H_1 : \mu = \mu_1 > \mu_0$ is a value of $\bar{x}$ half way between μ_0 and μ_1. In this case it follows from symmetry that $\alpha = \beta$. If $\pi_0 > \pi_1$ the borderline will shift to the right of this half-way point. The extent of this shift will depend upon the sizes of the constants σ^2, $\mu_1 - \mu_0$, and $\log \pi_0/\pi_1$ and upon the size of the sample n. Shifting the borderline to the right will decrease the size of the critical region and thereby increase the percentage of decisions in favor of H_0. This is certainly reasonable since $\pi_0 > \pi_1$ implies that H_0 is more likely to be true than H_1. As n becomes large, the second term in (21) approaches zero regardless of the relative sizes of π_0 and π_1, which means that the information supplied by the sample is swamping the information given by π_0 and π_1. In any given problem formula (21) can be used to calculate α and β as well as to determine the relative importance of the information supplied by the sample and the prior distribution.

EXERCISES

Sec. 1

1. Choosing $\alpha = .2$ and $\beta = .2$, test the hypothesis $H_0 : p = .5$ against $H_1 : p = .4$ sequentially by tossing a coin until a decision is reached. Here p is the probability of getting a head.

2. Choosing $\alpha = .1$ and $\beta = .2$, construct a sequential test for testing $H_0 : \sigma^2 = 8$ against $H_1 : \sigma^2 = 10$ for a normal variable with 0 mean.

3. Construct a sequential test for testing $H_0 : \theta = 3$ against $H_1 : \theta = 2$ for the density function $f(x; \theta) = \theta e^{-\theta x}$, $x \geq 0$. Choose $\alpha = .2$ and $\beta = .3$.

4. Using the value of n needed to reach a decision in problem 1, calculate by means of the normal approximation to the binomial what the value of β is for $\alpha = .2$ for a nonsequential test of H_0 based on this value of n.

5. Construct a sequential test for testing the hypothesis $\mu = \mu_0$ against the alternative $\mu = \mu_1 > \mu_0$ for a Poisson distribution.

6. By using random sampling numbers draw repeated samples from a Poisson distribution with $\mu = 2$. Use the test derived in problem 6 on these sample values to test $\mu = 2$ against $\mu = 3$ with $\alpha = .10$ and $\beta = .10$.

7. For the problem related to Table 1, calculate how large a fixed-size sample you would need to use to have α and β equal to the values used in that sequential test.

8. Given independent $X \sim \eta(\mu_X, 1)$ and $Y \sim \eta(\mu_Y, 1)$, construct a sequential

test for testing $H_0: \mu_Y = \mu_X$ against $H_1: \mu_Y = \mu_X + 1$ if pairs of samples (X_i, Y_i), $i = 1, 2, \cdots$ are taken at each stage. Use the variable $Z = Y - X$.

Sec. 1.1

9. Use formula (8) and the data for the problem used to illustrate it to verify that $E_1[n] \doteq 17$ for that problem.

10. Calculate $E_1[n]$ for the problem displayed in Table 1 and compare with the value of n actually realized in that problem.

11. Calculate $E_0[n]$ for problem 3.

Sec. 2

12. Let three single-variable normal subpopulations possess unit variances and means -1, 0, 1, respectively. Find what the best classification regions are if the population proportions are given by (a) $p_1 = p_2 = p_3 = \frac{1}{3}$, (b) $p_1 = 2p_2 = 4p_3$.

13. Work problem 12 if the three subpopulations are 2-variable populations with independent variables with unit variances and means $(-1, 0)$, $(0, 0)$, $(1, 0)$, respectively. Comment on the results in these two problems.

14. Let three single-variable subpopulations possess the distributions $f(x; \theta) = \theta e^{-x\theta}$, $\theta = 1, 2$, and 3, respectively, and let $p_1 = p_2 = p_3$. Find the best classification regions.

15. Apply the best classification technique to two 2-variable normal subpopulations whose means are $(-1, 0)$ and $(1, 0)$, whose correlation coefficients are both equal to $\frac{1}{2}$, and all of whose variances are equal to 1. Assume $p_1 = p_2$.

16. Compare the method of best classification when there are two normal subpopulations with the method of linear discriminant functions. Assume $p_1 = p_2$ and that the sample is so large that sample estimates used in the discriminant function may be treated as population values. Assume $\rho = 0$.

Sec. 3

17. Using $W(\theta, t) = |t - \theta|$, $f(x; \theta) = 1/\theta$, $0 \le x \le \theta$, and $\lambda(\theta) = e^{-\theta}$, $\theta > 0$, (a) calculate the risk function and the mean risk for the estimate $t = X$. (b) Compare this mean risk with that for the estimate $t = \frac{3}{2}X$.

18. Using $W(\theta, t) = (t - \theta)^2$, $f(x; \theta) = \theta e^{-x\theta}$, $x > 0$, and $\lambda(\theta) = \frac{2}{3}\theta$, $1 \le \theta \le 2$, calculate (a) the risk functions for the estimates $t_1 = X$ and $t_2 = X - 1$ and determine which is the minimax estimate with respect to these two estimates only. (b) Calculate the mean risk for estimate t_1. (c) Find the Bayes solution with respect to the estimates t_1 and t_2.

19. Given $W(\theta, t) = (t - \theta)^2$, $f(x; \theta) = e^{-\frac{1}{2}(x-\theta)^2}/\sqrt{2\pi}$, and $\lambda(\theta) = e^{-\frac{\theta^2}{2\sigma^2}}/\sqrt{2\pi}\,\sigma$, (a) calculate the risk function for the estimate $t = \overline{X}$, (b) calculate the mean risk for this estimate.

20. Given that X is a binomial variable with parameters n and p and that p possesses the beta distribution

$$\lambda(p) = \frac{\Gamma(\alpha + \beta)}{\Gamma(\alpha)\Gamma(\beta)} p^{\alpha-1}(1 - p)^{\beta-1}$$

calculate the Bayes solution for estimating p.

21. Given that X is a Poisson variable with parameter μ and that μ possesses the gamma distribution

$$\lambda(\mu) = \frac{1}{\Gamma(\alpha)} \mu^{\alpha-1} e^{-\mu}$$

calculate the Bayes solution for estimating μ.

22. Given that X possesses the gamma distribution

$$f(x \mid \theta) = \frac{\theta^{\alpha}}{\Gamma(\alpha)} x^{\alpha-1} e^{-\theta}$$

and that θ possesses the gamma distribution

$$\lambda(\theta) = \frac{\beta^a}{\Gamma(a)} \theta^{a-1} e^{-\beta\theta}$$

calculate the Bayes solution for estimating θ.

23. Suppose experience has shown that 20 percent of shipments of a certain chemical are of superior quality, with a mean of 65 and a standard deviation of 5 for this quality characteristic. If the remaining 80 percent have a mean of 60 and a standard deviation of 5, and a sample of 20 packages from a shipment yielded a mean of 63, test the hypothesis that the shipment is of superior quality. Assume normality.

24. Suppose 10 percent of certain automobile parts in a warehouse are from manufacturer A and 90 percent from manufacturer B. It is known that boxes of these parts from A contain 20 percent defectives and from B only 10 percent defectives. Test the hypothesis that a box, containing 200 parts, which is found to have 35 defectives is from A.

25. Calculate the values of α and β for problem 23.

Appendix

1 OUTLINE OF CENTRAL LIMIT THEOREM PROOF

The proof is based on using the characteristic function $\phi(\theta)$, which is the complex variable version of the moment generating function $M(\theta)$. The relationship is given by $\phi(\theta) = M(i\theta)$. The characteristic function possesses all the desirable properties of the moment generating function and in addition some highly useful properties not possessed by the moment generating function; therefore it is the proper tool to use in proofs.

Let X have the mean μ and the variance σ^2 and let $X_1, \cdots, X_n$ denote a random sample of size n of X. Write $Z_i = (X_i - \mu)/\sigma$. The $Z_i, i = 1, \cdots, n$, are a set of independent identically distributed random variables with mean 0 and variance 1. Let $Z = (\bar{X} - \mu)\sqrt{n}/\sigma$; then

$$Z = \frac{1}{\sqrt{n}} \sum_{i=1}^{n} Z_i \,.$$

Applying familiar moment generating function properties,

$$\phi_Z(\theta) = \phi_{\Sigma Z_i}\left(\frac{\theta}{\sqrt{n}}\right) = \phi_{Z_1}\left(\frac{\theta}{\sqrt{n}}\right) \cdots \phi_{Z_n}\left(\frac{\theta}{\sqrt{n}}\right) = \phi_{Z_1}{}^n\left(\frac{\theta}{\sqrt{n}}\right) \,.$$

Now apply the following Taylor expansion formula, which is valid for any function $G(z)$ that has a continuous second derivative. The variable z may be real or complex.

$$G(z) = G(0) + zG'(0) + \frac{z^2}{2} G''(0) + \frac{z^2}{2} g(z) \,,$$

where $\lim_{z \to 0} g(z) = 0$. Choose $G(z) = \phi(\theta)$, where $\phi(\theta)$ is the characteristic function

$$\phi(\theta) = \int_{-\infty}^{\infty} e^{i\theta x} f(x) \, dx \,.$$

359

Then $\phi(0) = 1$, $\phi'(0) = i\mu_1'$, $\phi''(0) = -\mu_2'$; consequently if it is assumed that $\phi''(\theta)$ is continuous,

$$\phi(\theta) = 1 + \theta i\mu_1' - \frac{\theta^2}{2}\mu_2' + \frac{\theta^2}{2}g(\theta)$$

where $\lim_{\theta \to 0} g(\theta) = 0$. Apply this formula to $\phi_{Z_1}(\theta/\sqrt{n})$. Since Z_1 is a variable with mean 0 and variance 1, $\mu_1' = 0$ and $\mu_2' = 1$; hence

$$\phi_{Z_1}\left(\frac{\theta}{\sqrt{n}}\right) = 1 - \frac{\theta^2}{2n} + \frac{\theta^2}{2n}g\left(\frac{\theta}{\sqrt{n}}\right).$$

Therefore

$$\phi_Z(\theta) = \left[1 - \frac{\theta^2}{2n} + \frac{\theta^2}{2n}g\left(\frac{\theta}{\sqrt{n}}\right)\right]^n.$$

Now apply the following limit formula from calculus, namely,

$$\lim_{n \to \infty}\left[1 + \frac{a}{n} + \frac{h(n)}{n}\right]^n = e^a$$

where $h(n)$ is any function for which $\lim_{n \to \infty} h(n) = 0$. Choosing $a = -\theta^2/2$ and $h(n) = g(\theta/\sqrt{n})\theta^2/2$, it follows from this formula and the expression for $\phi_Z(\theta)$ that

$$\lim_{n \to \infty} \phi_Z(\theta) = e^{-\frac{\theta^2}{2}}.$$

This is the characteristic function of a standard normal variable. To complete the proof it is necessary to use two theorems of advanced probability, which may be stated as follows:

THEOREM A: *To each characteristic function $\phi(\theta)$ there corresponds a unique distribution function $F(x)$ having $\phi(\theta)$ as its characteristic function.*

THEOREM B: *If a sequence of characteristic functions $\phi_n(\theta)$, $n = 1, 2, \ldots$, converges to a characteristic function $\phi(\theta)$, then the sequence of distribution functions $F_n(x)$ corresponding to the sequence $\phi_n(\theta)$ will converge to the distribution function determined by $\phi(\theta)$.*

Although the dependence of $\phi_Z(\theta)$ on n is not explicitly displayed, it is obviously a function of n. The preceding limit result shows that the sequence of characteristic functions represented by $\phi_Z(\theta)$ converges to the characteristic function $e^{-\frac{\theta^2}{2}}$. Theorem B guarantees that the sequence of distributions $F_n(x)$ corresponding to $\phi_Z(\theta)$ will converge to the distribution function determined by $e^{-\frac{\theta^2}{2}}$. But Theorem A guarantees that $e^{-\frac{\theta^2}{2}}$ uniquely determines a distribution function. Therefore since it is known that $e^{-\frac{\theta^2}{2}}$ is the characteristic function of a standard normal variable, it follows that the random variable

Z has a distribution converging to the distribution of a standard normal variable. These calculations and arguments constitute the outline of a proof of the central limit theorem.

In addition to relying upon Theorems A and B of advanced probability, it was also necessary to use a limit formula from calculus and a Taylor expansion formula from calculus. The limit formula, although rather uncommon, is readily demonstrated by beginning with a more familiar calculus inequality, namely,

$$x - \frac{x^2}{2} \leq \log(1 + x) \leq x, \qquad 0 < x < 1$$

and applying it to $x = [a + h(n)]/n$. Application of the Taylor formula to $\phi(\theta)$ requires that $\phi''(\theta)$ be continuous. It is easy to show that $\phi''(\theta)$ is continuous if μ'_3 exists; however by using more advanced techniques of analysis it is possible to show that $\phi''(\theta)$ is continuous if only μ'_2 exists. Thus, the central limit theorem is applicable to any random variable that possesses a second moment.

Theorem A is the uniqueness theorem that was referred to in earlier chapters with respect to the moment generating function. The only reason for using moment generating functions rather than characteristic functions in those chapters is that students with only an elementary calculus background are not expected to be familiar with complex numbers and functions.

2 PROPERTIES OF r

The purpose of this section is to prove that $|r| \leq 1$ and that $r = \pm 1$ if, and only if, all sample points lie on a straight line.

Let $a_i = x_i - \bar{x}$ and $b_i = y_i - \bar{y}$. Then r will assume the form

$$(1) \qquad r = \frac{\sum a_i b_i}{\sqrt{\sum a_i^2 \sum b_i^2}}.$$

In order to avoid trivial cases, which can easily be treated separately, it will be assumed that the x's are not all equal and that the y's are not all equal. This assumption prevents the denominator in (1) from having the value zero.

Now consider the inequality

$$(2) \qquad \sum (za_i + b_i)^2 > 0$$

where z is any real number. Since the left side is the sum of only squared terms, this inequality will be satisfied for all values of z if, and only if, a number z_0 does not exist such that

$$(3) \qquad z_0 a_i + b_i = 0, \qquad i = 1, 2, \cdots, n.$$

Assume for the present that no such number exists. Then squaring and summing in (2) will produce the inequality

(4) $$z^2 \sum a_i^2 + 2z \sum a_i b_i + \sum b_i^2 > 0 .$$

The left side is a quadratic function in z, which is everywhere positive; consequently the corresponding quadratic equation must have imaginary roots, which in turn implies that the discriminant of the quadratic must be negative. Thus it is necessary that

$$(2 \sum a_i b_i)^2 - 4(\sum a_i^2)(\sum b_i^2) < 0 ,$$

or that

$$(\sum a_i b_i)^2 < \sum a_i^2 \sum b_i^2 .$$

In view of (1), this shows that $r^2 < 1$, provided that a number z_0 satisfying (3) does not exist.

Now suppose a number z_0 does exist such that (3) holds. Then

$$\sum (z_0 a_i + b_i)^2 = 0 .$$

Squaring and summing will yield

(5) $$z_0^2 \sum a_i^2 + 2z_0 \sum a_i b_i + \sum b_i^2 = 0 .$$

This says that z_0 is a root of the corresponding quadratic equation. Since there obviously cannot be two different values of z_0 satisfying (3), z_0 must be a double root in (5). As a result, the discriminant will be equal to zero; hence

$$(\sum a_i b_i)^2 = \sum a_i^2 \sum b_i^2 .$$

In view of (1), this shows that if a number z_0 satisfying (3) exists, then $r^2 = 1$. Since it was shown earlier that $r^2 < 1$ unless such a number did exist, it follows that $r^2 = 1$ if, and only if, a number z_0 satisfying (3) exists. In terms of the original variables, (3) can be written in the form

$$y_i - \bar{y} = -z_0(x_i - \bar{x}), \qquad i = 1, 2, \cdots, n .$$

In geometrical language, it therefore follows that $r^2 \leq 1$ and that $r^2 = 1$ if, and only if, the points (x_i, y_i) lie on some straight line.

Students who have seen the inequality of Schwartz will recognize that $r^2 \leq 1$ is merely a version of that inequality.

3   CRAMER-RAO INEQUALITY

Consider the problem of how to find the best unbiased estimate of the parameter θ in the continuous frequency function $f(x; \theta)$. The solution of the problem lies in obtaining an inequality for the variance of any unbiased estimator $t = t(X_1, X_2, \cdots, X_n)$ of θ. This inequality is derived in the following manner.

Since $X_1, X_2, \cdots, X_n$ is a random sample from $f(x; \theta)$, its density function, which for brevity of notation is denoted by L, is given by

$$L = \prod_{i=1}^{n} f(x_i; \theta).$$

It therefore follows that

(1) $$\int \cdots \int L \, dx_1 \cdots dx_n = 1.$$

Since $t = t(X_1, X_2, \cdots, X_n)$ is assumed to be an unbiased estimator of θ, it follows that

(2) $$E[t] = \int \cdots \int t L \, dx_1 \cdots dx_n = \theta.$$

It is understood here that the integrand t is a function of the variables of integration, whereas t in $E[t]$ is a function of the corresponding random variables.

Formulas (1) and (2) are identities in θ; therefore they may be differentiated with respect to θ. In doing so, it will be assumed that it is permissible to differentiate under the integral sign and that the limits of integration do not depend on θ. Differentiation of (1) will give

(3) $$\int \cdots \int \frac{\partial L}{\partial \theta} \, dx_1 \cdots dx_n = 0.$$

Differentiation of (2) yields

(4) $$\int \cdots \int t \frac{\partial L}{\partial \theta} \, dx_1 \cdots dx_n = 1.$$

The value of $\partial L/\partial \theta$ is most easily obtained by calculating

$$\frac{\partial \log L}{\partial \theta} = \frac{1}{L} \frac{\partial L}{\partial \theta}.$$

Thus

$$\frac{\partial L}{\partial \theta} = L \sum_{i=1}^{n} \frac{\partial \log f(x_i; \theta)}{\partial \theta}.$$

To simplify the notation somewhat, let

(5) $$T = \sum_{i=1}^{n} \frac{\partial \log f(x_i; \theta)}{\partial \theta}.$$

Equation (3) can now be expressed as follows.

(6) $$0 = \int \cdots \int TL \, dx_1 \cdots dx_n = E[T]$$

Similarly, equation (4) will assume the form

(7) $$1 = \int \cdots \int tTL \, dx_1 \cdots dx_n = E[tT].$$

Next, consider the value of the correlation coefficient between the two random variables t and T. From formula (14), Chapter 6, it may be written as

$$\rho_{tT} = \frac{E[tT] - E[t]E[T]}{\sigma_t \sigma_T}.$$

In view of the results in (6) and (7), this will reduce to

(8) $$\rho_{tT} = \frac{1}{\sigma_t \sigma_T}.$$

Since any correlation coefficient satisfies the inequality $\rho^2 \leq 1$, it follows from (8) that σ_t and σ_T must satisfy the inequality

(9) $$\sigma_t^2 \geq \frac{1}{\sigma_T^2}.$$

In view of (5) and the independence of the terms in that sum, it follows that

(10) $$\sigma_T^2 = \sum_{i=1}^{n} \sigma_i^2$$

where σ_i^2 is the variance of $\partial \log f(X_i; \theta)/\partial \theta$. But from (5) and (6)

$$\sum_{i=1}^{n} E \frac{\partial \log f(X_i; \theta)}{\partial \theta} = 0.$$

Since the X_i possess the same distribution, the quantities $\partial \log f(X_i; \theta)/\partial \theta$, $i = 1, 2, \cdots, n$, must possess the same distribution, hence the same expected value. Since the sum of such expected values is zero, it follows that each expected value must be zero and therefore the variance σ_i^2 of $\partial \log f(X_i; \theta)/\partial \theta$ is equal to its second moment. Thus

$$\sigma_i^2 = E\left[\frac{\partial \log f(X_i; \theta)}{\partial \theta}\right]^2.$$

Consequently, from (10),

$$\sigma_T^2 = nE\left[\frac{\partial \log f(X; \theta)}{\partial \theta}\right]^2$$

because each X_i has the same distribution as the basic variable X. When this result is substituted in (9), the desired inequality will be obtained, namely

$$(11) \qquad \sigma_t^2 \geq \frac{1}{nE\left[\dfrac{\partial \log f(X;\theta)}{\partial \theta}\right]^2} .$$

Since a best unbiased estimate is by definition one with minimum variance, it follows that if one can find an unbiased estimate whose variance is equal to the quantity on the right of (11) he will have found a best unbiased estimate.

This formula can be used to show that $\bar{X}$ is a best unbiased estimate for the mean of a normal distribution. Toward this end, write

$$f(x;\mu) = \frac{e^{-\frac{1}{2}\left(\frac{x-\mu}{\sigma}\right)^2}}{\sqrt{2\pi}\,\sigma}$$

and assume that the value of σ is known. Then

$$\log f(x;\mu) = -\log \sqrt{2\pi}\,\sigma - \frac{1}{2}\left(\frac{x-\mu}{\sigma}\right)^2$$

and

$$\frac{\partial \log f(x;\mu)}{\partial \mu} = \frac{x-\mu}{\sigma^2} .$$

Hence

$$E\left[\frac{\partial \log f(x;\mu)}{\partial \mu}\right]^2 = \frac{1}{\sigma^4} E(x-\mu)^2 = \frac{1}{\sigma^2} .$$

Substituting this result in (11) will yield

$$\sigma_t^2 \geq \frac{\sigma^2}{n} .$$

But it is known that $\sigma_{\bar{X}}^2 = \sigma^2/n$; therefore $\bar{X}$ must be a best unbiased estimate of μ for a normal distribution.

4 LIKELIHOOD RATIO TEST FOR GOODNESS OF FIT

Consider k cells with probabilities $p_1, p_2, \cdots, p_k$, where $\sum_1^k p_i = 1$. Let $n_1, n_2, \cdots, n_k$, with $\sum_1^k n_i = n$, be the observed frequencies in those cells in n trials. Then the likelihood function is

$$(1) \qquad L(p) = p_1^{n_1} p_2^{n_2} \cdots p_k^{n_k}$$

Since $\sum p_i = 1$, there are only $k - 1$ independent parameters here; consequently in maximizing $L(p)$ by calculus methods it is necessary to keep this fact in mind. In this connection, it is convenient to choose p_k as the parameter to be expressed in terms of the remaining parameters. Taking logarithms and differentiating with respect to p_i will yield

$$\log L(p) = n_1 \log p_1 + n_2 \log p_2 + \cdots + n_k \log p_k$$

$$\frac{\partial \log L(p)}{\partial p_i} = \frac{n_i}{p_i} + \frac{n_k}{p_k} \frac{\partial p_k}{\partial p_i} = \frac{n_i}{p_i} - \frac{n_k}{p_k}.$$

For a maximum it is necessary that all $k - 1$ partial derivatives vanish; hence it is necessary that

(2)
$$\frac{n_i}{p_i} - \frac{n_k}{p_k} = 0, \qquad i = 1, 2, \cdots, k - 1.$$

If the maximum likelihood estimate of p_i is denoted by $\hat{p}_i$, it follows from (2) that

(3)
$$\hat{p}_i = \frac{\hat{p}_k}{n_k} n_i, \qquad i = 1, 2, \cdots, k,$$

where $\hat{p}_k = 1 - \sum_1^{k-1} \hat{p}_i$. Summing both sides of (3) will yield

$$1 = \sum_1^k \hat{p}_i = \frac{\hat{p}_k}{n_k} \sum_1^k n_i = \frac{\hat{p}_k}{n_k} n.$$

If this result is applied to (3), it will follow that

(4)
$$\hat{p}_i = \frac{n_i}{n}, \qquad i = 1, 2, \cdots, k.$$

Now consider the likelihood ratio test for testing the hypothesis

$$H_0 : p_i = p_{i0}, \qquad i = 1, 2, \cdots, k.$$

Since there are no unspecified parameters remaining when H_0 is true, it follows from (1) and (4) that the likelihood ratio here is given by

$$\lambda = \frac{L_0(\hat{p})}{L(\hat{p})} = \frac{p_{10}^{n_1} p_{20}^{n_2} \cdots p_{k0}^{n_k}}{\left(\dfrac{n_1}{n}\right)^{n_1} \left(\dfrac{n_2}{n}\right)^{n_2} \cdots \left(\dfrac{n_k}{n}\right)^{n_k}}$$

$$= \left(\frac{n p_{10}}{n_1}\right)^{n_1} \left(\frac{n p_{20}}{n_2}\right)^{n_2} \cdots \left(\frac{n p_{k0}}{n_k}\right)^{n_k}$$

$$= \left(\frac{e_1}{n_1}\right)^{n_1} \left(\frac{e_2}{n_2}\right)^{n_2} \cdots \left(\frac{e_k}{n_k}\right)^{n_k}$$

where $e_i = np_{i0}$. As a result

(5)
$$-2 \log \lambda = -2 \sum_1^k n_i \log \frac{e_i}{n_i}.$$

Now let $x_i = n_i - e_i$, which is the difference between the observed frequency and the expected frequency in the ith cell. Then (5) may be expressed in the following form.

$$-2 \log \lambda = -2 \sum (e_i + x_i) \log \frac{e_i}{e_i + x_i}$$

$$= 2 \sum (x_i + e_i) \log \frac{e_i + x_i}{e_i}$$

$$= 2 \sum (x_i + e_i) \log \left(1 + \frac{x_i}{e_i}\right)$$

$$= 2 \sum (x_i + e_i) \left[\frac{x_i}{e_i} - \frac{1}{2}\left(\frac{x_i}{e_i}\right)^2 + \frac{1}{3}\left(\frac{x_i}{e_i}\right)^3 - \cdots\right]$$

$$= 2 \left(\sum \frac{x_i^2}{e_i} - \frac{1}{2} \sum \frac{x_i^3}{e_i^2} + \frac{1}{3} \sum \frac{x_i^4}{e_i^3} - \cdots \right.$$

$$\left. + \sum x_i - \frac{1}{2} \sum \frac{x_i^2}{e_i} + \frac{1}{3} \sum \frac{x_i^3}{e_i^2} - \cdots\right)$$

(6)
$$= \sum \frac{x_i^2}{e_i} - \frac{1}{3} \sum \frac{x_i^3}{e_i^2} + \cdots.$$

The variable n_i is a binomial variable with mean $\mu_i = np_{i0} = e_i$ and variance $\sigma_i^2 = np_{i0}(1 - p_{i0}) = e_i(1 - p_{i0})$; consequently the variable x_i/e_i may be expressed in the form

(7)
$$\frac{x_i}{e_i} = \frac{n_i - e_i}{e_i} = \frac{n_i - \mu_i}{\sigma_i} \sqrt{\frac{1 - p_{i0}}{np_{i0}}}.$$

Now from Theorem 2, Chapter 5, the variable $(n_i - \mu_i)/\sigma_i$ has a distribution approaching that of a standard normal variable as $n \to \infty$, whereas the square root factor in (7) approaches zero at the same rate as $1/\sqrt{n}$. Thus, for large n, x_i/e_i will almost certainly be very small and of the order of $1/\sqrt{n}$; consequently the successive terms in the above expansion will be of order $1/\sqrt{n}$ times the preceding term. As a result, the large sample approximate value of $-2 \log \lambda$ is given by the first term in (6). Thus

$$-2 \log \lambda \sim \sum_{i=1}^k \frac{x_i^2}{e_i} = \sum_{i=1}^k \frac{(n_i - e_i)^2}{e_i}.$$

Since from (25), Chapter 8, it is known that $-2 \log \lambda$ possesses an approximate χ^2 distribution, this derivation shows that the quantity $\sum (n_i - e_i)^2/e_i$ possesses an approximate χ^2 distribution. With a little more attention to details the preceding derivation can be made to yield a mathematical theorem, which essentially states in the language of limiting distributions what has been said here concerning approximate distributions.

5   TRANSFORMATIONS AND JACOBIANS

Geometrically, the functions $u = u(x, y)$ and $v = v(x, y)$ represent a transformation from the coordinate system x, y to the coordinate system u, v. Now there exists a calculus formula that enables one to evaluate the integral of the function $f(x, y)$ over a region R in the x, y plane by means of the proper integral over a corresponding region R' in the u, v plane. This formula is

(1)
$$\iint_R f(x, y) \, dx \, dy = \iint_{R'} f(x, y) \, |J| \, du \, dv$$

where the quantity J, called the Jacobian of the transformation, is given by the formula

(2)
$$\frac{1}{J} = \begin{vmatrix} \dfrac{\partial u}{\partial x} & \dfrac{\partial u}{\partial y} \\ \dfrac{\partial v}{\partial x} & \dfrac{\partial v}{\partial y} \end{vmatrix}.$$

The region of integration R' on the right is the region in the u, v plane that corresponds to the region R in the x, y plane. It is understood that the variables x and y in the right integrand of (1) are to be replaced by their values in terms of u and v by solving the relations $u = u(x, y)$ and $v = v(x, y)$ for X and Y. It will be assumed that these functions are such that each point in the x, y plane corresponds to exactly one point in the u, v plane, and conversely.

The integral on the left of (1) yields the probability that the sample point X, Y will lie in the region R. Because of the one-to-one correspondence between points in the two coordinate systems, this can occur if, and only if, the sample point U, V lies in the corresponding region R'; consequently the integral on the right must yield the probability that the sample point U, V will lie in the region R'. Now formula (1) holds for all possible regions R in the x, y plane, hence for all possible regions R' in the u, v plane; consequently if $f(x, y)$ is a continuous function, the integrand in the integral on the right side of (1) must be the density function of the random variables U and V. Thus, denoting this function by $g(u, v)$, it follows that

(3) $$g(u, v) = f(x, y) |J|$$

where J is given by (2).

As an illustration, consider the problem solved earlier in (6), Chapter 10. Here it suffices to choose $u = z = t(x, y)$ and $v = x$. Then (2) becomes

$$\frac{1}{J} = \begin{vmatrix} \dfrac{\partial z}{\partial x} & \dfrac{\partial z}{\partial y} \\ 1 & 0 \end{vmatrix} = -\frac{\partial z}{\partial y}.$$

Application of (3) then yields

$$g(u, v) = \frac{f(x, y)}{\left| \dfrac{\partial z}{\partial y} \right|}$$

which is the result given earlier.

The method that has just been explained for finding the density function of two transformed variables U and V can be generalized to any number of variables. The formula that results is a direct consequence of probability considerations applied to the formula for evaluating a multiple integral by means of a new coordinate system. The following theorem, in which the functions u_i and f are assumed to satisfy certain regularity conditions, yields the desired general result for k variables.

THEOREM: *If the continuous variables $X_1, X_2, \cdots, X_k$ possess the frequency function $f(x_1, x_2, \cdots, x_k)$ and the transformed variables $U_i = u_i(X_1, X_2, \cdots, X_k)$, $i = 1, 2, \cdots, k$, yield a one-to-one transformation of the two coordinate systems, the density function of the U's will be given by the formula*

(4) $$g(u_1, u_2, \cdots, u_k) = f(x_1, x_2, \cdots, x_k) |J|$$

where

$$\frac{1}{J} = \begin{vmatrix} \dfrac{\partial u_1}{\partial x_1} & \cdots & \dfrac{\partial u_1}{\partial x_k} \\ \cdots\cdots\cdots \\ \dfrac{\partial u_k}{\partial x_1} & \cdots & \dfrac{\partial u_k}{\partial x_k} \end{vmatrix}$$

and where the x's on the right of (4) are to be replaced by their values in terms of the u's by solving the relations $u_i = u_i(x_1, x_2, \cdots, x_k)$ for the x's.

6 INDEPENDENCE OF $\bar{X}$ AND S^2 FOR NORMAL DISTRIBUTIONS

Consider the n independent normal variables $X_1, X_2, \cdots, X_n$ with means $\mu_1, \mu_2, \cdots, \mu_n$ and the common variance σ^2. Their joint density function is given by

$$(1) \qquad f(x_1, x_2, \cdots, x_n) = \frac{e^{-\frac{1}{2\sigma^2}\sum_{i=1}^{n}(x_i-\mu_i)^2}}{(2\pi\sigma^2)^{\frac{n}{2}}} .$$

Now $\sum (x_i - \mu_i)^2 = c^2$ is the equation of a sphere in n dimensions with center at the point $(\mu_1, \mu_2, \cdots, \mu_n)$ and with radius c. As a consequence, the geometrical interpretation of (1) is that the probability density is constant on the surface of any sphere with center at $(\mu_1, \mu_2, \cdots, \mu_n)$ and the magnitude of the density for any point on such a sphere is given by replacing $\sum (x_i - \mu_i)^2$ in (1) by the square of the radius of the sphere. These two geometrical properties completely determine the distribution of $X_1, X_2, \cdots, X_n$. A sketch illustrating these properties is shown in Fig. 1 where x and μ denote the sample point $(x_1, x_2, \cdots, x_n)$ and the mean point $(\mu_1, \mu_2, \cdots, \mu_n)$, respectively.

Now suppose one rotates the axes of this coordinate system in any desired manner. If the new axes are denoted by $y_1, y_2, \cdots, y_n$, as indicated in Fig. 1, the equation of the sphere sketched there will become $\sum (y_i - \nu_i)^2 = c^2$, where $(\nu_1, \nu_2, \cdots, \nu_n)$ denotes the coordinates of the mean point μ in terms

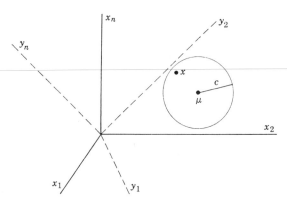

Fig. 1. Distribution of n independent normal variables.

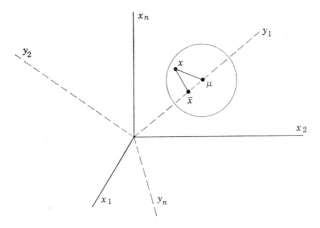

Fig. 2.   Transformation of coordinates.

of the new coordinate system. The typical sample point $x = (x_1, x_2, \cdots, x_n)$ becomes the point $y = (y_1, y_2, \cdots, y_n)$ in the new system. The y's represent the coordinates of a random point in the same sample space as for the X's. Since the only effect of rotating axes is to change the coordinates of the mean of the distribution, the geometrical properties of the distribution being considered show that the distribution of the new variables $Y_1, Y_2, \cdots, Y_n$ must be given by the frequency function

$$(2) \qquad g(y_1, y_2, \cdots, y_n) = \frac{e^{-\frac{1}{2\sigma^2}\sum_{i=1}^{n}(y_i-v_i)^2}}{(2\pi\sigma^2)^{\frac{n}{2}}}.$$

The preceding discussion will now be specialized to the case in which $\mu_1 = \mu_2 = \cdots = \mu_n$ and in which one rotates the axes in such a way that the y_1 axis becomes the line that makes equal angles with the positive x axes, as shown in Fig. 2.

In the new coordinate system the mean point μ will be on the y_1 axis, and therefore its coordinates with respect to the new axes are given by $(v, 0, \cdots, 0)$, where v is the distance from the origin to the point μ in the old coordinate system. It is clear from Figs. 1 and 2 that this distance is given by $v = \sqrt{\mu_1^2 + \mu_2^2 + \cdots + \mu_n^2} = \mu\sqrt{n}$, where μ here denotes the common numerical value of the equal-valued μ_i. As a result, formula (2) may be applied to give

$$(3) \qquad g(y_1, y_2, \cdots, y_n) = \frac{e^{-\frac{1}{2\sigma^2}[(y_1-\mu\sqrt{n})^2+y_2^2+\cdots+y_n^2]}}{(2\pi\sigma^2)^{\frac{n}{2}}}.$$

Thus the Y's are independent normal variables with a common variance σ^2, and all have zero means except Y_1, which has a mean of $\mu\sqrt{n}$.

Now consider the geometrical meaning of the equation defining the sample mean, namely

(4) $$x_1 + x_2 + \cdots + x_n = n\bar{x}.$$

This is the equation of a plane in n dimensions. Since the coefficients of the variables $x_1, x_2, \cdots, x_n$ are direction numbers of a normal (perpendicular) to the plane and since all these coefficients are equal, it follows that the y_1 axis is a normal, hence perpendicular, to this plane because the y_1 axis makes equal angles with the positive x axes. Further, since the coordinates of the point $(\bar{x}, \bar{x}, \cdots, \bar{x})$ satisfy equation (4) and since this point lies on the y_1 axis, it follows that the point of intersection of this plane with the y_1 axis is the point $(\bar{x}, \bar{x}, \cdots, \bar{x})$, which has been labeled $\bar{x}$ in Fig. 2. The point x lies in this plane also.

Now $ns^2 = \sum (x_i - \bar{x})^2$ is the square of the distance between the points labeled x and $\bar{x}$ in Fig. 2. In the new coordinate system the square of this distance is given by $y_2^2 + y_3^2 + \cdots + y_n^2$ because the points x and $\bar{x}$ possess the coordinates $(y_1, y_2, \cdots, y_n)$ and $(y_1, 0, \cdots, 0)$, respectively, in the new coordinate system. Thus

$$ns^2 = y_2^2 + y_3^2 + \cdots + y_n^2.$$

Furthermore, since the distance from the origin to the plane given by (4) is $\sqrt{n}\bar{x}$ and y_1, respectively, in the two coordinate systems, it follows that

$$\bar{x} = \frac{y_1}{\sqrt{n}}.$$

This shows that $\bar{X}$ is a function of the variable Y_1 only and that S^2 is a function of the variables $Y_2, Y_3, \cdots, Y_n$ only. Since the Y's are independent random variables, it therefore follows that $\bar{X}$ and S^2 must be independent random variables.

Answers

Numerical answers often depend on the order of operations and the extent of rounding off; hence a student's answers may differ slightly from those given here.

CHAPTER 2

1. $\frac{1}{9}$.

3. (a) Twenty points with probabilities $\frac{1}{20}$. (b) Three points with probability $\frac{3}{10}$ and one point with probability $\frac{1}{10}$.

5. $\frac{1}{3}, \frac{2}{3} \cdot \frac{1}{3}, (\frac{2}{3})^2 \cdot \frac{1}{3}, \cdots$.

7. (a) $\frac{1}{2}$, (b) $\frac{9}{20}$, (c) $\frac{11}{30}$.

9. (a) $\frac{1}{16}$, (b) $\frac{1}{32}$, (c) $\frac{5}{16}$.

11. (a) $\frac{1}{2}$, (b) $\frac{2}{3}$.

13. $\frac{5}{9}$.

15. $\frac{3}{4}$.

17. $\frac{7}{11}$.

19. $\frac{1}{4}$.

21. .28.

23. $\frac{45}{1081}$.

25. $\dfrac{94}{54,145}$.

27. $\dbinom{40}{7} \Big/ \dbinom{50}{7}$.

29. $\dfrac{10n - 30}{n^2 - n}$.

31. $2 \dbinom{4}{3} \dbinom{22}{10} \Big/ \dbinom{26}{13}$.

33. $\dfrac{\dbinom{24}{5} - \dbinom{12}{5}\dbinom{12}{0} - \dbinom{12}{4}\dbinom{12}{1} - \dbinom{12}{3}\dbinom{12}{2}}{\dbinom{52}{5} - \dbinom{40}{5} - \dbinom{40}{4}\dbinom{12}{1} - \dbinom{40}{3}\dbinom{12}{2}} = \dfrac{161}{1456}$.

35. (a) $\frac{1}{2}$, (b) $\frac{8}{11}$.

37. (a) .184, (b) .736, (c) $\displaystyle\sum_{x=0}^{\infty} \frac{1}{x!} = e$.

39. (a) $\frac{3}{8}$, (b) $f(x) = \dfrac{3}{2x!\,(4-x)!}$, (c) $f(0) = \frac{1}{16}, f(1) = \frac{4}{16}, f(2) = \frac{6}{16}, \cdots$.

41. Use $F(x) = 0, x < 0, F(0) = \frac{1}{16}, F(1) = \frac{5}{16}, F(2) = \frac{11}{16}, F(3) = \frac{15}{16}, F(x) = 1,$ $x \geq 4, \cdots$.

43. $f(x) = \binom{4}{x}\binom{2}{2-x} / \binom{6}{2}$, $F(x) = 0$, $x < 0$, $F(x) = \frac{1}{15}$, $0 \le x < 1$, $F(x) =$
$\frac{9}{15}$, $1 \le x < 2$, $F(x) = 1$, $x \ge 2$.

45. $f(x) = \frac{3}{4}(\frac{1}{4})^{x-1}$.   47. $f(x, y) = \frac{1}{4}$.

49. $f(x, y) = 1280/81x!\, y!\, (5 - x - y)!\, 4^{x+y}$, $0 \le x + y \le 5$.

51. $f(x, y) = \binom{13}{x}\binom{13}{y}\binom{26}{13-x-y} / \binom{52}{13}$.

53. $f(0) = \frac{2}{5}$, $f(y \mid 0) = \frac{1}{6}\binom{2}{y}\binom{2}{2-y}$.   55. $g(x \mid y) = 1/2\, x!\, (2-x)!$.

57. $3.09/y!\, (3 - y)!\, 4^y$.

59. (a) $c = \frac{1}{13}$, (b) $f(1) = \frac{2}{13}, f(2) = \frac{7}{13}, f(3) = \frac{4}{13}$, (c) $f(y \mid x) = 1$ if $x = 1$ or 3
    and $y = 1$, $f(y \mid 2) = (2 + y)/7$ if $y = 1$ or 2.

61. Find $f(x, y)$, then sum with respect to x from y to ∞ by letting $t = x - y$ and
    summing t from 0 to ∞.

63. (a) $c = \frac{1}{2}$, (b) .5, (c) .75.

65. (a) $F(x) = 0$, $x < 0$, $F(x) = \dfrac{x}{2}$, $0 \le x \le 2$, $F(x) = 1$, $x > 2$, (b) $F(x) = 0$,

$x < 0$, $F(x) = \dfrac{x^2}{2}$, $0 \le x \le 1$, $F(x) = -\dfrac{x^2}{2} + 2x - 1$, $1 < x \le 2$, $F(x) = 1$,

$x > 2$, (c) $F(x) = \dfrac{1}{\pi} \tan^{-1} x + \frac{1}{2}$.

67. (a) 100, (b) $F(x) = 0$, $x < 100$, $F(x) = 1 - \dfrac{100}{x}$, $x \ge 100$, (c) $\frac{1}{5}$.

69. .25.   71. $2e^{-1} - 4e^{-2} = .19$.

73. $f(x, y) = \frac{1}{2}$ for $\begin{Bmatrix} -1 < x < 0 \\ -1 < y < 0 \end{Bmatrix}$ and $\begin{Bmatrix} 0 < x < 1 \\ 0 < y < 1 \end{Bmatrix}$.

CHAPTER 3

1. (a) $E[X] = \frac{7}{3}$, (b) $E[X] = \frac{14}{3}$.   3. $\frac{6}{5}$.

5. $E[X] = \frac{7}{2}$, $V[X] = \frac{35}{12}$.   7. 7.

9. $\frac{1}{6}$ and $\frac{5}{6}$.   11. 0 and $\frac{3}{2}$.

13. $\sum [-p(x - 1)(1 - p)^{x-2} + (1 - p)^{x-1}] = 0$ yields
    $-\mu/(1 - p) + 1/(1 - p) + 1/p = 0$.

15. (a) $\frac{160}{729} = .22$, (b) $\frac{656}{729} = .90$.   17. $\displaystyle\sum_{x=0}^{6} \frac{12!}{x!\,(12-x)!}\left(\frac{1}{2}\right)^x\left(\frac{1}{2}\right)^{12-x} = .61$.

19. (a) .24.   21. (a) $\frac{9}{16}$, (b) $\frac{27}{128}$.

23. $p = \frac{1}{6}$, $n = 72$.

27. Using .2175 for estimating p, frequencies are 7, 19, 24, 18, 9, 3, 1, 0, $\cdots$.

29. .125.   31. (a) .018, (b) .029.

33. (a) .0067, (b) 41.   35. .26.

37. (a) .18, (b) $e^{-n/2}\left(\dfrac{n}{2}\right)^{x} \Big/ x!$.

39. Using $\bar{x} = .61$ for estimating μ, frequencies are 109, 66, 20, 4, 1, $\cdots$.

41. Integer between $\mu - 1$ and μ. Either will do if μ is an integer.

43. $\frac{14}{55}$.   45. $\frac{2}{21}$.

47. (a) $\frac{5}{6}$, (b) $\frac{1}{12}$.   49. .984.

51. (a) $\mu = \frac{1}{2}$, $\sigma^2 = \frac{1}{12}$, (b) $\mu = \frac{1}{3}$, $\sigma^2 = \frac{4}{45}$.

53. (a) $c = \frac{1}{2}$, (b) $2^{k+1}/(k + 2)$, (c) $(2\theta e^{2\theta} - e^{2\theta} + 1)/2\theta^2$.

55. $c = 1/\alpha!$, (b) $(\alpha + k)!/\alpha!$, (c) $(1 - \theta)^{-(\alpha+1)}$.

59. $\frac{7}{16}$.   61. (a) .001, (b) .499.

63. 15.6.   65. (a) .186, (b) .392.

67. 19.8.   69. .03.

71. $\hat{p} = .0255$; hence $.0255 \pm .0150$. Out on 18, 22, and 38.

73. .0004.   77. .43.

79. 11.

CHAPTER 4

1. Class marks: 156, 159, $\cdots$.   3. 4.43.

5. 71 and 96.   7. $\sigma \doteq 2\frac{1}{2}$ inches.

9. Possibly a long right tail.   13. (a) $\alpha = \frac{1}{2}$, $\beta = \frac{1}{2}$, (b) $\alpha = \frac{3}{4}$, $\beta = \frac{1}{4}$.

15. Choose $x > 5$, then $\beta = \frac{1}{2}$.   17. (a) $\frac{1}{64}$, (b) $\frac{1}{256}$.

19. $\alpha = \frac{1}{35}$, $\beta = \frac{31}{35}$.   21. (a) No, (b) conclude differently.

CHAPTER 5

1. Density not the same as for original population.

3. (a) 2, (b) 6, (c) $(1 - \theta)^{-2}$.

5. $E[Z] = (n - 1)/2$, $V[Z] = (n + 1)/12$.

9. (a) .401, (b) .159.   11. 171.

15. Use $E[X] = \beta$ and $V[X] = \beta^2$ and central limit theorem on $\bar{X}$.

19. Use moment generating function.   21. $nf(z)F^{n-1}(z)$.

23. ν and 2ν.   25. $n\sigma^2\left(\dfrac{1}{a} - \dfrac{1}{b}\right)$.

27. $z = -1.80$; hence not justified.

29. X and Y are not independent variables.

31. (a) No, (b) either X or Y or both are not taking random samples because $\bar{X} - \bar{Y}$ is not compatible with $\mu_X = \mu_Y$.

33. $z = -20/\sqrt{80}$; hence reject $\mu_1 = \mu_2$.

35. $z = 1.43$; hence accept $p_1 = p_2$.

37. $z = 1.55$; hence accept $p_1 - p_2 = .08$ rather than $p_1 - p_2 > .08$.

CHAPTER 6

1. (a) $\frac{7}{8}$, (b) $\dfrac{\pi}{4}$, (c) $\frac{6}{8}$.

3. (a) .24, (b) .6, (c) .5, (d) 0, (e) .6, (f) .8, (g) 1, (h) $\dfrac{\pi}{4}$.

5. (a) $c = 6/(6e - 5)$, (b) $f(x) = 6\left(\dfrac{x}{3} + e^x\right) \Big/ (6e - 5)$, (c) $f(y \mid x)$ depends on x; hence not independent.

7. (a) $1 - \log 2$, (b) $\dfrac{x}{2}$, (c) $(y - 1)/\log y$.

9. Find $f(x, y)$, then integrate with respect to x from 0 to ∞.

11. (a) $f(x) = e^{-x}$, $g(y) = (y + 1)^{-2}$,
    (b) $f(y \mid x) = xe^{-xy}$, $f(x \mid y) = (y + 1)^2 xe^{-x(y+1)}$,
    (c) $\mu_{Y \mid x} = \dfrac{1}{x}$.

13. (a) $(x + y)/(x + \frac{1}{2})$, (b) $\left(\dfrac{x}{2} + \dfrac{1}{3}\right) \Big/ (x + \frac{1}{2})$.

15. $\dfrac{2}{9\pi} \sqrt{9 - x^2}$, (b) $1/2 \sqrt{9 - x^2}$, (c) 0.   17. $\mu_{Y \mid x} = \mu_Y$.

23. (a) $\mu'_{pq} = 1/(p + 1)(q + 1)$, (b) $\rho = 0$, (c) $\mu_{Y \mid x} = \frac{1}{2}$.

25. (a) $c = 1$, (b) 0, (c) $y = -\dfrac{x}{2}$, $-1 < x < 0$, $y = \dfrac{x}{2}$, $0 < x < 1$.

27. $E[X] = np$, $E[Y] = n(n + 1)p/2$, $E[XY] = [1 + (n - 1)p]pn(n + 1)/2$,
    $\mu_{11} = n(n + 1)p(1 - p)/2$.

29. .06.

33. (a) $\mu_X = 1$, $\mu_Y = -2$, $\sigma_X{}^2 = \frac{20}{64}$, $\sigma_Y{}^2 = \frac{5}{64}$, $\rho = .6$,
    (b) $f(x) = \dfrac{4}{\sqrt{10\pi}} \exp\left[-\frac{8}{5}(x + 1)^2\right]$, (c) $\mu_{Y \mid x} = -2 + .3(x - 1)$.

35. $\frac{15}{4}$ as compared to 4.

CHAPTER 7

1. .92.
3. $|r| > .37$.

7. Maximum traffic occurs around 8–9 A.M. and near 5 P.M. Maximum tides may occur around 8 A.M. and 8 P.M.; hence maxima and minima occur fairly close together for the two variables.

9. $s^2 = (n^2 - 1)/12$.
11. $y' = .87 + .8x$.

13.
$$\begin{cases} \alpha \sum x^2 + \beta n = \sum yx \\ \alpha n + \beta \sum \dfrac{1}{x^2} = \sum \dfrac{y}{x}. \end{cases}$$

17. Calculate $\sum (y'_i - \bar{y})(y_i - \bar{y})$; then calculate $\sum (y'_i - \bar{y})^2$; and finally obtain r.

19. $y' = .39x_1 + .25x_2$.
21. $\alpha = .23, \beta = 3.3$.

23. $a_i = \sum_{k=1}^{n} y_k P_i(x_k)$.
25. .53.

CHAPTER 8

1. (a) 11.6, (b) $E = \frac{151}{175}\sigma^2$; hence biased.

5. Calculate the probability of $x - 1$ successes in $N - 1$ trials, followed by a success.

9. Ratio determines relative size samples needed for equality of variances.

11. $\dfrac{1}{2} - \dfrac{1}{2}\sqrt{1 - 1\Big/\Big(2 + \dfrac{1}{n}\Big)} < p < \dfrac{1}{2} + \dfrac{1}{2}\sqrt{1 - 1\Big/\Big(2 - \dfrac{1}{n}\Big)}$

13. $\frac{20}{3}$.
15. $\hat{\theta} = \sqrt{\sum x_i^2/n}$.

17. $f(x; p) = \dbinom{8p}{x}\dbinom{8 - 8p}{2 - x}\Big/\dbinom{8}{2}$. For $x = 0$, $p = 0$ maximizes, for $x = 1$, $p = \frac{1}{2}$ maximizes, for $x = 2, p = 1$ maximizes.

19. $\hat{p} = \dfrac{x}{n}$.
21. $\sum x_i/(\sum x_i + n)$.

23. $\hat{\mu} = \bar{x}, \hat{\sigma} = \sqrt{\sum (x_i - \bar{x})^2/n}$.
25. 10.1-11.9.

29. .62.

33. (a) Choose $x = 1, 3, 7$, and 8 as critical region, (b) no, because preceding test is not best for H_3.

35. 14.
37. $\sum \log x_i \leq$ constant.

39. $\sum x_i \geq$ constant. Yes.
41. 80.

43. $\lambda = \Big(\dfrac{e}{n}\Big)^{n/2}(\sum x_i^2)^{n/2}e^{-\frac{1}{2}\sum x_i^2}$.
45. $\lambda = (e\bar{x}\,e^{-\bar{x}})^n$.

47. $(\pi x_i)^{\theta_0 - \theta}/(1 + \theta)^n < \lambda_0$, where $\theta = -(1 + n/\sum \log x_i)$.

51. $P(\mu) = 1 - \dfrac{1}{\sqrt{2\pi}} \displaystyle\int_{-2-3\mu}^{2-3\mu} e^{-\frac{1}{2}z^2} \, dz.$ 53. $P(\mu) = \dfrac{1}{\sqrt{2\pi}} \displaystyle\int_{81.28-4\mu}^{\infty} e^{-\frac{1}{2}z^2} \, dz.$

55. $\displaystyle\int_{47.5}^{\infty} \dfrac{e^{-\frac{(x-100p)^2}{200p(1-p)}}}{\sqrt{200\pi p(1-p)}} \, dx.$

CHAPTER 9

1. $\frac{1}{36}$.

3. $\chi^2 = 2.4$, $\chi_0^2 = 3.84$; hence compatible.

5. $\chi^2 = 59$, $\chi_0^2 = 7.8$; hence reject compatibility.

9. (a) $\chi^2 = 8.2$, $\chi_0^2 = 5.99$; hence not compatible,
   (b) $\chi^2 = .5$, $\chi_0^2 = 3.84$; hence compatible.

13. (a) 3, (b) 4, (c) 4.

15. $\chi^2 = 2.9$ with 3 degrees of freedom; hence satisfactory.

17. $\chi^2 = 4.9$, $\chi_0^2 = 3.84$; hence reject hypothesis of no effect.

19. Obviously not independent since the contribution to χ^2 from the first cell alone is larger than critical χ^2 for 12 degrees of freedom.

21. Let $a + b + c + d = n$, $a + c = n_1$, $b + d = n_2$ and choose $\tilde{p} = (a + b)/n$ as estimate of p. Show that χ^2 will assume the form $(\tilde{p}_1 - \tilde{p}_2)^2/\tilde{p}\tilde{q}\left(\dfrac{1}{n_1} + \dfrac{1}{n_2}\right)$.

23. $\chi^2 = 9.8$, $\chi_0^2 = 9.5$; hence not comparable.

CHAPTER 10

1. (a) $e^{-1/y}/y^2$, (b) $\exp[y - e^y]$.

5. $g(y) = e^{-y/2}/2$.   7. $g(z) = \dfrac{\sigma}{\theta} e^{-(\sigma z + \mu)/\theta}$, $z > -\dfrac{\mu}{\sigma}$.

11. $\frac{7}{8}$.

13. $g(z) = 2z/(1 + z)^3$, $z \geq 0$.

17. (a) $g(0) = \frac{1}{3}$, $g(1) = \frac{2}{3}$, (b) $f(-1, 1) = \frac{1}{3}$, $f(0, 0) = \frac{1}{3}$, $f(1, 1) = \frac{1}{3}$,
    (c) Cov $(X, Y) = 0$.

19. $\dfrac{1}{2\sqrt{1 - z}\, z^{\frac{3}{2}}} \displaystyle\int_0^\infty x[f(x) + f(-x)] \left[f\left(x\sqrt{\dfrac{1-z}{z}}\right) + f\left(-x\sqrt{\dfrac{1-z}{z}}\right) \right] dx.$

21. $M_z(\theta) = (1 - 2\theta)^{-n}$; hence result follows.

23. Use the relation $2\nu = V\left[\dfrac{nS^2}{\sigma^2}\right] = \dfrac{n^2}{\sigma^4} V[S^2].$

25. $k = 1/(n + 1)$.

27. (a) $\chi^2 = 9$; hence accept H_0, (b) $21 < \sigma^2 < 84$.

29. $1.3 < \sigma^2 < 8.3$.   31. $3.0 < \sum (x_i - \bar{x})^2 < 133$.

33. Write $\left(1 + \dfrac{t^2}{\nu}\right)^{-(\nu+1)/2} = \left[\left(1 + \dfrac{t^2}{\nu}\right)^{\nu/t^2}\right]^{-t^2/2}\left[1 + \dfrac{t^2}{\nu}\right]^{-\frac{1}{2}}$ and let $\nu \to \infty$.

35. $17.8 < \mu < 22.2$.   37. $9.9 < \mu < 12.1$.

39. (a) $t = -1.2$; hence accept H_0, (b) $-7.3 < \mu_1 - \mu_2 < 1.3$.

41. $t = .44$; hence no effect indicated.   43. 871.

47. Normal with mean $\alpha + \beta(x - \bar{x})$, variance $\left(1 + \dfrac{(x - \bar{x})^2}{s_x^2}\right)\sigma^2/n$.

53. $F = 1.9$, $F_0 = 2.97$; hence accept H_0.

55. $\dfrac{n_1 s_1^2(n_2 - 1)}{n_2 s_2^2(n_1 - 1)} \dfrac{1}{F_0} < \dfrac{\sigma_1^2}{\sigma_2^2} < \dfrac{n_1 s_1^2(n_2 - 1)}{n_2 s_2^2(n_1 - 1)} F_0$.

57. $F_0 = \sqrt{2}$; hence $n \doteq 92$.   61. $(n - 1)e^{-R}(1 - e^{-R})^{n-2}$.

63. 18.   65. 71.6 ± 34.4.

CHAPTER 11

1. $F = 2.66$, $F_0 = 2.77$; hence accept H_0.

3. $F_r = 5.76$, $F_0 = 5.14$; hence rations differ. $F_r = 6.22$, $F_0 = 4.76$; hence types differ.

5. 86.   7. 12.8.

9. $\sigma_{\bar{x}_R}^2/\sigma_{\bar{x}}^2 = \frac{7}{16}$

13. Assume the number of groups of plants per unit is a Poisson variable with mean m; then μm is the number of plants per sampling unit. Since $P(0) = e^{-m}$, $1 - p = e^{-m}$, from which the solution follows.

15. For $c = 2$, $n = 106$ and $I = 283$; for $c = 3$, $n = 133$ and $I = 207$; for $c = 4$, $n = 160$ and $I = 213$; hence $c = 3$ and $n = 133$ minimize I.

17. $n = 120$, $c = 3$, AOQL $= .011$.

CHAPTER 12

1. $\tau = -2.2$; hence inferior.

3. $\tau = 2.0$; hence reject hypothesis that $\mu = 0$ in favor of $\mu > 0$.

5. $t = 2.93$; hence reject $\mu = 0$.

7. $\tau = -.9$; hence accept H_0.

11. $\tau = 2.2$; hence reject randomness.

15. .92.   17. 29.

19. $1/(n + 1)$.

21. Add and subtract .19 to the empirical distribution function.

CHAPTER 13

3. Boundaries for $\sum x_i$ are $\log \frac{3}{8} + m \log \frac{3}{2}$ and $\log 7 + m \log \frac{3}{2}$.

5. Boundaries for $\sum x_i$ are $\dfrac{m(\mu_1 - \mu_0) + \log \beta/(1 - \alpha)}{\log \mu_1 - \log \mu_0}$ and

$\dfrac{m(\mu_1 - \mu_0) + \log (1 - \beta)/\alpha}{\log \mu_1 - \log \mu_0}$.

7. 34.

11. $E_0(n) \doteq 7.4$.

13. Same solutions as for problem 12; hence Y is of no help.

15. Boundary is the line $y = 2x$.

17. (a) risk $= \dfrac{\theta}{2}$, mean risk $= \frac{1}{2}$, (b) mean risk $= \frac{5}{12}$; hence second estimate is better.

19. (a) $\dfrac{1}{n}$, (b) $\dfrac{1}{n}$.   21. $E(\mu \mid x) = x + \alpha$.

23. Critical region is $e^{250 - 4\bar{x}} \geq \frac{1}{4}$. Since $\bar{x} = 63$ does not satisfy this inequality, accept the hypothesis.

25. $\alpha = P\{\bar{X} < 62.85 \mid \mu = 65\} = P\{Z < -1.93\} = .03$,
$\beta = P\{\bar{X} > 62.85 \mid \mu = 60\} = P\{Z > 2.54\} = .01$.

Tables

TABLE I. Squares and Square Roots

N	N^2	$\sqrt{N}$	$\sqrt{10N}$	N	N^2	$\sqrt{N}$	$\sqrt{10N}$
1.00	1.0000	1.00000	3.16228	**1.50**	2.2500	1.22474	3.87298
1.01	1.0201	1.00499	3.17805	1.51	2.2801	1.22882	3.88587
1.02	1.0404	1.00995	3.19374	1.52	2.3104	1.23288	3.89872
1.03	1.0609	1.01489	3.20936	1.53	2.3409	1.23693	3.91152
1.04	1.0816	1.01980	3.22490	1.54	2.3716	1.24097	3.92428
1.05	1.1025	1.02470	3.24037	1.55	2.4025	1.24499	3.93700
1.06	1.1236	1.02956	3.25576	1.56	2.4336	1.24900	3.94968
1.07	1.1449	1.03441	3.27109	1.57	2.4649	1.25300	3.96232
1.08	1.1664	1.03923	3.28634	1.58	2.4964	1.25698	3.97492
1.09	1.1881	1.04403	3.30151	1.59	2.5281	1.26095	3.98748
1.10	1.2100	1.04881	3.31662	**1.60**	2.5600	1.26491	4.00000
1.11	1.2321	1.05357	3.33167	1.61	2.5921	1.26886	4.01248
1.12	1.2544	1.05830	3.34664	1.62	2.6244	1.27279	4.02492
1.13	1.2769	1.06301	3.36155	1.63	2.6569	1.27671	4.03733
1.14	1.2996	1.06771	3.37639	1.64	2.6896	1.28062	4.04969
1.15	1.3225	1.07238	3.39116	1.65	2.7225	1.28452	4.06202
1.16	1.3456	1.07703	3.40588	1.66	2.7556	1.28841	4.07431
1.17	1.3689	1.08167	3.42053	1.67	2.7889	1.29228	4.08656
1.18	1.3924	1.08628	3.43511	1.68	2.8224	1.29615	4.09878
1.19	1.4161	1.09087	3.44964	1.69	2.8561	1.30000	4.11096
1.20	1.4400	1.09545	3.46410	**1.70**	2.8900	1.30384	4.12311
1.21	1.4641	1.10000	3.47851	1.71	2.9241	1.30767	4.13521
1.22	1.4884	1.10454	3.49285	1.72	2.9584	1.31149	4.14729
1.23	1.5129	1.10905	3.50714	1.73	2.9929	1.31529	4.15933
1.24	1.5376	1.11355	3.52136	1.74	3.0276	1.31909	4.17133
1.25	1.5625	1.11803	3.53553	1.75	3.0625	1.32288	4.18330
1.26	1.5876	1.12250	3.54965	1.76	3.0976	1.32665	4.19524
1.27	1.6129	1.12694	3.56371	1.77	3.1329	1.33041	4.20714
1.28	1.6384	1.13137	3.57771	1.78	3.1684	1.33417	4.21900
1.29	1.6641	1.13578	3.59166	1.79	3.2041	1.33791	4.23084
1.30	1.6900	1.14018	3.60555	**1.80**	3.2400	1.34164	4.24264
1.31	1.7161	1.14455	3.61939	1.81	3.2761	1.34536	4.25441
1.32	1.7424	1.14891	3.63318	1.82	3.3124	1.34907	4.26615
1.33	1.7689	1.15326	3.64692	1.83	3.3489	1.35277	4.27785
1.34	1.7956	1.15758	3.66060	1.84	3.3856	1.35647	4.28952
1.35	1.8225	1.16190	3.67423	1.85	3.4225	1.36015	4.30116
1.36	1.8496	1.16619	3.68782	1.86	3.4596	1.36382	4.31277
1.37	1.8769	1.17047	3.70135	1.87	3.4969	1.36748	4.32435
1.38	1.9044	1.17473	3.71484	1.88	3.5344	1.37113	4.33590
1.39	1.9321	1.17898	3.72827	1.89	3.5721	1.37477	4.34741
1.40	1.9600	1.18322	3.74166	**1.90**	3.6100	1.37840	4.35890
1.41	1.9881	1.18743	3.75500	1.91	3.6481	1.38203	4.37035
1.42	2.0164	1.19164	3.76829	1.92	3.6864	1.38564	4.38178
1.43	2.0449	1.19583	3.78153	1.93	3.7249	1.38924	4.39318
1.44	2.0736	1.20000	3.79473	1.94	3.7636	1.39284	4.40454
1.45	2.1025	1.20416	3.80789	1.95	3.8025	1.39642	4.41588
1.46	2.1316	1.20830	3.82099	1.96	3.8416	1.40000	4.42719
1.47	2.1609	1.21244	3.83406	1.97	3.8809	1.40357	4.43847
1.48	2.1904	1.21655	3.84708	1.98	3.9204	1.40712	4.44972
1.49	2.2201	1.22066	3.86005	1.99	3.9601	1.41067	4.46094
1.50	2.2500	1.22474	3.87298	**2.00**	4.0000	1.41421	4.47214
N	N^2	$\sqrt{N}$	$\sqrt{10N}$	N	N^2	$\sqrt{N}$	$\sqrt{10N}$

TABLE I. (*Continued*)

N	N²	√N	√10N	N	N²	√N	√10N
2.00	4.0000	1.41421	4.47214	**2.50**	6.2500	1.58114	5.00000
2.01	4.0401	1.41774	4.48330	2.51	6.3001	1.58430	5.00999
2.02	4.0804	1.42127	4.49444	2.52	6.3504	1.58745	5.01996
2.03	4.1209	1.42478	4.50555	2.53	6.4009	1.59060	5.02991
2.04	4.1616	1.42829	4.51664	2.54	6.4516	1.59374	5.03984
2.05	4.2025	1.43178	4.52769	2.55	6.5025	1.59687	5.04975
2.06	4.2436	1.43527	4.53872	2.56	6.5536	1.60000	5.05964
2.07	4.2849	1.43875	4.54973	2.57	6.6049	1.60312	5.06952
2.08	4.3264	1.44222	4.56070	2.58	6.6564	1.60624	5.07937
2.09	4.3681	1.44568	4.57165	2.59	6.7081	1.60935	5.08920
2.10	4.4100	1.44914	4.58258	**2.60**	6.7600	1.61245	5.09902
2.11	4.4521	1.45258	4.59347	2.61	6.8121	1.61555	5.10882
2.12	4.4944	1.45602	4.60435	2.62	6.8644	1.61864	5.11859
2.13	4.5369	1.45945	4.61519	2.63	6.9169	1.62173	5.12835
2.14	4.5796	1.46287	4.62601	2.64	6.9696	1.62481	5.13809
2.15	4.6225	1.46629	4.63681	2.65	7.0225	1.62788	5.14782
2.16	4.6656	1.46969	4.64758	2.66	7.0756	1.63095	5.15752
2.17	4.7089	1.47309	4.65833	2.67	7.1289	1.63401	5.16720
2.18	4.7524	1.47648	4.66905	2.68	7.1824	1.63707	5.17687
2.19	4.7961	1.47986	4.67974	2.69	7.2361	1.64012	5.18652
2.20	4.8400	1.48324	4.69042	**2.70**	7.2900	1.64317	5.19615
2.21	4.8841	1.48661	4.70106	2.71	7.3441	1.64621	5.20577
2.22	4.9284	1.48997	4.71169	2.72	7.3984	1.64924	5.21536
2.23	4.9729	1.49332	4.72229	2.73	7.4529	1.65227	5.22494
2.24	5.0176	1.49666	4.73286	2.74	7.5076	1.65529	5.23450
2.25	5.0625	1.50000	4.74342	2.75	7.5625	1.65831	5.24404
2.26	5.1076	1.50333	4.75395	2.76	7.6176	1.66132	5.25357
2.27	5.1529	1.50665	4.76445	2.77	7.6729	1.66433	5.26308
2.28	5.1984	1.50997	4.77493	2.78	7.7284	1.66733	5.27257
2.29	5.2441	1.51327	4.78539	2.79	7.7841	1.67033	5.28205
2.30	5.2900	1.51658	4.79583	**2.80**	7.8400	1.67332	5.29150
2.31	5.3361	1.51987	4.80625	2.81	7.8961	1.67631	5.30094
2.32	5.3824	1.52315	4.81664	2.82	7.9524	1.67929	5.31037
2.33	5.4289	1.52643	4.82701	2.83	8.0089	1.68226	5.31977
2.34	5.4756	1.52971	4.83735	2.84	8.0656	1.68523	5.32917
2.35	5.5225	1.53297	4.84768	2.85	8.1225	1.68819	5.33854
2.36	5.5696	1.53623	4.85798	2.86	8.1796	1.69115	5.34790
2.37	5.6169	1.53948	4.86826	2.87	8.2369	1.69411	5.35724
2.38	5.6644	1.54272	4.87852	2.88	8.2944	1.69706	5.36656
2.39	5.7121	1.54596	4.88876	2.89	8.3521	1.70000	5.37587
2.40	5.7600	1.54919	4.89898	**2.90**	8.4100	1.70294	5.38516
2.41	5.8081	1.55252	4.90918	2.91	8.4681	1.70587	5.39444
2.42	5.8564	1.55563	4.91935	2.92	8.5264	1.70880	5.40370
2.43	5.9049	1.55885	4.92950	2.93	8.5849	1.71172	5.41295
2.44	5.9536	1.56205	4.93964	2.94	8.6436	1.71464	5.42218
2.45	6.0025	1.56525	4.94975	2.95	8.7025	1.71756	5.43139
2.46	6.0516	1.56844	4.95984	2.96	8.7616	1.72047	5.44059
2.47	6.1009	1.57162	4.96991	2.97	8.8209	1.72337	5.44977
2.48	6.1054	1.57480	4.97996	2.98	8.8804	1.72627	5.45894
2.49	6.2001	1.57797	4.98999	2.99	8.9401	1.72916	5.46809
2.50	6.2500	1.58114	5.00000	**3.00**	9.0000	1.73205	5.47723
N	N²	√N	√10N	N	N²	√N	√10N

TABLE I. (*Continued*)

N	N²	√N	√10N		N	N²	√N	√10N
3.00	9.0000	1.73205	5.47723		**3.50**	12.2500	1.87083	5.91608
3.01	9.0601	1.73494	5.48635		3.51	12.3201	1.87350	5.92453
3.02	9.1204	1.73781	5.49545		3.52	12.3904	1.87617	5.93296
3.03	9.1809	1.74069	5.50454		3.53	12.4609	1.87883	5.94138
3.04	9.2416	1.74356	5.51362		3.54	12.5316	1.88149	5.94979
3.05	9.3025	1.74642	5.52268		3.55	12.6025	1.88414	5.95819
3.06	9.3636	1.74929	5.53173		3.56	12.6736	1.88680	5.96657
3.07	9.4249	1.75214	5.54076		3.57	12.7449	1.88944	5.97495
3.08	9.4864	1.75499	5.54977		3.58	12.8164	1.89209	5.98331
3.09	9.5481	1.75784	5.55878		3.59	12.8881	1.89473	5.99166
3.10	9.6100	1.76068	5.56776		**3.60**	12.9600	1.89737	6.00000
3.11	9.6721	1.76352	5.57674		3.61	13.0321	1.90000	6.00833
3.12	9.7344	1.76635	5.58570		3.62	13.1044	1.90263	6.01664
3.13	9.7969	1.76918	5.59464		3.63	13.1769	1.90526	6.02495
3.14	9.8596	1.77200	5.60357		3.64	13.2496	1.90788	6.03324
3.15	9.9225	1.77482	5.61249		3.65	13.3225	1.91050	6.04152
3.16	9.9856	1.77764	5.62139		3.66	13.3956	1.91311	6.04949
3.17	10.0489	1.78045	5.63028		3.67	13.4689	1.91572	6.05805
3.18	10.1124	1.78326	5.63915		3.68	13.5424	1.91833	6.06630
3.19	10.1761	1.78606	5.64801		3.69	13.6161	1.92094	6.07454
3.20	10.2400	1.78885	5.65685		**3.70**	13.6900	1.92354	6.08276
3.21	10.3041	1.79165	5.66569		3.71	13.7641	1.92614	6.09098
3.22	10.3684	1.79444	5.67450		3.72	13.8384	1.92873	6.09918
3.23	10.4329	1.79722	5.68331		3.73	13.9129	1.93132	6.10737
3.24	10.4976	1.80000	5.69210		3.74	13.9876	1.93391	6.11555
3.25	10.5625	1.80278	5.70088		3.75	14.0625	1.93649	6.12372
3.26	10.6276	1.80555	5.70964		3.76	14.1376	1.93907	6.13188
3.27	10.6929	1.80831	5.71839		3.77	14.2129	1.94165	6.14003
3.28	10.7584	1.81108	5.72713		3.78	14.2884	1.94422	6.14817
3.29	10.8241	1.81384	5.73585		3.79	14.3641	1.94679	6.15630
3.30	10.8900	1.81659	5.74456		**3.80**	14.4400	1.94936	6.16441
3.31	10.9561	1.81934	5.75326		3.81	14.5161	1.95192	6.17252
3.32	10.0224	1.82209	5.76194		3.82	14.5924	1.95448	6.18061
3.33	11.0889	1.82483	5.77062		3.83	14.6689	1.95704	6.18870
3.34	11.1556	1.82757	5.77927		3.84	14.7456	1.95959	6.19677
3.35	11.2225	1.83030	5.78792		3.85	14.8225	1.96214	6.20484
3.36	11.2896	1.83303	5.79655		3.86	14.8996	1.96469	6.21289
3.37	11.3569	1.83576	5.80517		3.87	14.9769	1.96723	6.22093
3.38	11.4244	1.83848	5.81378		3.88	15.0544	1.96977	6.22896
3.39	11.4921	1.84120	5.82237		3.89	15.1321	1.97231	6.23699
3.40	11.5600	1.84391	5.83095		**3.90**	51.2100	1.97484	6.24500
3.41	11.6281	1.84662	5.83952		3.91	15.2881	1.97737	6.25300
3.42	11.6964	1.84932	5.84808		3.92	15.3664	1.97990	6.26099
3.43	11.7649	1.85203	5.85662		3.93	15.4449	1.98242	6.26897
3.44	11.8336	1.85472	5.86515		3.94	15.5236	1.98494	6.27694
3.45	11.9025	1.85742	5.87367		3.95	15.6025	1.98746	6.28490
3.46	11.9716	1.86011	5.88218		3.96	15.6816	1.98997	6.29285
3.47	12.0409	1.86279	5.89067		3.97	15.7609	1.99249	6.30079
3.48	12.1104	1.86548	5.89915		3.98	15.8404	1.99499	6.30872
3.49	12.1801	1.86815	5.90762		3.99	15.9201	1.99750	6.31644
3.50	12.2500	1.87083	5.91608		**4.00**	16.0000	2.00000	6.32456
N	N²	√N	√10N		N	N²	√N	√10N

TABLE I. (*Continued*)

N	N^2	$\sqrt{N}$	$\sqrt{10N}$	N	N^2	$\sqrt{N}$	$\sqrt{10N}$
4.00	16.0000	2.00000	6.32456	**4.50**	20.2500	2.12132	6.70820
4.01	16.0801	2.00250	6.33246	4.51	20.3401	2.12368	6.71565
4.02	16.1604	2.00499	6.34035	4.52	20.4304	2.12603	6.72309
4.03	16.2409	2.00749	6.34823	4.53	20.5209	2.12838	6.73053
4.04	16.3216	2.00998	6.35610	4.54	20.6116	2.13073	6.73795
4.05	16.4025	2.01246	6.36396	4.55	20.7025	2.13307	6.74537
4.06	16.4836	2.01494	6.37181	4.56	20.7936	2.13542	6.75278
4.07	16.5649	2.01742	6.37966	4.57	20.8849	2.13776	6.76018
4.08	16.6464	2.01990	6.38749	4.58	20.9764	2.14009	6.76757
4.09	16.7281	2.02237	6.39531	4.59	21.0681	2.14243	6.77495
4.10	16.8100	2.02485	6.40312	**4.60**	21.1600	2.14476	6.78233
4.11	16.8921	2.02731	6.41093	4.61	21.2521	2.14709	6.78970
4.12	16.9744	2.02978	6.41872	4.62	21.3444	2.14942	6.79706
4.13	17.0569	2.03224	6.42651	4.63	21.4369	2.15174	6.80441
4.14	17.1396	2.03470	6.43428	4.64	21.5296	2.15407	6.81175
4.15	17.2225	2.03715	6.44205	4.65	21.6225	2.15639	6.81909
4.16	17.3056	2.03961	6.44981	4.66	21.7156	2.15870	6.82642
4.17	17.3889	2.04206	6.45755	4.67	21.8089	2.16102	6.83374
4.18	17.4724	2.04450	6.46529	4.68	21.9024	2.16333	6.84105
4.19	17.5561	2.04695	6.47302	4.69	21.9961	2.16564	6.84836
4.20	17.6400	2.04939	6.48074	**4.70**	22.0900	2.16795	6.85565
4.21	17.7241	2.05183	6.48845	4.71	22.1841	2.17025	6.86294
4.22	17.8084	2.05426	6.49615	4.72	22.2784	2.17256	6.87023
4.23	17.8929	2.05670	6.50384	4.73	22.3729	2.17486	6.87750
4.24	17.9776	2.05913	6.51153	4.74	22.4676	2.17715	6.88477
4.25	18.0625	2.06155	6.51920	4.75	22.5625	2.17945	6.89202
4.26	18.1476	2.06398	6.52687	4.76	22.6576	2.18174	6.89928
4.27	18.2329	2.06640	6.53452	4.77	22.7529	2.18403	6.90652
4.28	18.3184	2.06882	6.54217	4.78	22.8484	2.18632	6.91375
4.29	18.4041	2.07123	6.54981	4.79	22.9441	2.18861	6.92098
4.30	18.4900	2.07364	6.55744	**4.80**	23.0400	2.19089	6.92820
4.31	18.5761	2.07605	6.66506	4.81	23.1361	2.19317	6.93542
4.32	18.6624	2.07846	6.57267	4.82	23.2324	2.19545	6.94262
4.33	18.7489	2.08087	6.58027	4.83	23.3289	2.19773	6.94982
4.34	18.8356	2.08327	6.58787	4.84	23.4256	2.20000	6.95701
4.35	18.9225	2.08567	6.59545	4.85	23.5225	2.20227	6.96419
4.36	19.0096	2.08806	6.60303	4.86	23.6196	2.20454	6.97137
4.37	19.0969	2.09045	6.61060	4.87	23.7169	2.20681	6.97854
4.38	19.1844	2.09284	6.61816	4.88	23.8144	2.20907	6.98570
4.39	19.2721	2.09523	6.62571	4.89	23.9121	2.21133	6.99285
4.40	19.3600	2.09762	6.63325	**4.90**	24.0100	2.21359	7.00000
4.41	19.4481	2.10000	6.64078	4.91	24.1081	2.21585	7.00714
4.42	19.5364	2.10238	6.64831	4.92	24.2064	2.21811	7.01427
4.43	19.6249	2.10476	6.65582	4.93	24.3049	2.22036	7.02140
4.44	19.7136	2.10713	6.66333	4.94	24.4036	2.22261	7.02851
4.45	19.8025	2.10950	6.67083	4.95	24.5025	2.22486	7.03562
4.46	19.8916	2.11187	6.67832	4.96	24.6016	2.22711	7.04273
4.47	19.9809	2.11424	6.68581	4.97	24.7009	2.22935	7.04982
4.48	20.0704	2.11660	6.69328	4.98	24.8004	2.23159	7.05691
4.49	20.1601	2.11896	6.70075	4.99	24.9001	2.23383	7.06399
4.50	20.2500	2.12132	6.70820	**5.00**	25.0000	2.23607	7.07107
N	N^2	$\sqrt{N}$	$\sqrt{10N}$	N	N^2	$\sqrt{N}$	$\sqrt{10N}$

TABLE I. (*Continued*)

N	N²	√N	√10N		N	N²	√10	√10N
5.00	25.0000	2.23607	7.07107		**5.50**	30.2500	2.34521	7.41620
5.01	25.1001	2.23830	7.07814		5.51	30.3601	2.34734	7.42294
5.02	25.2004	2.24054	7.08520		5.52	30.4704	2.34947	7.42967
5.03	25.3009	2.24277	7.09225		5.53	30.5809	2.35160	7.43640
5.04	25.4016	2.24499	7.09930		5.54	30.6916	2.35372	7.44312
5.05	25.5025	2.24722	7.10634		5.55	30.8025	2.35584	7.44983
5.06	25.6036	2.24944	7.11337		5.56	30.9136	2.35797	7.45654
5.07	25.7049	2.25167	7.12039		5.57	31.0249	2.36008	7.46324
5.08	25.8064	2.25389	7.12741		5.58	31.1364	2.36220	7.46994
5.09	25.9081	2.25610	7.13442		5.59	31.2481	2.36432	7.47663
5.10	26.0100	2.25832	7.14143		**5.60**	31.3600	2.36643	7.48331
5.11	26.1121	2.26053	7.14843		5.61	31.4721	2.36854	7.48999
5.12	26.2144	2.26274	7.15542		5.62	31.5844	2.37065	7.49667
5.13	26.3169	2.26495	7.16240		5.63	31.6969	2.37276	7.50333
5.14	26.4196	2.26716	7.16938		5.64	31.8096	2.37487	7.50999
5.15	26.5225	2.26936	7.17635		5.65	31.9225	2.37697	7.51665
5.16	26.6256	2.27156	7.18331		5.66	32.0356	2.37908	7.52330
5.17	26.7289	2.27376	7.19027		5.67	32.1489	2.38118	7.52994
5.18	26.8324	2.27596	7.19722		5.68	32.2624	2.38328	7.53658
5.19	26.9361	2.27816	7.20417		5.68	32.3761	2.38537	7.54321
5.20	27.0400	2.28035	7.21110		**5.70**	32.4900	2.38747	7.54983
5.21	27.1441	2.28254	7.21803		5.71	32.6041	2.38956	7.55645
5.22	27.2484	2.28473	7.22496		5.72	32.7184	2.39165	7.56307
5.23	27.3529	2.28692	7.23187		5.73	32.8329	2.39374	7.56968
5.24	27.4576	2.28910	7.23838		5.74	32.9476	2.39583	7.57628
5.25	27.5625	2.29129	7.24569		5.75	33.0625	2.39792	7.58288
5.26	27.6676	2.29347	7.25259		5.76	33.1776	2.40000	7.58947
5.27	27.7729	2.29565	7.25948		5.77	33.2929	2.40208	7.59605
5.28	27.8784	2.29783	7.26636		5.78	33.4084	2.40416	7.60263
5.29	27.9841	2.30000	7.27324		5.79	33.5241	2.40624	7.60920
5.30	28.0900	2.30217	7.28011		**5.80**	33.6400	2.40832	7.61577
5.31	28.1961	2.30434	7.28697		5.81	33.7561	2.41039	7.62234
5.32	28.3024	2.30651	7.29383		5.82	33.8724	2.41247	7.62889
5.33	28.4089	2.30868	7.30068		5.83	33.9889	2.41454	7.63544
5.34	28.5156	2.31084	7.30753		5.84	34.1056	2.41661	7.64199
5.35	28.6225	2.31301	7.31437		5.85	34.2225	2.41868	7.64853
5.36	28.7296	2.31517	7.32120		5.86	34.3396	2.42074	7.65506
5.37	28.8369	2.31733	7.32803		5.87	34.4569	2.42281	7.66159
5.38	28.9444	2.31948	7.33485		5.88	34.5744	2.42487	7.66812
5.39	29.0521	2.32164	7.34166		5.89	34.6921	2.42693	7.67463
5.40	29.1600	2.32379	7.34847		**5.90**	34.8100	2.42899	7.68115
5.41	29.2681	2.32594	7.35527		5.91	34.9281	2.43105	7.68765
5.42	29.3764	2.32809	7.36206		5.92	35.0464	2.43311	7.69415
5.43	29.4849	2.33024	7.36885		5.93	35.1649	2.43516	7.70065
5.44	29.5936	2.33238	7.37564		5.94	35.2836	2.43721	7.70714
5.45	29.7025	2.33452	7.38241		5.95	35.4025	2.43926	7.71362
5.46	29.8116	2.33666	7.38918		5.96	35.5216	2.44131	7.72010
5.47	29.9209	2.33880	7.39594		5.97	35.6409	2.44336	7.72658
5.48	30.0304	2.34094	7.40270		5.98	35.7604	2.44540	7.73305
5.49	30.1401	2.34307	7.40945		5.99	35.8801	2.44745	7.73951
5.50	30.2500	2.34521	7.41620		**6.00**	36.0000	2.44949	7.74597
N	N²	√N	√10N		N	N²	√N	√10N

TABLE I. (*Continued*)

N	N²	√N̄	√10N̄	N	N²	√N̄	√10N̄
6.00	36.0000	2.44949	7.74597	**6.50**	42.2500	2.54951	8.06226
6.01	36.1201	2.45153	7.75242	6.51	42.3801	2.55147	8.06846
6.02	36.2404	2.45357	7.75887	6.52	42.5104	2.55343	8.07465
6.03	36.3609	2.45561	7.76531	6.53	42.6409	2.55539	8.08084
6.04	36.4816	2.45764	7.77174	6.54	42.7716	2.55734	8.08703
6.05	36.6025	2.45967	7.77817	6.55	42.9025	2.55930	8.09321
6.06	36.7236	2.46171	7.78460	6.56	43.0336	2.56125	8.09938
6.07	36.8449	2.46374	7.79102	6.57	43.1649	2.56320	8.10555
6.08	36.9664	2.46577	7.79744	6.58	43.2964	2.56515	8.11172
6.09	37.0881	2.46779	7.80385	6.59	43.4281	2.56710	8.11788
6.10	37.2100	2.46982	7.81025	**6.60**	43.5600	2.56905	8.12404
6.11	37.3321	2.47184	7.81665	6.61	43.6921	2.57099	8.13019
6.12	37.4544	2.47386	7.82304	6.62	43.8244	2.57294	8.13634
6.13	37.5769	2.47588	7.82943	6.63	43.9569	2.57488	8.14248
6.14	37.6996	2.47790	7.83582	6.64	44.0896	2.57682	8.14862
6.15	37.8225	2.47992	7.84219	6.65	44.2225	2.57876	8.15475
6.16	37.9456	2.48193	7.84857	6.66	44.3556	2.58070	8.16088
6.17	38.0689	2.48395	7.85493	6.67	44.4889	2.58263	8.16701
6.18	38.1924	2.48596	7.86130	6.68	44.6224	2.58457	8.17313
6.19	38.3161	2.48797	7.86766	6.69	44.7561	2.58650	8.17924
6.20	38.4400	2.48998	7.87401	**6.70**	44.8900	2.58844	8.18535
6.21	38.5641	2.49199	7.88036	6.71	45.0241	2.59037	8.19146
6.22	38.6884	2.49399	7.88670	6.72	45.1584	2.59230	8.19756
6.23	38.8129	2.49600	7.89303	6.73	45.2929	2.59422	8.20366
6.24	38.9376	2.49800	7.89937	6.74	45.4276	2.59615	8.20975
6.25	39.0625	2.50000	7.90569	6.75	45.5625	2.59808	8.21584
6.26	39.1876	2.50200	7.91202	6.76	45.6976	2.60000	8.22192
6.27	39.3129	2.50400	7.91833	6.77	45.8329	2.60192	8.22800
6.28	39.4384	2.50599	7.92465	6.78	45.9684	2.60384	8.23408
6.29	39.5641	2.50799	7.93095	6.79	46.1041	2.60576	8.24015
6.30	39.6900	2.50998	7.93725	**6.80**	46.2400	2.60768	8.24621
6.31	39.8161	2.51197	7.94355	6.81	46.3761	2.60960	8.25227
6.32	39.9424	2.51396	7.94984	6.82	46.5124	2.61151	8.25833
6.33	40.0689	2.51595	7.95613	6.83	46.6489	2.61343	8.26438
6.34	40.1956	2.51794	7.96241	6.84	46.7856	2.61534	8.27043
6.35	40.3225	2.51992	7.96869	6.85	46.9225	2.61725	8.27647
6.36	40.4496	2.52190	7.97496	6.86	47.0596	2.61916	8.28251
6.37	40.5769	2.52389	7.98123	6.87	47.1969	2.62107	8.28855
6.38	40.7044	2.52587	7.98749	6.88	47.3344	2.62298	8.29458
6.39	40.8321	2.52784	7.99375	6.89	47.4721	2.62488	8.30060
6.40	40.9600	2.52982	8.00000	**6.90**	47.6100	2.62679	8.30662
6.41	41.0881	2.53180	8.00625	6.91	47.7481	2.62869	8.31264
6.42	41.2164	2.53377	8.01249	6.92	47.8864	2.63059	8.31865
6.43	41.3449	2.53574	8.01873	6.93	48.0249	2.63249	8.32466
6.44	41.4736	2.53772	8.02496	6.94	48.1636	2.63439	8.33067
6.45	41.6025	2.53969	8.03119	6.95	48.3025	2.63629	8.33667
6.46	41.7316	2.54165	8.03741	6.96	48.4416	2.63818	8.34266
6.47	41.8609	2.54362	8.04363	6.97	48.5809	2.64008	8.34865
6.48	41.9904	2.54558	8.04984	6.98	48.7204	2.64197	8.35464
6.49	42.1201	2.54755	8.05605	6.99	48.8601	2.64386	8.36062
6.50	42.2500	2.54951	8.06226	**7.00**	49,0000	2.64575	8.36660
N	N²	√N̄	√10N̄	N	N²	√N̄	√10N̄

TABLE I. (*Continued*)

N	N²	√N̄	√10N̄		N	N²	√N̄	√10N̄
7.00	49.0000	2.64575	8.36660		**7.50**	56.2500	2.73861	8.66025
7.01	49.1401	2.64764	8.37257		7.51	56.4001	2.74044	8.66603
7.02	49.2804	2.64953	8.37854		7.52	56.5504	2.74226	8.67179
7.03	49.4209	2.65141	8.38451		7.53	56.7009	2.74408	8.67756
7.04	49.5616	2.65330	8.39047		7.54	56.8516	2.74591	8.68332
7.05	49.7025	2.65518	8.39643		7.55	57.0025	2.74773	8.68907
7.06	49.8436	2.65707	8.40238		7.56	57.1536	2.74955	8.69483
7.07	49.9849	2.65895	8.40833		7.57	57.3049	2.75136	8.70057
7.08	50.1264	2.66083	8.41427		7.58	57.4564	2.75318	8.70632
7.09	50.2681	2.66271	8.42021		7.59	57.6081	2.75500	8.71206
7.10	50.4100	2.66458	8.42615		**7.60**	57.7600	2.75681	8.71780
7.11	50.5521	2.66646	8.43208		7.61	57.9121	2.75862	8.72353
7.12	50.6944	2.66833	8.43801		7.62	58.0644	2.76043	8.72926
7.13	50.8369	2.67021	8.44393		7.63	58.2169	2.76225	8.73499
7.14	50.9796	2.67208	8.44985		7.64	58.3696	2.76405	8.74071
7.15	51.1225	2.67395	8.45577		7.65	58.5225	2.76586	8.74643
7.16	51.2656	2.67582	8.46168		7.66	58.6756	2.76767	8.75214
7.17	51.4089	2.67769	8.46759		7.67	58.8289	2.76948	8.75785
7.18	51.5524	2.67955	8.47349		7.68	58.9824	2.77128	8.76356
7.19	51.6961	2.68142	8.47939		7.69	59.1361	2.77308	8.76926
7.20	51.8400	2.68328	8.48528		**7.70**	59.2900	2.77489	8.77496
7.21	51.9841	2.68514	8.49117		7.71	59.4441	2.77669	8.78066
7.22	52.1284	2.68701	8.49706		7.72	59.5984	2.77849	8.78635
7.23	52.2729	2.68887	8.50294		7.73	59.7529	2.78029	8.79204
7.24	52.4176	2.69072	8.50882		7.74	59.9076	2.78209	8.79773
7.25	52.5625	2.69258	8.51469		7.75	60.0625	2.78388	8.80341
7.26	52.7076	2.69444	8.52056		7.76	60.2176	2.78568	8.80909
7.27	52.8529	2.69629	8.52643		7.77	60.3729	2.78747	8.81476
7.28	52.9984	2.69815	8.53229		7.78	60.5284	2.78927	8.82043
7.29	53.1441	2.70000	8.53815		7.79	60.6841	2.79106	8.82610
7.30	53.2900	2.70185	8.54400		**7.80**	60.8400	2.79285	8.83176
7.31	53.4361	2.70370	8.54985		7.81	60.9961	2.79464	8.83742
7.32	53.5824	2.70555	8.55570		7.82	61.1524	2.79643	8.84308
7.33	53.7289	2.70740	8.56154		7.83	61.3089	2.79821	8.84873
7.34	53.8756	2.70924	8.56738		7.84	61.4656	2.80000	8.85438
7.35	54.0225	2.71109	8.57321		7.85	61.6225	2.80179	8.86002
7.36	54.1696	2.71293	8.57904		7.86	61.7796	2.80357	8.86566
7.37	54.3169	2.71477	8.58487		7.87	61.9369	2.80535	8.87130
7.38	54.4644	2.71662	8.59069		7.88	62.0944	2.80713	8.87694
7.39	54.6121	2.71846	8.59651		7.89	62.2521	2.80891	8.88257
7.40	54.7600	2.72029	8.60233		**7.90**	62.4100	2.81069	8.88819
7.41	54.9081	2.72213	8.60814		7.91	62.5681	2.81247	8.89382
7.42	55.0564	2.72397	8.61394		7.92	62.7264	2.81425	8.89944
7.43	55.2049	2.72580	8.61974		7.93	62.8849	2.81603	8.90505
7.44	55.3536	2.72764	8.62554		7.94	63.0436	2.81780	8.91067
7.45	55.5025	2.72947	8.63134		7.95	63.2025	2.81957	8.91628
7.46	55.6516	2.73130	8.63713		7.96	63.3616	2.82135	8.92188
7.47	55.8009	2.73313	8.64292		7.97	63.5209	2.82312	8.92749
7.48	55.9504	2.73496	8.64870		7.98	63.6804	2.82489	8.93308
7.49	56.1001	2.73679	8.65448		7.99	63.8401	2.82666	8.93868
7.50	56.2500	2.73861	8.66025		**8.00**	64.0000	2.82843	8.94427
N	N²	√N̄	√10N̄		N	N²	√N̄	√10N̄

TABLE I. (*Continued*)

N	N²	√N̄	√1̄0̄N̄	N	N²	√N̄	√1̄0̄N̄
8.00	64.0000	2.82843	8.94427	**8.50**	72.2500	2.91548	9.21954
8.01	64.1601	2.83019	8.94986	8.51	72.4201	2.91719	9.22497
8.02	64.3204	2.83196	8.95545	8.52	72.5904	2.91890	9.23038
8.03	64.4809	2.83373	8.96103	8.53	72.7609	2.92062	9.23580
8.04	64.6416	2.83549	8.96660	8.54	72.9316	2.92233	9.24121
8.05	64.8025	2.83725	8.97218	8.55	73.1025	2.92404	9.24662
8.06	64.9636	2.83901	8.97775	8.56	73.2736	2.92575	9.25203
8.07	65.1249	2.84077	8.98332	8.57	73.4449	2.92746	9.25743
8.08	65.2864	2.84253	8.98888	8.58	73.6164	2.92916	9.26283
8.09	65.4481	2.84429	8.99444	8.59	73.7881	2.93087	9.26823
8.10	65.6100	2.84605	9.00000	**8.60**	73.9600	2.93258	9.27362
8.11	65.7721	2.84781	9.00555	8.61	74.1321	2.93428	9.27901
8.12	65.9344	2.84956	9.01110	8.62	74.3044	2.93598	9.28440
8.13	66.0969	2.85132	9.01665	8.63	74.4769	2.93769	9.28978
8.14	66.2596	2.85307	9.02219	8.64	74.6496	2.93939	9.29516
8.15	66.4225	2.85482	9.02774	8.65	74.8225	2.94109	9.30054
8.16	66.5856	2.85657	9.03327	8.66	74.9956	2.94279	9.30591
8.17	66.7489	2.85832	9.03881	8.67	75.1689	2.94449	9.31128
8.18	66.9124	2.86007	9.04434	8.68	75.3424	2.94618	9.31665
8.19	67.0761	2.86182	9.04986	8.69	75.5161	2.94788	9.32202
8.20	67.2400	2.86356	9.05539	**8.70**	75.6900	2.94958	9.32738
8.21	67.4041	2.86531	9.06091	8.71	75.8641	2.95127	9.33274
8.22	67.5684	2.86705	9.06642	8.72	76.0384	2.95296	9.33809
8.23	67.7329	2.86880	9.07193	8.73	76.2129	2.95466	9.34345
8.24	67.8976	2.87054	9.07744	8.74	76.3876	2.95635	9.34880
8.25	68.0625	2.87228	9.08295	8.75	76.5625	2.95804	9.35414
8.26	68.2276	2.87402	9.08845	8.76	76.7376	2.95973	9.35949
8.27	68.3929	2.87576	9.09395	8.77	76.9129	2.96142	9.36483
8.28	68.5584	2.87750	9.09945	8.78	77.0884	2.96311	9.37017
8.29	68.7241	2.87924	9.10494	8.79	77.2641	2.96479	9.37550
8.30	68.8900	2.88097	9.11045	**8.80**	77.4400	2.96648	9.38083
8.31	69.0561	2.88271	9.11592	8.81	77.6161	2.96816	9.38616
8.32	69.2224	2.88444	9.12140	8.82	77.7924	2.96985	9.39149
8.33	69.3889	2.88617	9.12688	8.83	77.9689	2.97153	9.39681
8.34	69.5556	2.88791	9.13236	8.84	78.1456	2.97321	9.40213
8.35	69.7225	2.88964	9.13783	8.85	78.3225	2.97489	9.40744
8.36	69.8896	2.89137	9.14330	8.86	78.4996	2.97658	9.41276
8.37	70.0569	2.89310	9.14877	8.87	78.6769	2.97825	9.41807
8.38	70.2244	2.89482	9.15423	8.88	78.8544	2.97993	9.42338
8.39	70.3921	2.89655	9.15969	8.89	79.0321	2.98161	9.42868
8.40	70.5600	2.89828	9.16515	**8.90**	79.2100	2.98329	9.43398
8.41	70.7281	2.90000	9.17061	8.91	79.3881	2.98496	9.43928
8.42	70.8964	2.90172	9.17606	8.92	79.5664	2.98664	9.44458
8.43	71.0649	2.90345	9.18150	8.93	79.7449	2.98831	9.44987
8.44	71.2336	2.90517	9.18695	8.94	79.9236	2.98998	9.45516
8.45	71.4025	2.90689	9.19239	8.95	80.1025	2.99166	9.46044
8.46	71.5716	2.90861	9.19783	8.96	80.2816	2.99333	9.46573
8.47	71.7409	2.91033	9.20326	8.97	80.4609	2.99500	9.47101
8.48	71.9104	2.91204	9.20869	8.98	80.6404	2.99666	9.47629
8.49	72.0801	2.91376	9.21412	8.99	80.8201	2.99833	9.48156
8.50	72.2500	2.91548	9.21954	**9.00**	81.0000	3.00000	9.48683
N	N²	√N̄	√1̄0̄N̄	N	N²	√N̄	√1̄0̄N̄

389

TABLE I. (*Continued*)

N	N²	√N	√10N	N	N²	√N	√10N
9.00	81.0000	3.00000	9.48683	**9.50**	90.2500	3.08221	9.74679
9.01	81.1801	3.00167	9.49210	9.51	90.4401	3.08383	9.75192
9.02	81.3604	3.00333	9.49737	9.52	90.6304	3.08545	9.75705
9.03	81.5409	3.00500	9.50263	9.53	90.8209	3.08707	9.76217
9.04	81.7216	3.00666	9.50789	9.54	91.0116	3.08869	9.76729
9.05	81.9025	3.00832	9.51315	9.55	91.2025	3.09031	9.77241
9.06	82.0836	3.00998	9.51840	9.56	91.3936	3.09192	9.77753
9.07	82.2649	3.01164	9.52365	9.57	91.5849	3.09354	9.78264
9.08	82.4464	3.01330	9.52890	9.58	91.7764	3.09516	9.78775
9.09	82.6281	3.01496	9.53415	9.59	91.9681	3.09677	9.79285
9.10	82.8100	3.01662	9.53939	**9.60**	92.1600	3.09839	9.79796
9.11	82.9921	3.01828	9.54463	9.61	92.3521	3.10000	9.80306
9.12	83.1744	3.01993	9.54987	9.62	92.5444	3.10161	9.80816
9.13	83.3569	3.02159	9.55510	9.63	92.7369	3.10322	9.81326
9.14	83.5396	3.02324	9.56033	9.64	92.9296	3.10483	9.81835
9.15	83.7225	3.02490	9.56556	9.65	93.1225	3.10644	9.82344
9.16	83.9056	3.02655	9.57079	9.66	93.3156	3.10805	9.82853
9.17	84.0889	3.02820	9.57601	9.67	93.5089	3.10966	9.83362
9.18	84.2724	3.02985	9.58123	9.68	93.7024	3.11127	9.83870
9.19	84.4561	3.03150	9.58645	9.69	93.8961	3.11288	9.84378
9.20	84.6400	3.03315	9.59166	**9.70**	94.0900	3.11448	9.84886
9.21	84.8241	3.03480	9.59687	9.71	94.2841	3.11609	9.85393
9.22	85.0084	3.03645	9.60208	9.72	94.4784	3.11769	9.85901
9.23	85.1929	3.03809	9.60729	9.73	94.6729	3.11929	9.86408
9.24	85.3776	3.03974	9.61249	9.74	94.8676	3.12090	9.86914
9.25	85.5625	3.04138	9.61769	9.75	95.0625	3.12250	9.87421
9.26	85.7476	3.04302	9.62289	9.76	95.2576	3.12410	9.87927
9.27	85.9329	3.04467	9.62808	9.77	95.4529	3.12570	9.88433
9.28	86.1184	3.04631	9.63328	9.78	95.6484	3.12730	9.88939
9.29	86.3041	3.04795	9.63846	9.79	95.8441	3.12890	9.89444
9.30	86.4900	3.04959	9.64365	**9.80**	96.0400	3.13050	9.89949
9.31	86.6761	3.05123	9.64883	9.81	96.2361	3.13209	9.90454
9.32	86.8624	3.05287	9.65401	9.82	96.4324	3.13369	9.90959
9.33	87.0489	3.05450	9.65919	9.83	96.6289	3.13528	9.91464
9.34	87.2356	3.05614	9.66437	9.84	96.8256	3.13688	9.91968
9.35	87.4225	3.05778	9.66954	9.85	97.0225	3.13847	9.92472
9.36	87.6096	3.05941	9.67471	9.86	97.2196	3.14006	9.92975
9.37	87.7969	3.06105	9.67988	9.87	97.4169	3.14166	9.93479
9.38	87.9844	3.06268	9.68504	9.88	97.6144	3.14325	9.93982
9.39	88.1721	3.06431	9.69020	9.89	97.8121	3.14484	9.94485
9.40	88.3600	3.06594	9.69536	**9.90**	98.0100	3.14643	9.94987
9.41	88.5481	3.06757	9.70052	9.91	98.2081	3.14802	9.95490
9.42	88.7364	3.06920	9.70567	9.92	98.4064	3.14960	9.95992
9.43	88.9249	3.07083	9.71082	9.93	98.6049	3.15119	9.96494
9.44	89.1136	3.07246	9.71597	9.94	98.8036	3.15278	9.96995
9.45	89.3025	3.07409	9.72111	9.95	99.0025	3.15436	9.97497
9.46	89.4916	3.07571	9.72625	9.96	99.2016	3.15595	9.97998
9.47	89.6809	3.07734	9.73139	9.97	99.4009	3.15753	9.98499
9.48	89.8704	3.07896	9.73653	9.98	99.6004	3.15911	9.98999
9.49	90.0601	3.08058	9.74166	9.99	99.8001	3.16070	9.99500
9.50	90.2500	3.08221	9.74679	**10.0**	100.000	3.16228	10.0000
N	N²	√N	√10N	N	N²	√N	√10N

TABLE II.   Areas of a Standard Normal Distribution

An entry in the table is the proportion under the
entire curve which is between $z = 0$ and a positive
value of z. Areas for negative values of z are
obtained by symmetry.

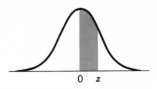

z	.00	.01	.02	.03	.04	.05	.06	.07	.08	.09
0.0	.0000	.0040	.0080	.0120	.0160	.0199	.0239	.0279	.0319	.0359
0.1	.0398	.0438	.0478	.0517	.0557	.0596	.0636	.0675	.0714	.0753
0.2	.0793	.0832	.0871	.0910	.0948	.0987	.1026	.1064	.1103	.1141
0.3	.1179	.1217	.1255	.1293	.1331	.1368	.1406	.1443	.1480	.1517
0.4	.1554	.1591	.1628	.1664	.1700	.1736	.1772	.1808	.1844	.1879
0.5	.1915	.1950	.1985	.2019	.2054	.2088	.2123	.2157	.2190	.2224
0.6	.2257	.2291	.2324	.2357	.2389	.2422	.2454	.2486	.2517	.2549
0.7	.2580	.2611	.2642	.2673	.2703	.2734	.2764	.2794	.2823	.2852
0.8	.2881	.2910	.2939	.2967	.2995	.3023	.3051	.3078	.3106	.3133
0.9	.3159	.3186	.3212	.3238	.3264	.3289	.3315	.3340	.3365	.3389
1.0	.3413	.3438	.3461	.3485	.3508	.3531	.3554	.3577	.3599	.3621
1.1	.3643	.3665	.3686	.3708	.3729	.3749	.3770	.3790	.3810	.3830
1.2	.3849	.3869	.3888	.3907	.3925	.3944	.3962	.3980	.3997	.4015
1.3	.4032	.4049	.4066	.4082	.4099	.4115	.4131	.4147	.4162	.4177
1.4	.4192	.4207	.4222	.4236	.4251	.4265	.4279	.4292	.4306	.4319
1.5	.4332	.4345	.4357	.4370	.4382	.4394	.4406	.4418	.4429	.4441
1.6	.4452	.4463	.4474	.4484	.4495	.4505	.4515	.4525	.4535	.4545
1.7	.4554	.4564	.4573	.4582	.4591	.4599	.4608	.4616	.4625	.4633
1.8	.4641	.4649	.4656	.4664	.4671	.4678	.4686	.4693	.4699	.4706
1.9	.4713	.4719	.4726	.4732	.4738	.4744	.4750	.4756	.4761	.4767
2.0	.4772	.4778	.4783	.4788	.4793	.4798	.4803	.4808	.4812	.4817
2.1	.4821	.4826	.4830	.4834	.4838	.4842	.4846	.4850	.4854	.4857
2.2	.4861	.4864	.4868	.4871	.4875	.4878	.4881	.4884	.4887	.4890
2.3	.4893	.4896	.4898	.4901	.4904	.4906	.4909	.4911	.4913	.4916
2.4	.4918	.4920	.4922	.4925	.4927	.4929	.4931	.4932	.4934	.4936
2.5	.4938	.4940	.4941	.4943	.4945	.4946	.4948	.4949	.4951	.4952
2.6	.4953	.4955	.4956	.4957	.4959	.4960	.4961	.4962	.4963	.4964
2.7	.4965	.4966	.4967	.4968	.4969	.4970	.4971	.4972	.4973	.4974
2.8	.4974	.4975	.4976	.4977	.4977	.4978	.4979	.4979	.4980	.4981
2.9	.4981	.4982	.4982	.4983	.4984	.4984	.4985	.4985	.4986	.4986
3.0	.4987	.4987	.4987	.4988	.4988	.4989	.4989.	.4989	.4990	.4990

TABLE III. χ^2 Distribution

Degrees of freedom	$P = 0.99$	0.98	0.95	0.90	0.80	0.70	0.50	0.30	0.20	0.10	0.05	0.02	0.01
1	0.000157	0.000628	0.00393	0.0158	0.0642	0.148	0.455	1.074	1.642	2.706	3.841	5.412	6.635
2	0.0201	0.0404	0.103	0.211	0.446	0.713	1.386	2.408	3.219	4.605	5.991	7.824	9.210
3	0.115	0.185	0.352	0.584	1.005	1.424	2.366	3.665	4.642	6.251	7.815	9.837	11.341
4	0.297	0.429	0.711	1.064	1.649	2.195	3.357	4.878	5.989	7.779	9.488	11.668	13.277
5	0.554	0.752	1.145	1.610	2.343	3.000	4.351	6.064	7.289	9.236	11.070	13.388	15.086
6	0.872	1.134	1.635	2.204	3.070	3.828	5.348	7.231	8.558	10.645	12.592	15.033	16.812
7	1.239	1.564	2.167	2.833	3.822	4.671	6.346	8.383	9.803	12.017	14.067	16.622	18.475
8	1.646	2.032	2.733	3.490	4.594	5.527	7.344	9.524	11.030	13.362	15.507	18.168	20.090
9	2.088	2.532	3.325	4.168	5.380	6.393	8.343	10.656	12.242	14.684	16.919	19.679	21.666
10	2.558	3.059	3.940	4.865	6.179	7.267	9.342	11.781	13.442	15.987	18.307	21.161	23.209
11	3.053	3.609	4.575	5.578	6.989	8.148	10.341	12.899	14.631	17.275	19.675	22.618	24.725
12	3.571	4.178	5.226	6.304	7.807	9.034	11.340	14.011	15.812	18.549	21.026	24.054	26.217
13	4.107	4.765	5.892	7.042	8.634	9.926	12.340	15.119	16.985	19.812	22.362	25.472	27.688
14	4.660	5.368	6.571	7.790	9.467	10.821	13.339	16.222	18.151	21.064	23.685	26.873	29.141
15	5.229	5.985	7.261	8.547	10.307	11.721	14.339	17.322	19.311	22.307	24.996	28.259	30.578
16	5.812	6.614	7.962	9.312	11.152	12.624	15.338	18.418	20.465	23.542	26.296	29.633	32.000
17	6.408	7.255	8.672	10.085	12.002	13.531	16.338	19.511	21.615	24.769	27.587	30.995	33.409
18	7.015	7.906	9.390	10.865	12.857	14.440	17.338	20.601	22.760	25.989	28.869	32.346	34.805
19	7.633	8.567	10.117	11.651	13.716	15.352	18.338	21.689	23.900	27.204	30.144	33.687	36.191
20	8.260	9.237	10.851	12.443	14.578	16.266	19.337	22.775	25.038	28.412	31.410	35.020	37.566
21	8.897	9.915	11.591	13.240	15.445	17.182	20.337	23.858	26.171	29.615	32.671	36.343	38.932
22	9.542	10.600	12.338	14.041	16.314	18.101	21.337	24.939	27.301	30.813	33.924	37.659	40.289
23	10.196	11.293	13.091	14.848	17.187	19.021	22.337	26.018	28.429	32.007	35.172	38.968	41.638
24	10.856	11.992	13.848	15.659	18.062	19.943	23.337	27.096	29.553	33.196	36.415	40.270	42.980
25	11.524	12.697	14.611	16.473	18.940	20.867	24.337	28.172	30.675	34.382	37.652	41.566	44.314
26	12.198	13.409	15.379	17.292	19.820	21.792	25.336	29.246	31.795	35.563	38.885	42.856	45.642
27	12.879	14.125	16.151	18.114	20.703	22.719	26.336	30.319	32.912	36.741	40.113	44.140	46.963
28	13.565	14.847	16.928	18.939	21.588	23.647	27.336	31.391	34.027	37.916	41.337	45.419	48.278
29	14.256	15.574	17.708	19.768	22.475	24.577	28.336	32.461	35.139	39.087	42.557	46.693	49.588
30	14.953	16.306	18.493	20.599	23.364	25.508	29.336	33.530	36.250	40.256	43.773	47.962	50.892

For degrees of freedom greater than 30, the expression $\sqrt{2\chi^2} - \sqrt{2n' - 1}$ may be used as a normal deviate with unit variance, where n' is the number of degrees of freedom. Reproduced from *Statistical Methods for Research Workers*, 6th ed., with the permission of the author, R. A. Fisher, and his publisher, Oliver and Boyd, Edinburgh.

TABLE IV. Student's t Distribution

The first column lists the number of degrees of freedom (ν). The headings of the other columns give probabilities (P) for t to exceed the entry value. Use symmetry for negative t values.

ν \ P	.10	.05	.025	.01	.005
1	3.078	6.314	12.706	31.821	63.657
2	1.886	2.920	4.303	6.965	9.925
3	1.638	2.353	3.182	4.541	5.841
4	1.533	2.132	2.776	3.747	4.604
5	1.476	2.015	2.571	3.365	4.032
6	1.440	1.943	2.447	3.143	3.707
7	1.415	1.895	2.365	2.998	3.499
8	1.397	1.860	2.306	2.896	3.355
9	1.383	1.833	2.262	2.821	3.250
10	1.372	1.812	2.228	2.764	3.169
11	1.363	1.796	2.201	2.718	3.106
12	1.356	1.782	2.179	2.681	3.055
13	1.350	1.771	2.160	2.650	3.012
14	1.345	1.761	2.145	2.624	2.977
15	1.341	1.753	2.131	2.602	2.947
16	1.337	1.746	2.120	2.583	2.921
17	1.333	1.740	2.110	2.567	2.898
18	1.330	1.734	2.101	2.552	2.878
19	1.328	1.729	2.093	2.539	2.861
20	1.325	1.725	2.086	2.528	2.845
21	1.323	1.721	2.080	2.518	2.831
22	1.321	1.717	2.074	2.508	2.819
23	1.319	1.714	2.069	2.500	2.807
24	1.318	1.711	2.064	2.492	2.797
25	1.316	1.708	2.060	2.485	2.787
26	1.315	1.706	2.056	2.479	2.779
27	1.314	1.703	2.052	2.473	2.771
28	1.313	1.701	2.048	2.467	2.763
29	1.311	1.699	2.045	2.462	2.756
30	1.310	1.697	2.042	2.457	2.750
40	1.303	1.684	2.021	2.423	2.704
60	1.296	1.671	2.000	2.390	2.660
120	1.289	1.658	1.980	2.358	2.617
∞	1.282	1.645	1.960	2.326	2.576

TABLE V. F Distribution*

5% (Roman Type) and 1% (Bold-Face Type) Points for the Distribution of F

Degrees of freedom for denominator (ν_2)	\	Degrees of freedom for numerator (ν_1)																						
	1	2	3	4	5	6	7	8	9	10	11	12	14	16	20	24	30	40	50	75	100	200	500	∞
1	161 / **4052**	200 / **4999**	216 / **5403**	225 / **5625**	230 / **5764**	234 / **5859**	237 / **5928**	239 / **5981**	241 / **6022**	242 / **6056**	243 / **6082**	244 / **6106**	245 / **6142**	246 / **6169**	248 / **6208**	249 / **6234**	250 / **6258**	251 / **6286**	252 / **6302**	253 / **6323**	253 / **6334**	254 / **6352**	254 / **6361**	254 / **6366**
2	18.51 / **98.49**	19.00 / **99.01**	19.16 / **99.17**	19.25 / **99.25**	19.30 / **99.30**	19.33 / **99.33**	19.36 / **99.34**	19.37 / **99.36**	19.38 / **99.38**	19.39 / **99.40**	19.40 / **99.41**	19.41 / **99.42**	19.42 / **99.43**	19.43 / **99.44**	19.44 / **99.45**	19.45 / **99.46**	19.46 / **99.47**	19.47 / **99.48**	19.47 / **99.48**	19.48 / **99.49**	19.49 / **99.49**	19.49 / **99.49**	19.50 / **99.50**	19.50 / **99.50**
3	10.13 / **34.12**	9.55 / **30.81**	9.28 / **29.46**	9.12 / **28.71**	9.01 / **28.24**	8.94 / **27.91**	8.88 / **27.67**	8.84 / **27.49**	8.81 / **27.34**	8.78 / **27.23**	8.76 / **27.13**	8.74 / **27.05**	8.71 / **26.92**	8.69 / **26.83**	8.66 / **26.69**	8.64 / **26.60**	8.62 / **26.50**	8.60 / **26.41**	8.58 / **26.30**	8.57 / **26.27**	8.56 / **26.23**	8.54 / **26.18**	8.54 / **26.14**	8.53 / **26.12**
4	7.71 / **21.20**	6.94 / **18.00**	6.59 / **16.69**	6.39 / **15.98**	6.26 / **15.52**	6.16 / **15.21**	6.09 / **14.98**	6.04 / **14.80**	6.00 / **14.66**	5.96 / **14.54**	5.93 / **14.45**	5.91 / **14.37**	5.87 / **14.24**	5.84 / **14.15**	5.80 / **14.02**	5.77 / **13.93**	5.74 / **13.83**	5.71 / **13.74**	5.70 / **13.69**	5.68 / **13.61**	5.66 / **13.57**	5.65 / **13.52**	5.64 / **13.48**	5.63 / **13.46**
5	6.61 / **16.26**	5.79 / **13.27**	5.41 / **12.06**	5.19 / **11.39**	5.05 / **10.97**	4.95 / **10.67**	4.88 / **10.45**	4.82 / **10.27**	4.78 / **10.15**	4.74 / **10.05**	4.70 / **9.96**	4.68 / **9.89**	4.64 / **9.77**	4.60 / **9.68**	4.56 / **9.55**	4.53 / **9.47**	4.50 / **9.38**	4.46 / **9.29**	4.44 / **9.24**	4.42 / **9.17**	4.40 / **9.13**	4.38 / **9.07**	4.37 / **9.04**	4.36 / **9.02**
6	5.99 / **13.74**	5.14 / **10.92**	4.76 / **9.78**	4.53 / **9.15**	4.39 / **8.75**	4.28 / **8.47**	4.21 / **8.26**	4.15 / **8.10**	4.10 / **7.98**	4.06 / **7.87**	4.03 / **7.79**	4.00 / **7.72**	3.96 / **7.60**	3.92 / **7.52**	3.87 / **7.39**	3.84 / **7.31**	3.81 / **7.23**	3.77 / **7.14**	3.75 / **7.09**	3.72 / **7.02**	3.71 / **6.99**	3.69 / **6.94**	3.68 / **6.90**	3.67 / **6.88**
7	5.59 / **12.25**	4.74 / **9.55**	4.35 / **8.45**	4.12 / **7.85**	3.97 / **7.46**	3.87 / **7.19**	3.79 / **7.00**	3.73 / **6.84**	3.68 / **6.71**	3.63 / **6.62**	3.60 / **6.54**	3.57 / **6.47**	3.52 / **6.35**	3.49 / **6.27**	3.44 / **6.15**	3.41 / **6.07**	3.38 / **5.98**	3.34 / **5.90**	3.32 / **5.85**	3.29 / **5.78**	3.28 / **5.75**	3.25 / **5.70**	3.24 / **5.67**	3.23 / **5.65**
8	5.32 / **11.26**	4.46 / **8.65**	4.07 / **7.59**	3.84 / **7.01**	3.69 / **6.63**	3.58 / **6.37**	3.50 / **6.19**	3.44 / **6.03**	3.39 / **5.91**	3.34 / **5.82**	3.31 / **5.74**	3.28 / **5.67**	3.23 / **5.56**	3.20 / **5.48**	3.15 / **5.36**	3.12 / **5.28**	3.08 / **5.20**	3.05 / **5.11**	3.03 / **5.06**	3.00 / **5.00**	2.98 / **4.96**	2.96 / **4.91**	2.94 / **4.88**	2.93 / **4.86**
9	5.12 / **10.56**	4.26 / **8.02**	3.86 / **6.99**	3.63 / **6.42**	3.48 / **6.06**	3.37 / **5.80**	3.29 / **5.62**	3.23 / **5.47**	3.18 / **5.35**	3.13 / **5.26**	3.10 / **5.18**	3.07 / **5.11**	3.02 / **5.00**	2.98 / **4.92**	2.93 / **4.80**	2.90 / **4.73**	2.86 / **4.64**	2.82 / **4.56**	2.80 / **4.51**	2.77 / **4.45**	2.76 / **4.41**	2.73 / **4.36**	2.72 / **4.33**	2.71 / **4.31**

*Reprinted, by permission, from Snedecor, *Statistical Methods*, Collegiate Press, Iowa State College, Ames.

TABLE V. (Continued)

10	2.54 / 3.91	2.55 / 3.93	2.56 / 3.96	2.59 / 4.01	2.61 / 4.05	2.64 / 4.12	2.67 / 4.17	2.70 / 4.25	2.74 / 4.33	2.77 / 4.41	2.82 / 4.52	2.86 / 4.60	2.91 / 4.71	2.94 / 4.78	2.97 / 4.85	3.02 / 4.95	3.07 / 5.06	3.14 / 5.21	3.22 / 5.39	3.33 / 5.64	3.48 / 5.99	3.71 / 6.55	4.10 / 7.56	4.96 / 10.04
11	2.40 / 3.60	2.41 / 3.62	2.42 / 3.66	2.45 / 3.70	2.47 / 3.74	2.50 / 3.80	2.53 / 3.86	2.57 / 3.94	2.61 / 4.02	2.65 / 4.10	2.70 / 4.21	2.74 / 4.29	2.79 / 4.40	2.82 / 4.46	2.86 / 4.54	2.90 / 4.63	2.95 / 4.74	3.01 / 4.88	3.09 / 5.07	3.20 / 5.32	3.36 / 5.67	3.59 / 6.22	3.98 / 7.20	4.84 / 9.65
12	2.30 / 3.36	2.31 / 3.38	2.32 / 3.41	2.35 / 3.46	2.36 / 3.49	2.40 / 3.56	2.42 / 3.61	2.46 / 3.70	2.50 / 3.78	2.54 / 3.86	2.60 / 3.98	2.64 / 4.05	2.69 / 4.16	2.72 / 4.22	2.76 / 4.30	2.80 / 4.39	2.85 / 4.50	2.92 / 4.65	3.00 / 4.82	3.11 / 5.06	3.26 / 5.41	3.49 / 5.95	3.88 / 6.93	4.75 / 9.33
13	2.21 / 3.16	2.22 / 3.18	2.24 / 3.21	2.26 / 3.27	2.28 / 3.30	2.32 / 3.37	2.34 / 3.42	2.38 / 3.51	2.42 / 3.59	2.46 / 3.67	2.51 / 3.78	2.55 / 3.85	2.60 / 3.96	2.63 / 4.02	2.67 / 4.10	2.72 / 4.19	2.77 / 4.30	2.84 / 4.44	2.92 / 4.62	3.02 / 4.86	3.18 / 5.20	3.41 / 5.74	3.80 / 6.70	4.67 / 9.07
14	2.13 / 3.00	2.14 / 3.02	2.16 / 3.06	2.19 / 3.11	2.21 / 3.14	2.24 / 3.21	2.27 / 3.26	2.31 / 3.34	2.35 / 3.43	2.39 / 3.51	2.44 / 3.62	2.48 / 3.70	2.53 / 3.80	2.56 / 3.86	2.60 / 3.94	2.65 / 4.03	2.70 / 4.14	2.77 / 4.28	2.85 / 4.46	2.96 / 4.69	3.11 / 5.03	3.34 / 5.56	3.74 / 6.51	4.60 / 8.86
15	2.07 / 2.87	2.08 / 2.89	2.10 / 2.92	2.12 / 2.97	2.15 / 3.00	2.18 / 3.07	2.21 / 3.12	2.25 / 3.20	2.29 / 3.29	2.33 / 3.36	2.39 / 3.48	2.43 / 3.56	2.48 / 3.67	2.51 / 3.73	2.55 / 3.80	2.59 / 3.89	2.64 / 4.00	2.70 / 4.14	2.79 / 4.32	2.90 / 4.56	3.06 / 4.89	3.29 / 5.42	3.68 / 6.36	4.54 / 8.68
16	2.01 / 2.75	2.02 / 2.77	2.04 / 2.80	2.07 / 2.86	2.09 / 2.89	2.13 / 2.96	2.16 / 3.01	2.20 / 3.10	2.24 / 3.18	2.28 / 3.25	2.33 / 3.37	2.37 / 3.45	2.42 / 3.55	2.45 / 3.61	2.49 / 3.69	2.54 / 3.78	2.59 / 3.89	2.66 / 4.03	2.74 / 4.20	2.85 / 4.44	3.01 / 4.77	3.24 / 5.29	3.63 / 6.23	4.49 / 8.53
17	1.96 / 2.65	1.97 / 2.67	1.99 / 2.70	2.02 / 2.76	2.04 / 2.79	2.08 / 2.86	2.11 / 2.92	2.15 / 3.00	2.19 / 3.08	2.23 / 3.16	2.29 / 3.27	2.33 / 3.35	2.38 / 3.45	2.41 / 3.52	2.45 / 3.59	2.50 / 3.68	2.55 / 3.79	2.62 / 3.93	2.70 / 4.10	2.81 / 4.34	2.96 / 4.67	3.20 / 5.18	3.59 / 6.11	4.45 / 8.40
18	1.92 / 2.57	1.93 / 2.59	1.95 / 2.62	1.98 / 2.68	2.00 / 2.71	2.04 / 2.78	2.07 / 2.83	2.11 / 2.91	2.15 / 3.00	2.19 / 3.07	2.25 / 3.19	2.29 / 3.27	2.34 / 3.37	2.37 / 3.44	2.41 / 3.51	2.46 / 3.60	2.51 / 3.71	2.58 / 3.85	2.66 / 4.01	2.77 / 4.25	2.93 / 4.58	3.16 / 5.09	3.55 / 6.01	4.41 / 8.28
19	1.88 / 2.49	1.90 / 2.51	1.91 / 2.54	1.94 / 2.60	1.96 / 2.63	2.00 / 2.70	2.02 / 2.76	2.07 / 2.84	2.11 / 2.92	2.15 / 3.00	2.21 / 3.12	2.26 / 3.19	2.31 / 3.30	2.34 / 3.36	2.38 / 3.43	2.43 / 3.52	2.48 / 3.63	2.55 / 3.77	2.63 / 3.94	2.74 / 4.17	2.90 / 4.50	3.13 / 5.01	3.52 / 5.93	4.38 / 8.18
20	1.84 / 2.42	1.85 / 2.44	1.87 / 2.47	1.90 / 2.53	1.92 / 2.56	1.96 / 2.63	1.99 / 2.69	2.04 / 2.77	2.08 / 2.86	2.12 / 2.94	2.18 / 3.05	2.23 / 3.13	2.28 / 3.23	2.31 / 3.30	2.35 / 3.37	2.40 / 3.45	2.45 / 3.56	2.52 / 3.71	2.60 / 3.87	2.71 / 4.10	2.87 / 4.43	3.10 / 4.94	3.49 / 5.85	4.35 / 8.10
21	1.81 / 2.36	1.82 / 2.38	1.84 / 2.42	1.87 / 2.47	1.89 / 2.51	1.93 / 2.58	1.96 / 2.63	2.00 / 2.72	2.05 / 2.80	2.09 / 2.88	2.15 / 2.99	2.20 / 3.07	2.25 / 3.17	2.28 / 3.24	2.32 / 3.31	2.37 / 3.40	2.42 / 3.51	2.49 / 3.65	2.57 / 3.81	2.68 / 4.04	2.84 / 4.37	3.07 / 4.87	3.47 / 5.78	4.32 / 8.02
22	1.78 / 2.31	1.80 / 2.33	1.81 / 2.37	1.84 / 2.42	1.87 / 2.46	1.91 / 2.53	1.93 / 2.58	1.98 / 2.67	2.03 / 2.75	2.07 / 2.83	2.13 / 2.94	2.18 / 3.02	2.23 / 3.12	2.26 / 3.18	2.30 / 3.26	2.35 / 3.35	2.40 / 3.45	2.47 / 3.59	2.55 / 3.76	2.66 / 3.99	2.82 / 4.31	3.05 / 4.82	3.44 / 5.72	4.30 / 7.94
23	1.76 / 2.26	1.77 / 2.28	1.79 / 2.32	1.82 / 2.37	1.84 / 2.41	1.88 / 2.48	1.91 / 2.53	1.96 / 2.62	2.00 / 2.70	2.04 / 2.78	2.10 / 2.89	2.14 / 2.97	2.20 / 3.07	2.24 / 3.14	2.28 / 3.21	2.32 / 3.30	2.38 / 3.41	2.45 / 3.54	2.53 / 3.71	2.64 / 3.94	2.80 / 4.26	3.03 / 4.76	3.42 / 5.66	4.28 / 7.88
24	1.73 / 2.21	1.74 / 2.23	1.76 / 2.27	1.80 / 2.33	1.82 / 2.36	1.86 / 2.44	1.89 / 2.49	1.94 / 2.58	1.98 / 2.66	2.02 / 2.74	2.09 / 2.85	2.13 / 2.93	2.18 / 3.03	2.22 / 3.09	2.26 / 3.17	2.30 / 3.25	2.36 / 3.36	2.43 / 3.50	2.51 / 3.67	2.62 / 3.90	2.78 / 4.22	3.01 / 4.72	3.40 / 5.61	4.26 / 7.82
25	1.71 / 2.17	1.72 / 2.19	1.74 / 2.23	1.77 / 2.29	1.80 / 2.32	1.84 / 2.40	1.87 / 2.45	1.92 / 2.54	1.96 / 2.62	2.00 / 2.70	2.06 / 2.81	2.11 / 2.89	2.16 / 2.99	2.20 / 3.05	2.24 / 3.13	2.28 / 3.21	2.34 / 3.32	2.41 / 3.46	2.49 / 3.63	2.60 / 3.86	2.76 / 4.18	2.99 / 4.68	3.38 / 5.57	4.24 / 7.77

TABLE V. (Continued)

5% (Roman Type) and 1% (Bold-Face Type) Points for the Distribution of F

Values are shown as: 5% point (Roman) / **1% point (Bold-Face)**

Degrees of freedom for numerator (ν_1)

ν_2	1	2	3	4	5	6	7	8	9	10	11	12	14	16	20	24	30	40	50	75	100	200	500	∞
26	4.22 / **7.72**	3.37 / **5.53**	2.98 / **4.64**	2.74 / **4.14**	2.59 / **3.82**	2.47 / **3.59**	2.39 / **3.42**	2.32 / **3.29**	2.27 / **3.17**	2.22 / **3.09**	2.18 / **3.02**	2.15 / **2.96**	2.10 / **2.86**	2.05 / **2.77**	1.99 / **2.66**	1.95 / **2.58**	1.90 / **2.50**	1.85 / **2.41**	1.82 / **2.36**	1.78 / **2.28**	1.76 / **2.25**	1.72 / **2.19**	1.70 / **2.15**	1.69 / **2.13**
27	4.21 / **7.68**	3.35 / **5.49**	2.96 / **4.60**	2.73 / **4.11**	2.57 / **3.79**	2.46 / **3.56**	2.37 / **3.39**	2.30 / **3.26**	2.25 / **3.14**	2.20 / **3.06**	2.16 / **2.98**	2.13 / **2.93**	2.08 / **2.83**	2.03 / **2.74**	1.97 / **2.63**	1.93 / **2.55**	1.88 / **2.47**	1.84 / **2.38**	1.80 / **2.33**	1.76 / **2.25**	1.74 / **2.21**	1.71 / **2.16**	1.68 / **2.12**	1.67 / **2.10**
28	4.20 / **7.64**	3.34 / **5.45**	2.95 / **4.57**	2.71 / **4.07**	2.56 / **3.76**	2.44 / **3.53**	2.36 / **3.36**	2.29 / **3.23**	2.24 / **3.11**	2.19 / **3.03**	2.15 / **2.95**	2.12 / **2.90**	2.06 / **2.80**	2.02 / **2.71**	1.96 / **2.60**	1.91 / **2.52**	1.87 / **2.44**	1.81 / **2.35**	1.78 / **2.30**	1.75 / **2.22**	1.72 / **2.18**	1.69 / **2.13**	1.67 / **2.09**	1.65 / **2.06**
29	4.18 / **7.60**	3.33 / **5.42**	2.93 / **4.54**	2.70 / **4.04**	2.54 / **3.73**	2.43 / **3.50**	2.35 / **3.33**	2.28 / **3.20**	2.22 / **3.08**	2.18 / **3.00**	2.14 / **2.92**	2.10 / **2.87**	2.05 / **2.77**	2.00 / **2.68**	1.94 / **2.57**	1.90 / **2.49**	1.85 / **2.41**	1.80 / **2.32**	1.77 / **2.27**	1.73 / **2.19**	1.71 / **2.15**	1.68 / **2.10**	1.65 / **2.06**	1.64 / **2.03**
30	4.17 / **7.56**	3.32 / **5.39**	2.92 / **4.51**	2.69 / **4.02**	2.53 / **3.70**	2.42 / **3.47**	2.34 / **3.30**	2.27 / **3.17**	2.21 / **3.06**	2.16 / **2.98**	2.12 / **2.90**	2.09 / **2.84**	2.04 / **2.74**	1.99 / **2.66**	1.93 / **2.55**	1.89 / **2.47**	1.84 / **2.38**	1.79 / **2.29**	1.76 / **2.24**	1.72 / **2.16**	1.69 / **2.13**	1.66 / **2.07**	1.64 / **2.03**	1.62 / **2.01**
32	4.15 / **7.50**	3.30 / **5.34**	2.90 / **4.46**	2.67 / **3.97**	2.51 / **3.66**	2.40 / **3.42**	2.32 / **3.25**	2.25 / **3.12**	2.19 / **3.01**	2.14 / **2.94**	2.10 / **2.86**	2.07 / **2.80**	2.02 / **2.70**	1.97 / **2.62**	1.91 / **2.51**	1.86 / **2.42**	1.82 / **2.34**	1.76 / **2.25**	1.74 / **2.20**	1.69 / **2.12**	1.67 / **2.08**	1.64 / **2.02**	1.61 / **1.98**	1.59 / **1.96**
34	4.13 / **7.44**	3.28 / **5.29**	2.88 / **4.42**	2.65 / **3.93**	2.49 / **3.61**	2.38 / **3.38**	2.30 / **3.21**	2.23 / **3.08**	2.17 / **2.97**	2.12 / **2.89**	2.08 / **2.82**	2.05 / **2.76**	2.00 / **2.66**	1.95 / **2.58**	1.89 / **2.47**	1.84 / **2.38**	1.80 / **2.30**	1.74 / **2.21**	1.71 / **2.15**	1.67 / **2.08**	1.64 / **2.04**	1.61 / **1.98**	1.59 / **1.94**	1.57 / **1.91**
36	4.11 / **7.39**	3.26 / **5.25**	2.86 / **4.38**	2.63 / **3.89**	2.48 / **3.58**	2.36 / **3.35**	2.28 / **3.18**	2.21 / **3.04**	2.15 / **2.94**	2.10 / **2.86**	2.06 / **2.78**	2.03 / **2.72**	1.98 / **2.62**	1.93 / **2.54**	1.87 / **2.43**	1.82 / **2.35**	1.78 / **2.26**	1.72 / **2.17**	1.69 / **2.12**	1.65 / **2.04**	1.62 / **2.00**	1.59 / **1.94**	1.56 / **1.90**	1.55 / **1.87**
38	4.10 / **7.35**	3.25 / **5.21**	2.85 / **4.34**	2.62 / **3.86**	2.46 / **3.54**	2.35 / **3.32**	2.26 / **3.15**	2.19 / **3.02**	2.14 / **2.91**	2.09 / **2.82**	2.05 / **2.75**	2.02 / **2.69**	1.96 / **2.59**	1.92 / **2.51**	1.85 / **2.40**	1.80 / **2.32**	1.76 / **2.22**	1.71 / **2.14**	1.67 / **2.08**	1.63 / **2.00**	1.60 / **1.97**	1.57 / **1.90**	1.54 / **1.86**	1.53 / **1.84**
40	4.08 / **7.31**	3.23 / **5.18**	2.84 / **4.31**	2.61 / **3.83**	2.45 / **3.51**	2.34 / **3.29**	2.25 / **3.12**	2.18 / **2.99**	2.12 / **2.88**	2.07 / **2.80**	2.04 / **2.73**	2.00 / **2.66**	1.95 / **2.56**	1.90 / **2.49**	1.84 / **2.37**	1.79 / **2.29**	1.74 / **2.20**	1.69 / **2.11**	1.66 / **2.05**	1.61 / **1.97**	1.59 / **1.94**	1.55 / **1.88**	1.53 / **1.84**	1.51 / **1.81**
42	4.07 / **7.27**	3.22 / **5.15**	2.83 / **4.29**	2.59 / **3.80**	2.44 / **3.49**	2.32 / **3.26**	2.24 / **3.10**	2.17 / **2.96**	2.11 / **2.86**	2.06 / **2.77**	2.02 / **2.70**	1.99 / **2.64**	1.94 / **2.54**	1.89 / **2.46**	1.82 / **2.35**	1.78 / **2.26**	1.73 / **2.17**	1.68 / **2.08**	1.64 / **2.02**	1.60 / **1.94**	1.57 / **1.91**	1.54 / **1.85**	1.51 / **1.80**	1.49 / **1.78**
44	4.06 / **7.24**	3.21 / **5.12**	2.82 / **4.26**	2.58 / **3.78**	2.43 / **3.46**	2.31 / **3.24**	2.23 / **3.07**	2.16 / **2.94**	2.10 / **2.84**	2.05 / **2.75**	2.01 / **2.68**	1.98 / **2.62**	1.92 / **2.52**	1.88 / **2.44**	1.81 / **2.32**	1.76 / **2.24**	1.72 / **2.15**	1.66 / **2.06**	1.63 / **2.00**	1.58 / **1.92**	1.56 / **1.88**	1.52 / **1.82**	1.50 / **1.78**	1.48 / **1.75**
46	4.05 / **7.21**	3.20 / **5.10**	2.81 / **4.24**	2.57 / **3.76**	2.42 / **3.44**	2.30 / **3.22**	2.22 / **3.05**	2.14 / **2.92**	2.09 / **2.82**	2.04 / **2.73**	2.00 / **2.66**	1.97 / **2.60**	1.91 / **2.50**	1.87 / **2.42**	1.80 / **2.30**	1.75 / **2.22**	1.71 / **2.13**	1.65 / **2.04**	1.62 / **1.98**	1.57 / **1.90**	1.54 / **1.86**	1.51 / **1.80**	1.48 / **1.76**	1.46 / **1.72**
48	4.04 / **7.19**	3.19 / **5.08**	2.80 / **4.22**	2.56 / **3.74**	2.41 / **3.42**	2.30 / **3.20**	2.21 / **3.04**	2.14 / **2.90**	2.08 / **2.80**	2.03 / **2.71**	1.99 / **2.64**	1.96 / **2.58**	1.90 / **2.48**	1.86 / **2.40**	1.79 / **2.28**	1.74 / **2.20**	1.70 / **2.11**	1.64 / **2.02**	1.61 / **1.96**	1.56 / **1.88**	1.53 / **1.84**	1.50 / **1.78**	1.47 / **1.73**	1.45 / **1.70**

Degrees of freedom for denominator (ν_2)

TABLE V. (Continued)

50	4.03 / **7.17**	3.18 / **5.06**	2.79 / **4.20**	2.56 / **3.72**	2.40 / **3.41**	2.29 / **3.18**	2.20 / **3.02**	2.13 / **2.88**	2.07 / **2.78**	2.02 / **2.70**	1.98 / **2.62**	1.95 / **2.56**	1.90 / **2.46**	1.85 / **2.39**	1.78 / **2.26**	1.74 / **2.18**	1.69 / **2.10**	1.63 / **2.00**	1.60 / **1.94**	1.55 / **1.86**	1.52 / **1.82**	1.48 / **1.76**	1.46 / **1.71**	1.44 / **1.68**
55	4.02 / **7.12**	3.17 / **5.01**	2.78 / **4.16**	2.54 / **3.68**	2.38 / **3.37**	2.27 / **3.15**	2.18 / **2.98**	2.11 / **2.85**	2.05 / **2.75**	2.00 / **2.66**	1.97 / **2.59**	1.93 / **2.53**	1.88 / **2.43**	1.83 / **2.35**	1.76 / **2.23**	1.72 / **2.15**	1.67 / **2.06**	1.61 / **1.96**	1.58 / **1.90**	1.52 / **1.82**	1.50 / **1.78**	1.46 / **1.71**	1.43 / **1.66**	1.41 / **1.64**
60	4.00 / **7.08**	3.15 / **4.98**	2.76 / **4.13**	2.52 / **3.65**	2.37 / **3.34**	2.25 / **3.12**	2.17 / **2.95**	2.10 / **2.82**	2.04 / **2.72**	1.99 / **2.63**	1.95 / **2.56**	1.92 / **2.50**	1.86 / **2.40**	1.81 / **2.32**	1.75 / **2.20**	1.70 / **2.12**	1.65 / **2.03**	1.59 / **1.93**	1.56 / **1.87**	1.50 / **1.79**	1.48 / **1.74**	1.44 / **1.68**	1.41 / **1.63**	1.39 / **1.60**
65	3.99 / **7.04**	3.14 / **4.95**	2.75 / **4.10**	2.51 / **3.62**	2.36 / **3.31**	2.24 / **3.09**	2.15 / **2.93**	2.08 / **2.79**	2.02 / **2.70**	1.98 / **2.61**	1.94 / **2.54**	1.90 / **2.47**	1.85 / **2.37**	1.80 / **2.30**	1.73 / **2.18**	1.68 / **2.09**	1.63 / **2.00**	1.57 / **1.90**	1.54 / **1.84**	1.49 / **1.76**	1.46 / **1.71**	1.42 / **1.64**	1.39 / **1.60**	1.37 / **1.56**
70	3.98 / **7.01**	3.13 / **4.92**	2.74 / **4.08**	2.50 / **3.60**	2.35 / **3.29**	2.23 / **3.07**	2.14 / **2.91**	2.07 / **2.77**	2.01 / **2.67**	1.97 / **2.59**	1.93 / **2.51**	1.89 / **2.45**	1.84 / **2.35**	1.79 / **2.28**	1.72 / **2.15**	1.67 / **2.07**	1.62 / **1.98**	1.56 / **1.88**	1.53 / **1.82**	1.47 / **1.74**	1.45 / **1.69**	1.40 / **1.63**	1.37 / **1.56**	1.35 / **1.53**
80	3.96 / **6.96**	3.11 / **4.88**	2.72 / **4.04**	2.48 / **3.56**	2.33 / **3.25**	2.21 / **3.04**	2.12 / **2.87**	2.05 / **2.74**	1.99 / **2.64**	1.95 / **2.55**	1.91 / **2.48**	1.88 / **2.41**	1.82 / **2.32**	1.77 / **2.24**	1.70 / **2.11**	1.65 / **2.03**	1.60 / **1.94**	1.54 / **1.84**	1.51 / **1.78**	1.45 / **1.70**	1.42 / **1.65**	1.38 / **1.57**	1.35 / **1.52**	1.32 / **1.49**
100	3.94 / **6.90**	3.09 / **4.82**	2.70 / **3.98**	2.46 / **3.51**	2.30 / **3.20**	2.19 / **2.99**	2.10 / **2.82**	2.03 / **2.69**	1.97 / **2.59**	1.92 / **2.51**	1.88 / **2.43**	1.85 / **2.36**	1.79 / **2.26**	1.75 / **2.19**	1.68 / **2.06**	1.63 / **1.98**	1.57 / **1.89**	1.51 / **1.79**	1.48 / **1.73**	1.42 / **1.64**	1.39 / **1.59**	1.34 / **1.51**	1.30 / **1.46**	1.28 / **1.43**
125	3.92 / **6.84**	3.07 / **4.78**	2.68 / **3.94**	2.44 / **3.47**	2.29 / **3.17**	2.17 / **2.95**	2.08 / **2.79**	2.01 / **2.65**	1.95 / **2.56**	1.90 / **2.47**	1.86 / **2.40**	1.83 / **2.33**	1.77 / **2.23**	1.72 / **2.15**	1.65 / **2.03**	1.60 / **1.94**	1.55 / **1.85**	1.49 / **1.75**	1.45 / **1.68**	1.39 / **1.59**	1.36 / **1.54**	1.31 / **1.46**	1.27 / **1.40**	1.25 / **1.37**
150	3.91 / **6.81**	3.06 / **4.75**	2.67 / **3.91**	2.43 / **3.44**	2.27 / **3.13**	2.16 / **2.92**	2.07 / **2.76**	2.00 / **2.62**	1.94 / **2.53**	1.89 / **2.44**	1.85 / **2.37**	1.82 / **2.30**	1.76 / **2.20**	1.71 / **2.12**	1.64 / **2.00**	1.59 / **1.91**	1.54 / **1.83**	1.47 / **1.72**	1.44 / **1.66**	1.37 / **1.56**	1.34 / **1.51**	1.29 / **1.43**	1.25 / **1.37**	1.22 / **1.33**
200	3.89 / **6.76**	3.04 / **4.71**	2.65 / **3.88**	2.41 / **3.41**	2.26 / **3.11**	2.14 / **2.90**	2.05 / **2.73**	1.98 / **2.60**	1.92 / **2.50**	1.87 / **2.41**	1.83 / **2.34**	1.80 / **2.28**	1.74 / **2.17**	1.69 / **2.09**	1.62 / **1.97**	1.57 / **1.88**	1.52 / **1.79**	1.45 / **1.69**	1.42 / **1.62**	1.35 / **1.53**	1.32 / **1.48**	1.26 / **1.39**	1.22 / **1.33**	1.19 / **1.28**
400	3.86 / **6.70**	3.02 / **4.66**	2.62 / **3.83**	2.39 / **3.36**	2.23 / **3.06**	2.12 / **2.85**	2.03 / **2.69**	1.96 / **2.55**	1.90 / **2.46**	1.85 / **2.37**	1.81 / **2.29**	1.78 / **2.23**	1.72 / **2.12**	1.67 / **2.04**	1.60 / **1.92**	1.54 / **1.84**	1.49 / **1.74**	1.42 / **1.64**	1.38 / **1.57**	1.32 / **1.47**	1.28 / **1.42**	1.22 / **1.32**	1.16 / **1.24**	1.13 / **1.19**
1000	3.85 / **6.66**	3.00 / **4.62**	2.61 / **3.80**	2.38 / **3.34**	2.22 / **3.04**	2.10 / **2.82**	2.02 / **2.66**	1.95 / **2.53**	1.89 / **2.43**	1.84 / **2.34**	1.80 / **2.26**	1.76 / **2.20**	1.70 / **2.09**	1.65 / **2.01**	1.58 / **1.89**	1.53 / **1.81**	1.47 / **1.71**	1.41 / **1.61**	1.36 / **1.54**	1.30 / **1.44**	1.26 / **1.38**	1.19 / **1.28**	1.13 / **1.19**	1.08 / **1.11**
∞	3.84 / **6.64**	2.99 / **4.60**	2.60 / **3.78**	2.37 / **3.32**	2.21 / **3.02**	2.09 / **2.80**	2.01 / **2.64**	1.94 / **2.51**	1.88 / **2.41**	1.83 / **2.32**	1.79 / **2.24**	1.75 / **2.18**	1.69 / **2.07**	1.64 / **1.99**	1.57 / **1.87**	1.52 / **1.79**	1.46 / **1.69**	1.40 / **1.59**	1.35 / **1.52**	1.28 / **1.41**	1.24 / **1.36**	1.17 / **1.25**	1.11 / **1.15**	1.00 / **1.00**

TABLE VI.   Random Digits

03991	10461	93716	16894	98953	73231	39528	72484	82474	25593
38555	95554	32886	59780	09958	18065	81616	18711	53342	44276
17546	73704	92052	46215	15917	06253	07586	16120	82641	22820
32643	52861	95819	06831	19640	99413	90767	04235	13574	17200
69572	68777	39510	35905	85244	35159	40188	28193	29593	88627
24122	66591	27699	06494	03152	19121	34414	82157	86887	55087
61196	30231	92962	61773	22109	78508	63439	75363	44989	16822
30532	21704	10274	12202	94205	20380	67049	09070	93399	45547
03788	97599	75867	20717	82037	10268	79495	04146	52162	90286
48228	63379	85783	47619	87481	37220	91704	30552	04737	21031
88618	19161	41290	67312	71857	15957	48545	35247	18619	13674
71299	23853	05870	01119	92784	26340	75122	11724	74627	73707
27954	58909	82444	99005	04921	73701	92904	13141	32392	19763
80863	00514	20247	81759	45197	25332	69902	63742	78464	22501
33564	60780	48460	85558	15191	18782	94972	11598	62095	36787
90899	75754	60833	25983	01291	41349	19152	00023	12302	80783
78038	70267	43529	06318	38384	74761	36024	00867	76378	41605
55986	66485	88722	56736	66164	49431	94458	74284	05041	49807
87539	08823	94813	31900	54155	83436	54158	34243	46978	35482
16818	60311	74457	90561	72848	11834	75051	93029	47665	64382
34677	58300	74910	64345	19325	81549	60365	94653	35075	33949
45305	07521	61318	31855	14413	70951	83799	42402	56623	34442
59747	67277	76503	34513	39663	77544	32960	07405	36409	83232
16520	69676	11654	99893	02181	68161	19322	53845	57620	52606
68652	27376	92852	55866	88448	03584	11220	94747	07399	37408
79375	95220	01159	63267	10622	48391	31751	57260	68980	05339
33521	26665	55823	47641	86225	31704	88492	99382	14454	04504
59589	49067	66821	41575	49767	04037	30934	47744	07481	83828
20554	91409	96277	48257	50816	97616	22888	48893	27499	98748
59404	72059	43947	51680	43852	59693	78212	16993	35902	91386
42614	29297	01918	28316	25163	01889	70014	15021	68971	11403
34994	41374	70071	14736	65251	07629	37239	33295	18477	65622
99385	41600	11133	07586	36815	43625	18637	37509	14707	93997
66497	68646	78138	66559	64397	11692	05327	82162	83745	22567
48509	23929	27482	45476	04515	25624	95096	67946	16930	33361
15470	48355	88651	22596	83761	60873	43253	84145	20368	07126
20094	98977	74843	93413	14387	06345	80854	09279	41196	37480
73788	06533	28597	20405	51321	92246	80088	77074	66919	31678
60530	45128	74022	84617	72472	00008	80890	18002	35352	54131
44372	15486	65741	14014	05466	55306	93128	18464	79982	68416
18611	19241	66083	24653	84609	58232	41849	84547	46850	52326
58319	15997	08355	60860	29735	47762	46352	33049	69248	93460
61199	67940	55121	29281	59076	07936	11087	96294	14013	31792
18627	90872	00911	98936	76355	93779	52701	08337	56303	87315
00441	58997	14060	40619	29549	69616	57275	36898	81304	48585
32624	68691	14845	46672	61958	77100	20857	73156	70284	24326
65961	73488	41839	55382	17267	70943	15633	84924	90415	93614
20288	34060	39685	23309	10061	68829	92694	48297	39904	02115
59362	95938	74416	53166	35208	33374	77613	19019	88152	00080
99782	93478	53152	67433	35663	52972	38688	32486	45134	63545

398

TABLE VI. (*Continued*)

27767	43584	85301	88977	29490	69714	94015	64874	32444	48277
13025	14338	54066	15243	47724	66733	74108	88222	88570	74015
80217	36292	98525	24335	24432	24896	62880	87873	95160	59221
10875	62004	90391	61105	57411	06368	11748	12102	80580	41867
54127	57326	26629	19087	24472	88779	17944	05600	60478	03343
60311	42824	37301	42678	45990	43242	66067	42792	95043	52680
49739	71484	92003	98086	76668	73209	54244	91030	45547	70818
78626	51594	16453	94614	39014	97066	30945	57589	31732	57260
66692	13986	99837	00582	81232	44987	69170	37403	86995	90307
44071	28091	07362	97703	76447	42537	08345	88975	35841	85771
59820	96163	78851	16499	87064	13075	73035	41207	74699	09310
25704	91035	26313	77463	55387	72681	47431	43905	31048	56699
22304	90314	78438	66276	18396	73538	43277	58874	11466	16082
17710	59621	15292	76139	59526	52113	53856	30743	08670	84741
25852	58905	55018	56374	35824	71708	30540	27886	61732	75454
46780	56487	75211	10271	36633	68424	17374	52003	70707	70214
59849	96169	87195	46092	26787	60939	59202	11973	02902	33250
47670	07654	30342	40277	11049	72049	83012	09832	25571	77628
94304	71803	73465	09819	58869	35220	09504	96412	90193	79568
08105	59987	21437	36786	49226	77837	98524	97831	65704	09514
64281	61826	18555	64937	64654	25843	41145	42820	14924	39650
66847	70495	32350	02985	01755	14750	48968	38603	70312	05682
72461	33230	21529	53424	72877	17334	39283	04149	90850	64618
21032	91050	13058	16218	06554	07850	73950	79552	24781	89683
95362	67011	06651	16136	57216	39618	49856	99326	40902	05069
49712	97380	10404	55452	09971	59481	37006	22186	72682	07385
58275	61764	97586	54716	61459	21647	87417	17198	21443	41808
89514	11788	68224	23417	46376	25366	94746	49580	01176	28838
15472	50669	48139	36732	26825	05511	12459	91314	80582	71944
12120	86124	51247	44302	87112	21476	14713	71181	13177	55292
95294	00556	70481	06905	21785	41101	49386	54480	23604	23554
66986	34099	74474	20740	47458	64809	06312	88940	15995	69321
80620	51790	11436	38072	40405	68032	60942	00307	11897	92674
55411	85667	77535	99892	71209	92061	92329	98932	78284	46347
95083	06783	28102	57816	85561	29671	77936	63574	31384	51924
90726	57166	98884	08583	95889	57067	38101	77756	11657	13897
68984	83620	89747	98882	92613	89719	39641	69457	91339	22502
36421	16489	18059	51061	67667	60631	84054	40455	99396	63680
92638	40333	67054	16067	24700	71594	47468	03577	57649	63266
21036	82808	77501	97427	76479	68562	43321	31370	28977	23896
13173	33365	41468	85149	49554	17994	91178	10174	29420	90438
86716	38746	94559	37559	49678	53119	98189	81851	29651	84215
92581	02262	41615	70360	64114	58660	96717	54244	10701	41393
12470	56500	50273	93113	41794	86861	39448	93136	25722	08564
01016	00857	41396	80504	90670	08289	58137	17820	22751	36518
34030	60726	25807	24260	71529	78920	47648	13885	70669	93406
50259	46345	06170	97965	88302	98041	11947	56203	19324	20504
73959	76145	60808	54444	74412	81105	69181	96845	38525	11600
46874	37088	80940	44893	10408	36222	14004	23153	69249	05747
60883	52109	19516	90120	46759	71643	62342	07589	08899	05985

Table VII. Rank-Sum Critical Values*

The sample sizes are shown in parentheses (n_1, n_2). The probability associated with a pair of critical values is the probability that $T \leq$ smaller value, or equally, it is the probability that $T \geq$ larger value. These probabilities are the closest ones to .025 and .05 that exist for integer values of T. The approximate .025 values should be used for a two-sided test with $\alpha = .05$, and the approximate .05 values for a one-sided test.

(2, 4)			(4, 4)			(6, 7)		
3	11	.067	11	25	.029	28	56	.026
(2, 5)			12	24	.057	30	54	.051
3	13	.047	**(4, 5)**			**(6, 8)**		
(2, 6)			12	28	.032	29	61	.021
3	15	.036	13	27	.056	32	58	.054
4	14	.071	**(4, 6)**			**(6, 9)**		
(2, 7)			12	32	.019	31	65	.025
3	17	.028	14	30	.057	33	63	.044
4	16	.056	**(4, 7)**			**(6, 10)**		
(2, 8)			13	35	.021	33	69	.028
3	19	.022	15	33	.055	35	67	.047
4	18	.044	**(4, 8)**			**(7, 7)**		
(2, 9)			14	38	.024	37	68	.027
3	21	.018	16	36	.055	39	66	.049
4	20	.036	**(4, 9)**			**(7, 8)**		
(2, 10)			15	41	.025	39	73	.027
4	22	.030	17	39	.053	41	71	.047
5	21	.061	**(4, 10)**			**(7, 9)**		
(3, 3)			16	44	.026	41	78	.027
6	15	.050	18	42	.053	43	76	.045
(3, 4)			**(5, 5)**			**(7, 10)**		
6	18	.028	18	37	.028	43	83	.028
7	17	.057	19	36	.048	46	80	.054
(3, 5)			**(5, 6)**			**(8, 8)**		
6	21	.018	19	41	.026	49	87	.025
7	20	.036	20	40	.041	52	84	.052
(3, 6)			**(5, 7)**			**(8, 9)**		
7	23	.024	20	45	.024	51	93	.023
8	22	.048	22	43	.053	54	90	.046
(3, 7)			**(5, 8)**			**(8, 10)**		
8	25	.033	21	49	.023	54	98	.027
9	24	.058	23	47	.047	57	95	.051
(3, 8)			**(5, 9)**			**(9, 9)**		
8	28	.024	22	53	.021	63	108	.025
9	27	.042	25	50	.056	66	105	.047
(3, 9)			**(5, 10)**			**(9, 10)**		
9	30	.032	24	56	.028	66	114	.027
10	29	.050	26	54	.050	69	111	.047
(3, 10)			**(6, 6)**			**(10, 10)**		
9	33	.024	26	52	.021	79	131	.026
11	31	.056	28	50	.047	83	127	.053

*This table was extracted from a more complete table (A-20) in *Introduction to Statistical Analysis*, 2nd edition, by W. J. Dixon and F. J. Massey, with permission from the publishers, the McGraw-Hill Book Company.

TABLE VIII. Critical Values for D_α in the Kolmogorov-Smirnov Test

α \ n	.20	.10	.05	.01
5	.45	.51	.56	.67
10	.32	.37	.41	.49
15	.27	.30	.34	.40
20	.23	.26	.29	.36
25	.21	.24	.27	.32
30	.19	.22	.24	.29
35	.18	.20	.23	.27
40	.17	.19	.21	.25
45	.16	.18	.20	.24
50	.15	.17	.19	.23
Large Values	$\dfrac{1.07}{\sqrt{n}}$	$\dfrac{1.22}{\sqrt{n}}$	$\dfrac{1.36}{\sqrt{n}}$	$\dfrac{1.63}{\sqrt{n}}$

TABLE IX. Poisson Distribution Function:

$$F(t) = \sum_{x=0}^{[t]} \frac{\mu^x}{x!} e^{-\mu}$$

						μ				
[t]	.50	1.0	2.0	3.0	4.0	5.0	6.0	7.0	8.0	9.0
0	.607	.368	.135	.050	.018	.007	.002	.001	.000	.000
1	.910	.736	.406	.199	.092	.040	.017	.007	.003	.001
2	.986	.920	.677	.423	.238	.125	.062	.030	.014	.006
3	.998	.981	.857	.647	.433	.265	.151	.082	.042	.021
4	1.000	.996	.947	.815	.629	.440	.285	.173	.100	.055
5	1.000	.999	.983	.961	.785	.616	.446	.301	.191	.116
6	1.000	1.000	.995	.966	.889	.762	.606	.450	.313	.207
7	1.000	1.000	.999	.988	.949	.867	.744	.599	.453	.324
8	1.000	1.000	1.000	.996	.979	.932	.847	.729	.593	.456
9	1.000	1.000	1.000	.999	.992	.968	.916	.830	.717	.587
10	1.000	1.000	1.000	1.000	.997	.986	.957	.901	.816	.706
11	1.000	1.000	1.000	1.000	.999	.995	.980	.947	.888	.803
12	1.000	1.000	1.000	1.000	1.000	.998	.991	.973	.936	.876
13	1.000	1.000	1.000	1.000	1.000	.999	.996	.987	.966	.926
14	1.000	1.000	1.000	1.000	1.000	1.000	.999	.994	.983	.959
15	1.000	1.000	1.000	1.000	1.000	1.000	.999	.998	.992	.978
16	1.000	1.000	1.000	1.000	1.000	1.000	1.000	.999	.996	.989
17	1.000	1.000	1.000	1.000	1.000	1.000	1.000	1.000	.998	.995
18	1.000	1.000	1.000	1.000	1.000	1.000	1.000	1.000	.999	.998
19	1.000	1.000	1.000	1.000	1.000	1.000	1.000	1.000	1.000	.999
20	1.000	1.000	1.000	1.000	1.000	1.000	1.000	1.000	1.000	1.000

TABLE IX. *(Continued)*

[*t*]	μ					
	10.0	11.0	12.0	13.0	14.0	15.0
2	.003	.001	.001	.000	.000	.000
3	.010	.005	.002	.001	.000	.000
4	.029	.015	.008	.004	.002	.001
5	.067	.038	.020	.011	.006	.003
6	.130	.079	.046	.026	.014	.008
7	.220	.143	.090	.054	.032	.018
8	.333	.232	.155	.100	.062	.037
9	.458	.341	.242	.166	.109	.070
10	.583	.460	.347	.252	.176	.118
11	.697	.579	.462	.353	.260	.185
12	.792	.689	.576	.463	.358	.268
13	.864	.781	.682	.573	.464	.363
14	.917	.854	.772	.675	.570	.466
15	.951	.907	.844	.764	.669	.568
16	.973	.944	.899	.835	.756	.664
17	.986	.968	.937	.890	.827	.749
18	.993	.982	.963	.930	.883	.819
19	.997	.991	.979	.957	.923	.875
20	.998	.995	.988	.975	.952	.917
21	.999	.998	.994	.986	.971	.947
22	1.000	.999	.997	.992	.983	.967
23	1.000	1.000	.999	.996	.991	.981
24	1.000	1.000	.999	.998	.995	.989
25	1.000	1.000	1.000	.999	.997	.994
26	1.000	1.000	1.000	1.000	.999	.997
27	1.000	1.000	1.000	1.000	.999	.998
28	1.000	1.000	1.000	1.000	1.000	.999
29	1.000	1.000	1.000	1.000	1.000	1.000

Abridged with permission from E. C. Molina, *Poisson's Exponential Binomial Limit,* D. Van Nostrand, 1949.

Index